PROBLEM ANALYSIS IN SCIENCE AND ENGINEERING

Academic Press Rapid Manuscript Reproduction

PROBLEM ANALYSIS IN SCIENCE AND ENGINEERING

Edited by

F. H. Branin, Jr.

IBM Corporation
Kingston, New York

K. Huseyin

University of Waterloo
Waterloo, Ontario, Canada

ACADEMIC PRESS
New York San Francisco London 1977
A Subsidiary of Harcourt Brace Jovanovich, Publishers

ACADEMIC PRESS, INC.
111 Fifth Avenue, New York, New York 10003

United Kingdom Edition published by
ACADEMIC PRESS, INC. (LONDON) LTD.
24/28 Oval Road, London NW1

Library of Congress Cataloging in Publication Data

Main entry under title:

Problem analysis in science and engineering.

"Outgrowth of the Interdisciplinary Conference on Problem Analysis in Science and Engineering held at University of Waterloo, Waterloo, Canada, in May 1975."
1. Engineering mathematics—Congresses. 2. Problem solving—Congresses. I. Branin, F. H. II. Huseyin, K.
TA329.P76 620'.001'8 77-16043
ISBN 0-12-125550-6

PRINTED IN THE UNITED STATES OF AMERICA

CONTENTS

LIST OF CONTRIBUTORS

Numbers in parentheses indicate page on which authors' contributions begin.

A. Abd-Ali (113), Department of Electrical and Electronic Engineering, Queen Mary College, University of London, England

G. C. Andrews (1), Department of Mechanical Engineering, University of Waterloo, Waterloo, Ontario, Canada

F. H. Branin, Jr. (41), IBM Corporation, Kingston, New York. (Visiting Professor, Departments of Electrical Engineering and Systems Design, University of Waterloo, 1973–1975.)

D. A. Calahan (361), Department of Electrical and Computer Engineering, The University of Michigan, Ann Arbor, Michigan

T. C. Edens (283), Department of Electrical Engineering and Systems Science, Michigan State University, East Lansing, Michigan

F. J. Evans (113), Department of Electrical and Electronic Engineering, Queen Mary College, University of London, England

R. Fitz (199), Department of Biology, University of Dayton, Dayton, Ohio

O. I. Franksen (147), Electric Power Engineering Department, The Technical University of Denmark, Lyngby, Denmark

D. J. Hornbach (199), Department of Biology, University of Dayton, Dayton, Ohio

K. Huseyin (229), Department of Systems Design, University of Waterloo, Waterloo, Ontario, Canada

H. Kardestuncer (257), Department of Civil Engineering, University of Connecticut, Storrs, Connecticut

H. E. Koenig (283), Department of Electrical Engineering and Systems Science, Michigan State University, East Lansing, Michigan

N. C. Lind (341), Department of Civil Engineering, University of Waterloo, Waterloo, Ontario, Canada

N. Orlandea (361), Department of Electrical and Computer Engineering, The University of Michigan, Ann Arbor, Michigan

M. J. Sewell (391), Department of Mathematics, The University of Reading, Reading, England

G. H. Smerage (427), Institute of Food and Agricultural Sciences, University of Florida, Gainesville, Florida

E. Tonti (463), Istituto di Matematica, Politecnica di Milano, Milan, Italy

PREFACE

This book is the outgrowth of the Interdisciplinary Conference on Problem Analysis in Science and Engineering held at University of Waterloo, Waterloo, Ontario, Canada, in May 1975. This conference was convened to consider the common elements of problem formulation and solution in various fields of engineering, the physical sciences, economics, operations research, and so on. The objectives of the conference were as follows:

1. To identify and characterize fully the underlying mathematical features of problem formulation and solution that are common to these different disciplines, with major—though not exclusive—emphasis on the roles of graph theory and sparse matrix techniques.
2. To cross disciplinary boundaries in comparing and sharing methods of analysis that are common to different fields.
3. To stimulate research into new ways of taking advantage of problem *structure* in developing more efficient methods of solution.
4. To consider the possible use—or even the invention—of radically different forms of mathematics than are currently being used to solve these problems.

The first two of these objectives were achieved to a satisfying degree as the various chapters of this book demonstrate; the third objective is only beginning to show evidence of realizability; but the fourth appears still to be a dim and distant hope. Even so, these last two objectives are perhaps the most important to be pondered by readers of this book and others who are at work in the many overlapping areas addressed herein.

The Waterloo Conference, though organized independently, was in much the same spirit as the 1973 Colloquium held at Twente University of Technology, Enschede, The Netherlands, and documented in *Physical Structure in Systems Theory—Network Approaches to Engineering and Economics*, J. J. vanDixhoorn and F. J. Evans, Eds. (Academic Press, New York, 1974). Moreover, three of the participants in the Twente colloquium have also contributed to the present volume. As the list of contributors clearly shows, the Waterloo conference was indeed both interdisciplinary and international in scope. It is the hope of the editors and contributors that this volume will be of interest to a wide audience and that it will help to stimulate further advances by methods and techniques that we have scarcely begun to imagine.

The chapters of this book are arranged alphabetically by author rather than being organized by subject. Five chapters (1, 2, 10, 12, 13) deal with different aspects of *network theory* and its applications in engineering and the physical sciences. Two chapters (8, 9) deal with *economics*, three (5, 8, 12) with *ecology*, two (6, 11) with *catastrophe theory*, and three (3, 4, 7) with mathematical aspects of *problem structure* and *analysis tools*.

The major theme of a network analytic approach is carried out by Andrews in his paper (Chapter 1) on a restatement of the laws of dynamics in terms of graph theoretical concepts. In particular, he shows that a fundamental (and purely topological) relation in graph theory, often called "orthogonality," is a much more general concept than—and actually subsumes—the *principle of virtual work*. Andrews then describes a "vector network" technique and a program that uses this technique for compiling and solving the equations of mechanical systems automatically on a digital computer.

Orlandea and Calahan (Chapter 10) also treat the problem of automatically compiling and solving the equations for large mechanical systems where matrix sparsity is of major concern. They show in some detail how the effects of various *joint constraints* can be taken into account and they stress the importance of working with *all* of the system equations in order to take optimum advantage of matrix sparsity. (This is essentially the *tableau* approach now widely adopted in electrical engineering.) Finally, they illustrate the effectiveness of their sparse matrix techniques by solving some very large dynamical problems.

Smerage (Chapter 12) applies the ideas of network theory to *biological* and *ecological* systems. Using the concept of "complementary" (i.e., "through" and "across") variables, he shows how physicochemical systems may be treated, with *chemical potential* as the across variable and *flow rate* as the through variable. In addition, he derives network models for the storage, transportation, and transformation of chemical species and he applies these models to various processes—including *photosynthesis*—that take place in living plants. This work is a significant step toward enabling biologists and ecologists to take direct advantage of the techniques of and programs for network analysis developed particularly by electrical engineers.

Branin (Chapter 2) presents a treatment of the *network concept* not only in its usual context related to linear graph theory (or "l-networks") but also in relation to a higher dimensional homolog called the *3-network*. He shows that the algebraic structure of the 1-network problem can be extended to characterize the 3-network problem as well. Then, by introducing the concept of a *dual* 3-network, he demonstrates that the resulting algebraic structure is the progenitor of an isomorphic *operational* structure that characterizes the vector calculus. Finally, he illustrates the utility of this operational structure in arriving at an algebraic–topological treatment of Maxwell's equations.

Branin's paper is a prelude to and companion of Tonti's very remarkable work (Chapter 13) on a *formal treatment* of physical theories of an impressive variety. Using algebraic–topological concepts, together with the basic principles of physics, Tonti classifies numerous physical phenomena as 1-, 2-, 3-, and even 4-network problems. In this paper, and elsewhere, Tonti shows how a major portion of theoretical physics comes within the scope of this algebraic–topological approach, or what may in its broadest sense be called *network theory*.

Shifting to an entirely different topic, Koenig and Edens (Chapter 8) show that the principles of *thermodynamics*—particularly the second law—not only can but *must* be applied if the effective and proper management of energy resources, the ecology of the planet, and the human economic system are ever to be achieved. They point out that economists have generally overlooked the importance of the ecological and thermodynamic constraints that limit access to resources. As a result, serious inconsistencies relative to ecological and thermodynamic considerations have become deeply imbedded in current economic policies and planning.

In an effort to bring more scientific principles to bear on the global problems involved, Koenig and Edens liken our production and consumption system to a *heat engine* and point out the importance of giving attention to *thermodynamic efficiency*—of labor, use of resources, etc.—in the production of "economic substances." Finally, they seek to define suitable *control instruments,* such as marketing, pricing, and taxation policies, to bring about a more effective long-term solution to the problems of resource management.

Lind (Chapter 9) carries this theme into the analysis of *publicly regulated* economic activities—for example, the construction industry as influenced by the administration of building codes. By setting up two basic postulates involving such economic factors as investment and production rates, regulation parameters, socioeconomic loss functionals, safety, and security, Lind is able to show that there is a direct and useful analogy with an axiomatic statement of the laws of thermodynamics.

In a similar vein, Hornbach and Fitz (Chapter 5) apply the principles of *digraph theory* to the extremely complex problem of analyzing the ecology of the Sahel desert. Using a technique called *interpretive structural modeling* as a qualitative analysis tool, they characterize one of eight subsystems of the Sahelian ecology, namely, the quality of rangeland. This modeling activity, though in its early stages of development, has potential in guiding the decision making processes involved in dealing with such complicated problems as the Sahelian drought.

Although it is difficult to foresee the consequences of applying such unusual analytical tools as thermodynamics and digraph theory to economic and ecological problems, the authors of these three papers have shown a commendable degree of ingenuity and imagination in their assault on these challenging and complex problem areas.

Huseyin (Chapter 6) and Sewell (Chapter 11) discuss the subject of *catastrophe theory* and its application to various fields ranging from the buckling of structures and the beating of hearts to thermodynamic phase transitions and social behavior phenomena. In keeping with the objectives of the Waterloo conference, these authors have shown that novel mathematical tools can indeed be applied with success across many disciplinary boundaries.

Evans and Abd-Ali (Chapter 3) show that by the proper study of the *structure* of a system—even though highly nonlinear—some qualitative assessment of its behavior can often be gained without actually solving the equations that describe the system. They illustrate the value of this approach by treating a number of linear and nonlinear examples and finding stability characteristics from the structure of the equations themselves. Finally, they suggest that in the case of certain biological, economic, and social science problems, this qualitative analysis of system structure may provide the only hope of success.

Franksen (Chapter 4) contributes an ingenious description of the *group-theoretic foundations of measurement*—physical, economic, and social—and characterizes the algebraic properties thereof in terms of *digraphs*. After discoursing on digraph theory, he discusses the *Erlanger program of systems science* by which he means an extension of Klein's classification of geometrics to include both topology and tensor analysis. The purpose of this effort is to gain a more comprehensive overview of the structure of systems science and the interrelation of its various parts.

Kardestuncer (Chapter 7) advocates the use of *tensors*, rather than matrices, wherever they apply, not only in the formulation phase of problem analysis but also during the *solution process*. He suggests that the physics and geometry of the problem may be helpful in guiding the solution and in making it more efficient than mere "number crunching" might otherwise permit. He illustrates this idea by examples taken from discrete mechanics, showing that transforming coordinates to the *principal directions* can greatly simplify the problem. Although the idea is only in its early stages of development, it merits further study.

The editors of this book wish to acknowledge with thanks the efforts of the other members of the Waterloo Conference Committee, Professors N. C. Lind, H. K. Kesavan, and G. C. Andrews, for their cooperation in organizing the conference. On behalf of this committee, we also wish to express our gratitude to the University of Waterloo Research Grant Program, to the Solid Mechanics Division of University of Waterloo, and to the

National Research Council of Canada for their very generous financial support of the conference.

Our contributors have been most cooperative in preparing well-written papers and we are especially grateful to Mrs. Diane Hammar, who so carefully typed the majority of the final copy. Finally, we thank our publisher for its patience in shepherding the book through to completion.

A GENERAL RE-STATEMENT OF THE LAWS OF DYNAMICS BASED ON GRAPH THEORY

GORDON C. ANDREWS
Department of Mechanical Engineering
University of Waterloo
Waterloo, Ontario, Canada

SUMMARY

This paper describes how some concepts of graph theory can be incorporated into dynamics to yield a new and more symmetrical format for the basic laws of dynamics. Several new results are obtained by this approach: equations of motion can be formulated by a procedure so methodical that it can be used as an algorithm for a "self-formulating" computer program; a "dual" relationship relating force and displacement vectors is discovered; and a general "principle of orthogonality" can be derived for mechanical systems. This principle is observed to be a more general form of the principle of virtual work and may prove to be very useful.

1. INTRODUCTION

Scientific research is a continuing process of organizing and classifying scientific concepts into a coherent structure. As Henri Poincaré observed: "The scientist must make order; one builds science with facts, as a house is built with stones ..." With Poincaré's quotation in mind, this paper expands the traditional principles of dynamics to include some concepts from graph theory. The resulting structure of dynamics is slightly larger, stronger and more symmetrical.

The procedure for applying graph theory to dynamics, described in this paper, can be summarized very simply in a few sentences: given a dynamic mechanical system, a schematic diagram is constructed with displacement vectors, in which each vector represents an element of the system. This network of vectors is easily constructed, yet, by using some simple theorems from graph theory, a necessary and sufficient set of equations can be methodically created from the network to solve the dynamic system. Moreover, these equations can be manipulated to yield some unexpected new and useful results.

2. THE BASIS OF DYNAMICS

Newton's Laws of motion, as described in the *Principia* (1687), form the basis of all dynamics. These laws are repeated briefly below, in their modern format, in order to illustrate their relevance to the discussion which follows:

First Law: An object remains in a state of rest or in uniform motion in a straight line, unless acted upon by an unbalanced force.

Second Law: The time-rate of change of momentum of a given object is proportional to the force causing the change in momentum and in the same direction as the force.

Third Law: To every action, there is an equal and opposite reaction.

It is evident that the first law, although useful as a teaching example, is really a consequence of the second law, since an object which is *not* acted upon by an unbalanced force will have a constant momentum and therefore (assuming the mass to be constant) it will remain at rest or will continue in a straight line at constant speed.

The second law is by far the most important law in dynamics and it is retained (in the form proposed by D'Alembert) as one of the basic postulates (the vertex postulate) in this re-statement of the laws of dynamics.

The third law is uniquely satisfied by the procedures described in this paper: the application of graph theory to dynamics leads us to visualize the inter-connection of a system's elements as a set of line segments (vectors) connected to form a network. In this network, the equality of action and reaction are axiomatically satisfied, since the force associated with a given element is also associated with the line that represents the element and the force is therefore applied equally but oppositely at the two end-points of the line. In fact, it could be argued that the third law is applied more rigorously in the network approach than it is in traditional methods, which permit a force to be applied to a free-body diagram without specifying the reaction-point from which that force is generated.

Consequently, Newton's laws of motion are entirely satisfied by the graph theory methods described in this paper. However, some additional information is generated conveniently using graph theory. It should be observed that all of Newton's laws of motion refer to force relationships, yet there are certain displacement and velocity equations which are essential for the solution of a given system. A similar observation could be made of Lagrangian and Hamiltonian formulation techniques: the existence of a proper set of generalized coordinates is axiomatically assumed; it is further assumed that the potential and kinetic energies in the system can be expressed in terms of the generalized coordinates (or their time-derivatives), thus implying the existence of certain kinematic equations without stating the equations, themselves.

However, when concepts of graph theory are combined with Newton's laws, a procedure for writing the essential displacement and velocity equations is obtained. In fact, it is possible to tabulate the number of independent equations available to solve the system and compare it with the number of unknown variables, as is shown later in this paper.

3. SOME ESSENTIAL GRAPH THEORY CONCEPTS

3.1 Historical background

Leonhard Euler was the first to show the potential value of graph techniques when he used simple graphs in 1736 to solve the parlour-game known as the "Koenigsberg Bridge" problem [1,2]. Since that time, graph theory has evolved as part of the branch of Mathematics known as Combinatorics. Graph theory may be defined as the study of inter-connected systems of line segments.

The application of graph theory to physical systems began with electrical networks, understandably, since diagrams representing electrical networks are merely inter-connected systems of line segments. The extension of graph theory to other physical systems is usually associated with the work by Trent [3] in 1955 although the proposal of graph theory as a unifying concept (rather than an electrical analog) was made by Koenig and Blackwell [4,5] in 1960. Since then, there have been numerous publications on graph theory applications, including the comprehensive work by Koenig, Tokad and Kesavan [6], the basic texts by Roe [7] and by Shearer et al [8] as well as several more recent texts [9,10,11]. All of the basic graph theory concepts are defined in these earlier publications and only a few essential points are repeated here.

One fundamental new concept, not mentioned in earlier work, is the idea that displacement vectors should be used in the construction of the "system graph" which, in this application, is called the "vector-network diagram" [12,13,14]. The construction of this diagram is described in the following paragraph.

3.2 Vector-Network diagram

To derive the equations which make up the mathematical model of a mechanical system using graph theory, it is convenient to draw a schematic diagram of the system, in which the system elements are identified and their points of inter-connection are plotted as points in space. These points are usually called "nodes". Each element is then represented in the schematic diagram by a displacement vector which spans the two nodes which represent the element's end-points (or points of inter-connection to the system). Each of the five most commonly-encountered elements can be represented by a single vector and more complex elements can be represented by sets of vectors [12,13,15].

The resulting schematic diagram is a network of vectors and therefore it is usually called the "vector-network" diagram. It is a somewhat more general form of the "system graph" described in earlier texts. (It will be observed that earlier system graphs are scalar forms of vector-network diagram.)

As an example of the procedure for constructing the vector-network diagram, consider the simple system shown in Figure 2.1, composed of a particle mass at point p, supported by two springs and acted upon by a dashpot. The vector-network diagram for this

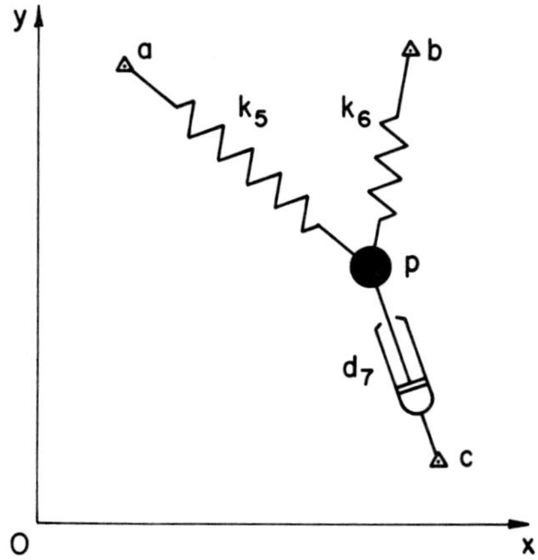

Fig. 2.1 - A Simple Mass-Spring-Dashpot System

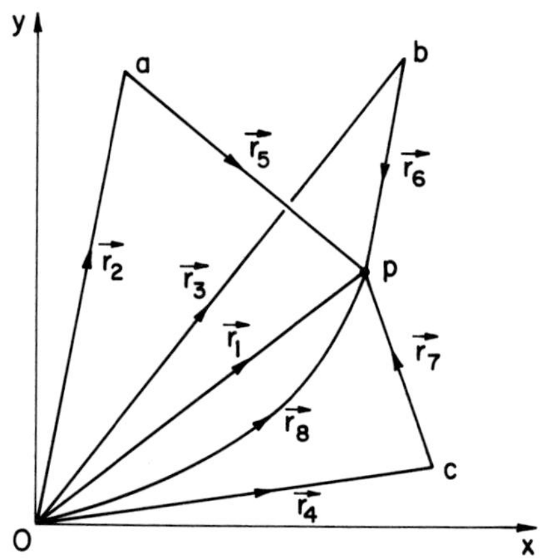

Fig. 2.2 - Vector-Network Diagram for the Simple System in Fig. 2.1

system is shown in Figure 2.2, where position vector $\vec{r}_1$ represents the mass; vectors $\vec{r}_2$, $\vec{r}_3$ and $\vec{r}_4$ define the location of points a, b and c, respectively, and are known as "displacement drivers"; vectors $\vec{r}_5$ and $\vec{r}_6$ represent the springs and vector $\vec{r}_7$ represents the dashpot. Position vector $\vec{r}_8$ is a "force driver" representing the weight of the particle. (Vector $\vec{r}_8$ is distorted for clarity in Figure 2.2, since $\vec{r}_8$ coincides with $\vec{r}_1$.) Mass elements and most driver elements are represented by position vectors radiating from the origin (or inertial node) since the physical characteristics of these elements are defined relative to the inertial reference frame.

When the vector-network diagram has been drawn, three sets of equations which form the mathematical model can be written: the "terminal" equations, which describe the characteristics of the system elements, and the "fundamental-cutset" and "fundamental-circuit" equations, which are applications of the basic postulates and depend on the way that the system elements are inter-connected.

3.3 Basic postulates

Unlike the traditional approach to dynamics in which Newton's second law is the basic postulate, in the vector-network method there are *two* equally-important fundamental postulates:

The Vertex Postulate: The "through" variables (forces) associated with the elements incident on a given node must sum to zero, when due account is taken of the directions of the variables.

This is a statement of d'Alembert's principle which, in turn, is an alternate form of Newton's second law. However, in this form it is more general, since it applies to any physical system; for example, in electrical systems, it is equivalent to Kirchhoff's current law. The second fundamental postulate is:

The Circuit Postulate: The "across" variables (displacements) associated with the elements in any closed loop must sum to zero, when due account is taken of the direction of the variables.

This is a statement of a basic axiom of vector algebra; all of the across variables in the vector-network diagram are vector displacements, therefore any closed polygon must sum to zero. In the above form, however, it applies to any physical system; for example, in electrical systems, it is equivalent to Kirchhoff's voltage law.

The importance of the circuit postulate is usually not recognized in traditional approaches to dynamics, because it does not involve force relationships and therefore it is kinematic, not dynamic, and is usually assumed to be obvious.

3.4 Fundamental-cutset equations

Each of the basic postulates yields a set of equations; however, if the vertex postulate is applied to each and every vertex, then one of the resulting equations will not be independent. Consequently, to ensure a useful independent set, a "tree" is identified in the vector-network diagram. A "tree" is a connected subset of vectors which touches all of the nodes but includes no closed loops. If there are N nodes in the system, then it can be shown, using graph theory, that there will be (N-1) vectors in the tree; these vectors are called tree "branches". The remaining vectors are called "chords" of the complement or "co-tree".

A "fundamental-cutset" can be defined for each branch, such that the branch plus a unique set of chords, when deleted, divide the network into two parts. Then, by applying the vertex postulate to the nodes isolated by each fundamental-cutset and summing, a set of (N-1) fundamental-cutset equations will result. (The fundamental-cutset is identical to a free-body diagram, since it isolates a portion of a system and a force equation results.) The fundamental-cutset equations are usually written in the form:

$$[U_t \ A] \begin{bmatrix} \vec{F}_t \\ \\ \vec{F}_c \end{bmatrix} = [0] \tag{3.1}$$

where the matrix $[U_t \ A]$ is called the "fundamental-cutset matrix" since each row identifies the vectors (and their orientation) in a fundamental-cutset. Equation (3.1) can be re-written:

$$[\vec{F}_t] = -[A][\vec{F}_c] \tag{3.2}$$

which shows that the forces associated with the tree branches can be written explicitly as the vector sum of the forces associated with the co-tree chords. The term "fundamental-cutset" is usually abbreviated, for convenience, to "f-cutset".

3.5 Fundamental-circuit equations

If the circuit postulate is applied to each and every circuit in a network, the resulting set of equations will usually contain redundant equations (summations of other equations). The previously defined "tree" is again useful, since a "fundamental-circuit" can be defined for each chord of the co-tree, such that the chord plus a unique set of branches form a closed loop. Then, by applying the circuit postulate to each fundamental-circuit, a set of fundamental-circuit equations will result. If the system contains E elements represented by vectors which are interconnected at N nodes, then there will be (E - N + 1) fundamental-circuit equations, which are usually written in the form:

$$[B \ U_c] \begin{bmatrix} \vec{r}_t \\ \\ \vec{r}_c \end{bmatrix} = [0] \tag{3.3}$$

where the matrix $[B \ U_c]$ is called the "fundamental-circuit matrix" since each row identifies the vectors (and their orientation) in a fundamental-circuit. Equation (3.3) can be re-written:

$$[\vec{r}_c] = -[B][\vec{r}_t] \tag{3.4}$$

which shows that the vector displacements associated with the co-tree chords can be written explicitly as the vector sum of the displacements associated with the tree branches.

It is convenient to treat displacement $\vec{r}$ and velocity $\vec{v}$ as independent variables; therefore, by differentiating equation (3.4) with respect to time we get:

$$[\vec{v}_c] = -[B][\vec{v}_t] \tag{3.5}$$

which is an independent set of fundamental-circuit equations in terms of velocities.

The term "fundamental-circuit" is usually abbreviated, for convenience, to "f-circuit".

3.6 Terminal equations

The "f-cutset" and "f-circuit" equations described earlier depend on the vertex and circuit postulates and the way in which the elements of the system are inter-connected. The physical characteristics of the elements do not enter into these equations and are defined by a third set of equations: the "terminal" equations.

The "terminal" equation of a given element expresses the experimentally-observed characteristics of the element, as they would be observed if the element were tested on a laboratory bench; for example, the characteristics of a spring are given by Hooke's law, which relates the spring force to its deformation. In a similar way, the characteristics of other system elements can be defined.

The terminal equations for the five most common types of element can be written in a general three-dimensional form using vector notation as follows:

Mass:	$\vec{F}_1 = - m \ddot{\vec{r}}_1$	(d'Alembert Force)	(3.6)
Spring:	$\vec{F}_2 = - k \{ \lvert\vec{r}_2\rvert - s \} \hat{r}_2$	(Hooke's Law)	(3.7)
Dashpot:	$\vec{F}_3 = - d \{ \vec{v}_3 \cdot \hat{r}_3 \} \hat{r}_3$	(Viscous Damping)	(3.8)
Force Driver:	$\vec{F}_4 = F_x \hat{i} + F_y \hat{j} + F_z \hat{k}$	(Specified Force)	(3.9)
Displacement Driver:	$\vec{r}_5 = r_x \hat{i} + r_y \hat{j} + r_z \hat{k}$	(Specified Displacement)	(3.10)

where $\vec{F}$ is the force associated with a given element and $\vec{r}$ is the vector between the element's two points of interconnection to the system. The terminal equations above describe only the common system elements, but the terminal equation(s) for any system element can be written if the physical characteristics of the element are completely known.

3.7 Creating **f**-cutset and **f**-circuit equations

It is convenient, when describing a network, to construct an "incidence" matrix, which has its columns numbered to correspond to the vectors in a network and its rows numbered to correspond to the nodes. The incidence matrix is composed of +1, -1 and zero entries depending whether a given vector enters, leaves, or does not touch a given node, respectively. The matrix is easily constructed and describes the inter-connection and orientation of the vectors in a network completely.

However, it can be shown, by a basic theorem in graph theory, that the f-cutset matrix can be obtained from the incidence matrix by simple row operations. Another basic theorem proves that the f-cutset and f-circuit matrices contain sub-matrices that are related. Sub-matrix [A] in equation (3.1) is related to sub-matrix [B] in equation (3.3) as follows:

$$[A] = - [B]^T \quad \text{and} \quad [B] = - [A]^T \tag{3.11}$$

Consequently, given the incidence matrix describing a system, it is possible, through matrix manipulations, to create the f-cutset and f-circuit equations for the system. This point was particularly useful in developing the computer algorithm described later in this paper.

Finally, it should be noted that the sign convention for forces and displacements, used in this paper, is the convention commonly used in dynamics and it differs somewhat from earlier publications which use a sign convention based on electrical network analysis.

4. A NEW BASIS FOR DYNAMICS

The previous paragraphs have defined some essential concepts and have described the procedure for constructing the vector-network diagram and writing the terminal equations. It is now possible to show the relationships between these basic concepts.

4.1 The re-organized structure

The basic concepts and their inter-relationship are summarized in the chart in Figure 4.1 which constitutes a new way of reorganizing the basic laws of dynamics. The vertex and circuit postulates are the two main axioms supporting this re-organized structure. These postulates apply throughout each system and are independent from the experimentally-observed characteristics of the elements in the system. In other words, unlike the traditional analysis based on Newton's second law, the characteristics of the elements are separated from the laws governing the system.

The existence of two basic postulates (rather than one) is observed: the vertex postulate applies to through-variables (force and its time-derivatives) and the circuit postulate applies to across-variables (displacement and its time-derivatives). Consequently, there is a symmetry in the theory which is lacking in the traditional approach. These postulates are also unifying concepts, since they apply to all physical systems.

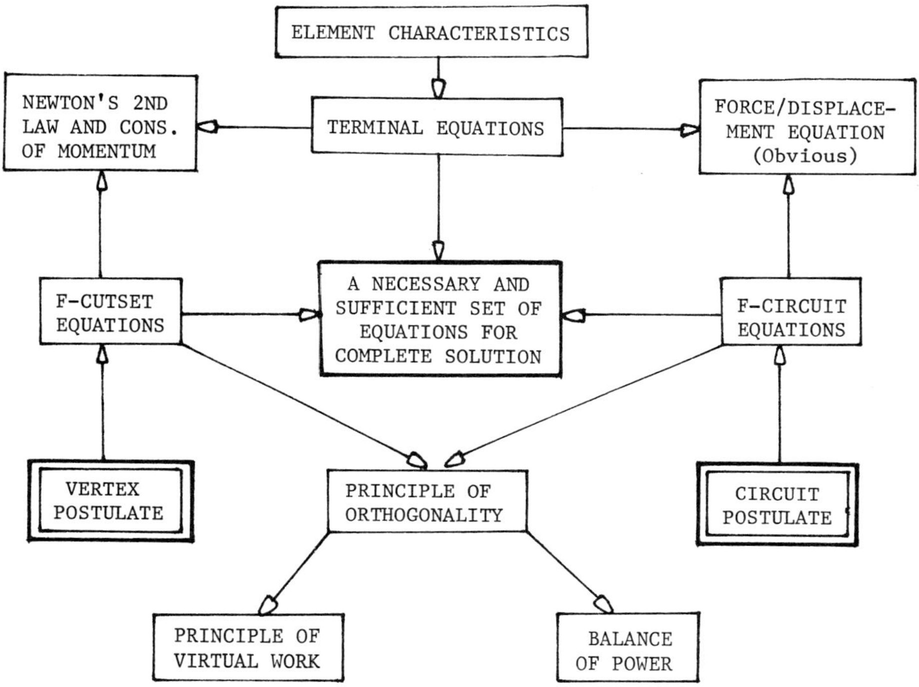

Fig. 4.1 - The Basic Postulates and their Relationship to Some Traditional Principles

4.2 Equations of motion

As illustrated in Figure 4.1, the two basic postulates, combined with the experimentally-observed element characteristics, are the sources for three sets of equations, the f-circuit, f-cutset and terminal equations, which form a necessary and sufficient set for complete solution of a given system. This formulation procedure (described in detail later in this paper) is much more methodical than traditional techniques. In fact, it is sufficiently methodical that it can be implemented on a digital computer to yield a computer program which produces the system response, given only a system description.

4.3 Principle of orthogonality

Because of the orthogonality of the matrices in the f-cutset and f-circuit equations, a new and general principle, called the "principle of orthogonality" can be derived. This is a fairly significant principle, since it can be shown that many traditional concepts, such as the principle of virtual work and the instantaneous balance of power are special cases of the principle of orthogonality. This general principle is derived and discussed later in this paper.

4.4 Additional observations

It will be observed that Newton's second law is classified, in the re-organized structure shown in Figure 4.1, as a combination of the mass terminal equation (3.6) and a f-cutset equation (3.2). This is true, since the f-cutset equation is equivalent to the force equation which would result from a free-body diagram of the mass and the characteristics of the mass are introduced by the mass terminal equation. It is obvious from Figure 4.1 that the f-circuit equations (3.4) and (3.5), which are displacement and velocity equations essential to the formulation procedure, are not generated by the traditional technique using free-body diagrams and Newton's second law. In the traditional techniques, it is assumed that these equations can be obtained by inspection.

It should also be pointed out that the law of conservation of momentum could also be derived using only a f-cutset equation and the mass terminal equation, as indicated in Figure 4.1, although to perform this derivation it would be necessary to use impulse as the through variable and velocity as the across variable, and to write the equations in terms of these variables.

For completeness, it can be observed that the combination of a f-circuit equation and a terminal equation is also possible. For example, a spring might be connected between two masses and the length of the spring could be expressed as a function of the positions of the masses using a f-circuit equation (3.4). If this equation is then substituted into the spring's terminal equation (3.7), then the force in the spring would be expressed as a function of the displacements of the masses. This force/displacement relationship is obvious and therefore it has not been accorded a name in the traditional study of dynamics.

5. FORMULATING EQUATIONS OF MOTION

As shown in Figure 4.1, the three sets of equations generated using this technique form a necessary and sufficient set for complete solution of the system. This will be illustrated in the paragraphs which follow.

5.1 Formulation tree

In selecting the tree, prior to formulating equations, it is essential that certain vectors should be included in the tree and others should be excluded. The decision depends on the element's terminal equation: if the terminal equation can be written with the across-variable (displacement, or its time-derivatives) on the left-hand side as an explicit function of time or other system variables, then the vector representing the element should be included in the tree; otherwise, the vector must be included in the co-tree. Experience has shown that the simplest formulation of equations of motion results when this criterion is satisfied. In the case of simple systems, the result is a "proper" tree composed solely of vectors representing masses or across (displacement) drivers and a co-tree composed solely of springs, dashpots and force drivers. In some cases, it may appear that a proper tree does not exist in a system; usually this occurs when two or more springs or dashpots are connected to a node which is assumed to have negligible mass. This problem can be overcome by assuming that the node has a small but finite mass, although, of course, this usually results in high-frequency oscillations occurring in the system. The result, however, is a more accurate model of the system, which can be easily simulated on the digital computer.

5.2 Necessary and sufficient equations

When the formulation tree has been identified, the f-cutset, f-circuit and terminal equations may be written. These equations, together, form a necessary and sufficient set for complete solution of the system, where "complete solution" is defined as the determination of the force, velocity and displacement associated with each and every element in the system as a function of the independent variable, time.

Consider a system for which the vector-network model has been constructed and a proper tree selected. Assume that the system has E elements connected at N nodes. If we wish to determine the complete solution of the system, we would want to know the force, displacement and velocity associated with each and every element. Since there are E elements and we want to know the values of 3 quantities associated with each element, then we have 3E unknowns to determine as a function of the independent variable, time.

Note that we have treated displacement and velocity as independent variables. Therefore, we may use the relationship that velocity is the derivative of displacement in solving for the unknowns:

$$\vec{v} = \dot{\vec{r}} \tag{5.1}$$

This gives us an extra terminal equation for each tree element, which will prove very useful. In addition, since the f-circuit equations are written in terms of displacement, we can differentiate these equations to get a second set of f-circuit equations in terms of velocities, as shown in equation (3.5).

At this point, we can sum the number of equations available to solve for the 3E unknown variables in the system:

f-cutset equations:	(N - 1)	eqs. (3.2)
f-circuit equations:	2(E - N + 1)	eqs. (3.4) and (3.5)
tree terminal equations:	2(N - 1)	eqs. (3.6), (3.10) and (5.1)
co-tree terminal equations:	(E - N + 1)	eqs. (3.7), (3.8) and (3.9)
TOTAL	3E	equations

Consequently, we have 3E equations in 3E variables, not including the independent variable, time. Some of these equations will be first-order differential equations. However, it will be demonstrated in the paragraph that follows that if a proper tree exists, these equations can always be arranged in normal form, ready for analytical or numerical integration, and the result after integration is a set of 3E independent equations in 3E unknowns. Consequently, the vector-network method yields a necessary and sufficient set of equations for complete solution.

5.3 Formulation procedure

If a proper tree exists in the vector-network diagram, then the equations of motion for the system can always be arranged in "normal form" as a set of first-order vector differential equations (with two equations for each mass element). The formulation of these equations of motion is a simple substitution procedure, illustrated as follows: Assume that the f-cutset equations have been written as shown in (3.2), the f-circuit equations as shown in (3.4) and (3.5) and the terminal equations have been written as shown in (3.6) through (3.10). Starting with the mass terminal equation (3.6) re-written as shown:

$$\dot{\vec{v}}_1 = \frac{-\vec{F}_1}{m_1} \tag{5.2}$$

substitute the f-cutset equations (3.2) for $\vec{F}_1$ to get (in functional notation):

$$\dot{\vec{v}}_1 = f_a\{\vec{F}_c\} \tag{5.3}$$

Then substitute the co-tree terminal equations (3.7) through (3.9) to get:

$$\dot{\vec{v}}_1 = f_b\{\vec{r}_c, \vec{v}_c, t\} \tag{5.4}$$

Then substitute the f-circuit equations (3.4) and (3.5) to eliminate $\vec{r}_c$ and $\vec{v}_c$:

$$\dot{\vec{v}}_1 = f_c\{\vec{r}_t, \vec{v}_t, t\} \tag{5.5}$$

And since the across drivers (3.10) are specified functions of time, then substituting these equations into (5.5) yields:

$$\vec{v}_1 = f_d\{\vec{r}_1, \vec{v}_1, t\} \tag{5.6}$$

which, when augmented by (5.1):

$$\dot{\vec{r}}_1 = \vec{v}_1 \tag{5.1}$$

gives two first-order vector differential equations for each mass element. (Obviously, in the above derivation, f_a, f_b, f_c and f_d are similar but different functions). Consequently, if a proper tree exists in the vector-network diagram, it is always possible to derive the equations of motion of the system. However, the analytical solution of the differential equations may not always be possible; nevertheless, the equations are in a suitable form for computer integration.

5.4 Example problem

As an example of the formulation procedure, consider the simple system shown earlier in Figure 2.1; the vector-network diagram for this system is shown in Figure 2.2 and a proper tree can be identified, composed of vectors $\vec{r}_1$, $\vec{r}_2$, $\vec{r}_3$ and $\vec{r}_4$. The three basic sets of equations can then be written. From a knowledge of the element types, the terminal equations are:

$$\dot{\vec{r}}_1 = \vec{v}_1 \quad ; \quad \dot{\vec{v}}_1 = \frac{-\vec{F}_1}{m} \tag{5.7}$$

$$\vec{r}_2 = a_x\hat{i} + a_y\hat{j} \quad ; \quad \vec{v}_2 = 0 \tag{5.8}$$

$$\vec{r}_3 = b_x\hat{i} + b_y\hat{j} \quad ; \quad \vec{v}_3 = 0 \tag{5.9}$$

$$\vec{r}_4 = c_x\hat{i} + c_y\hat{j} \quad ; \quad \vec{v}_4 = 0 \tag{5.10}$$

$$\vec{F}_5 = -k_5(|\vec{r}_5| - s_5)\,\hat{r}_5 \tag{5.11}$$

$$\vec{F}_6 = -k_6(|\vec{r}_6| - s_6)\,\hat{r}_6 \tag{5.12}$$

$$\vec{F}_7 = -d(\vec{v}_7 \cdot \hat{r}_7)\,\hat{r}_7 \tag{5.13}$$

$$\vec{F}_8 = -\,W\hat{j} \tag{5.14}$$

The f-cutset equations can be derived from the incidence matrix or written from inspection of Figure 2.2 in the form:

$$\begin{bmatrix} \vec{F}_1 \\ \vec{F}_2 \\ \vec{F}_3 \\ \vec{F}_4 \end{bmatrix} = -\begin{bmatrix} 1 & 1 & 1 & 1 \\ -1 & 0 & 0 & 0 \\ 0 & -1 & 0 & 0 \\ 0 & 0 & -1 & 0 \end{bmatrix} \begin{bmatrix} \vec{F}_5 \\ \vec{F}_6 \\ \vec{F}_7 \\ \vec{F}_8 \end{bmatrix} \tag{5.15}$$

The f-circuit equations can be derived from the f-cutset equations (5.15) or written by inspection of Figure 2.2 as follows:

$$\begin{bmatrix} \vec{r}_5 \\ \vec{r}_6 \\ \vec{r}_7 \\ \vec{r}_8 \end{bmatrix} = -\begin{bmatrix} -1 & 1 & 0 & 0 \\ -1 & 0 & 1 & 0 \\ -1 & 0 & 0 & 1 \\ -1 & 0 & 0 & 0 \end{bmatrix} \begin{bmatrix} \vec{r}_1 \\ \vec{r}_2 \\ \vec{r}_3 \\ \vec{r}_4 \end{bmatrix} \tag{5.16}$$

By differentiating (5.16) a second set of f-circuit equations is obtained with velocity as the variable:

$$\begin{bmatrix} \vec{v}_5 \\ \vec{v}_6 \\ \vec{v}_7 \\ \vec{v}_8 \end{bmatrix} = - \begin{bmatrix} -1 & 1 & 0 & 0 \\ -1 & 0 & 1 & 0 \\ -1 & 0 & 0 & 1 \\ -1 & 0 & 0 & 0 \end{bmatrix} \begin{bmatrix} \vec{v}_1 \\ \vec{v}_2 \\ \vec{v}_3 \\ \vec{v}_4 \end{bmatrix} \tag{5.17}$$

The above equations (5.7) to (5.17) are a necessary and sufficient set to solve the system. The formulation begins by substituting the f-cutset equation (5.15) for $\vec{F}_1$ into mass terminal equation (5.7) to get:

$$\dot{\vec{v}}_1 = \frac{1}{m} \{\vec{F}_5 + \vec{F}_6 + \vec{F}_7 + \vec{F}_8\} \tag{5.18}$$

This first step is immediately recognized as being equivalent to constructing a free-body diagram of the mass and applying Newton's second law. Next, the co-tree terminal equations (5.11) to (5.14) are substituted into (5.18):

$$\dot{\vec{v}}_1 = \frac{1}{m} \{-k_5(|\vec{r}_5| - s_5)\hat{r}_5 - k_6(|\vec{r}_6| - s_6)\hat{r}_6 - d(\vec{v}_7 \cdot \hat{r}_7)\, \hat{r}_7 - W\hat{j}\} \tag{5.19}$$

At this point, the f-circuit equations (5.16) and (5.17) can be substituted:

$$\dot{\vec{v}}_1 = \frac{1}{m} \{-k_5(|\vec{r}_1 - \vec{r}_2| - s_5) \frac{(\vec{r}_1 - \vec{r}_2)}{|\vec{r}_1 - \vec{r}_2|} - k_6(|\vec{r}_1 - \vec{r}_3| - s_6) \frac{(\vec{r}_1 - \vec{r}_3)}{|\vec{r}_1 - \vec{r}_3|} - d((\vec{v}_1 - \vec{v}_4) \cdot (\vec{r}_1 - \vec{r}_4)) \frac{(\vec{r}_1 - \vec{r}_4)}{|\vec{r}_1 - \vec{r}_4|^2} - W\hat{j}\} \tag{5.20}$$

Since vectors $\vec{r}_2$, $\vec{r}_3$, $\vec{r}_4$ and $\vec{v}_4$ are constants, then (5.20) is in a form similar to (5.6):

$$\dot{\vec{v}}_1 = f\,\{\vec{r}_1, \vec{v}_1\}$$

and it could be augmented by (5.7)

$$\dot{\vec{r}}_1 = \vec{v}_1 \tag{5.7}$$

to yield two first-order vector differential equations. Before these equations can be integrated, they must be written in scalar form. Equation (5.20), for example, is easily re-written as two scalar equations:

$$\ddot{x}_1 = \frac{1}{m}\{-k_5(r_5 - s_5)\frac{(x_1 - a_x)}{r_5} - k_6(r_6 - s_6)\frac{(x_1 - b_x)}{r_6} - d(v_d)\frac{(x_1 - c_x)}{r_7^2}\} \tag{5.21}$$

$$\ddot{y}_1 = \frac{1}{m}\{-k_5(r_5 - s_5)\frac{(y_1 - a_y)}{r_5} - k_6(r_6 - s_6)\frac{(y_1 - b_y)}{r_6} - d(v_d)\frac{(x_1 - c_y)}{r_7^2} - W\} \tag{5.22}$$

where:

$$\vec{r}_1 = x_1\hat{i} + y_1\hat{j}; \quad \vec{v}_1 = \dot{x}_1\hat{i} + \dot{y}_1\hat{j}; \quad \dot{\vec{v}}_1 = \ddot{x}_1\hat{i} + \ddot{y}_1\hat{j} \tag{5.23}$$

$$r_5 \equiv |\vec{r}_5| = \{(x_1 - a_x)^2 + (y_1 - a_y)^2\}^{1/2} \tag{5.24}$$

$$r_6 \equiv |\vec{r}_6| = \{(x_1 - b_x)^2 + (y_1 - b_y)^2\}^{1/2} \tag{5.25}$$

$$r_7 \equiv |\vec{r}_7| = \{(x_1 - c_x)^2 + (y_1 - c_y)^2\}^{1/2} \tag{5.26}$$

$$v_d = (\vec{v}_1 - \vec{v}_4)\cdot(\vec{r}_1 - \vec{r}_4) = \dot{x}_1(x_1 - c_x) + \dot{y}_1(y_1 - c_y) \tag{5.27}$$

Obviously, although this is a simple example, it is non-linear and the resulting differential equations of motion (5.20) or (5.21) and (5.22) are rather complicated and are not suitable for analytical integration. Nevertheless, it can be confirmed by traditional methods that these are the correct equations of motion.

6. SELF-FORMULATING COMPUTER PROGRAMS

The theoretical concepts described in this paper can be put to a very practical use as an algorithm for a "self-formulating" computer program for simulating the response of dynamic mechanical systems. The term "self-formulating" means that the computer program requires only the system description as input and the system response is generated as output, thereby giving the impression that the equations of motion have been formulated by the computer. Using the techniques described earlier in this paper, a computer program called VECNET (for *Vec*tor-*Net*work Simulation Program) has been written, tested and applied to the simulation of dynamic mechanical systems. [12, 13, 16] VECNET is capable of simulating the response of three-dimensional mass-spring-dashpot systems (in which the mass elements are particles) subject to both force and displacement drivers. The program consists of approximately 500 executable FORTRAN statements and utilizes a formulation procedure similar to that described in equations (5.2) to (5.6) except that the order of substitution is different (since the computer substitutes numerical values, not analytical expressions).

VECNET requires only the description of a system's vector-network diagram, element magnitudes and initial conditions as input and provides the system response as output. Each vector in the vector network is described on a data card which includes the element number, name and magnitude, as well as the numbers of the nodes between which the element is connected. The terminal equations for masses, springs and dashpots are permanently contained in the program and only the magnitude of these elements must be specified. Terminal equations for common drivers are included in the program, as well, although uncommon driving functions must be inserted in the program deck by writing a special subroutine.

The operation of the program in simulating the response of a dynamic system is illustrated in the flow-chart in Figure 6.1. The data cards describing the vector network are read and a sorting algorithm checks to ensure that a proper tree exists. An "incidence" matrix is created (composed of +1, -1 and zero entries) as a convenient method of recording the nodes upon which each and every vector is incident. Using a basic theorem from graph theory, the fundamental-cutset matrix can be obtained from the incidence matrix by a series of simple row operations, thus yielding the fundamental-cutset equations. A second basic theorem permits the fundamental-circuit matrix to be obtained by transposing a submatrix of the fundamental-cutset matrix, yielding the fundamental-circuit equations. Since the terminal equations for common elements are permanently contained in the program (and terminal equations for other elements may be inserted) and since a proper tree exists, then the computer contains sufficient information for complete solution.

The values of all system variables at time zero are determined by substitution of the f-circuit, f-cutset and terminal equations. The values of these variables can be printed out. The x, y and z components of the accelerations of the mass elements are calculated from the (d'Alembert) forces associated with the masses using the mass terminal equations and integrated numerically, yielding the positions and velocities of the masses at the next time-step. The computation and print-out cycle can then be repeated, as shown in Figure 6.1, until the response has been determined over some given time interval. The output from the program consists of the force, velocity and displacement associated with each system element as a function of the independent variable, time.

Computer graphics subroutines were combined with the VECNET

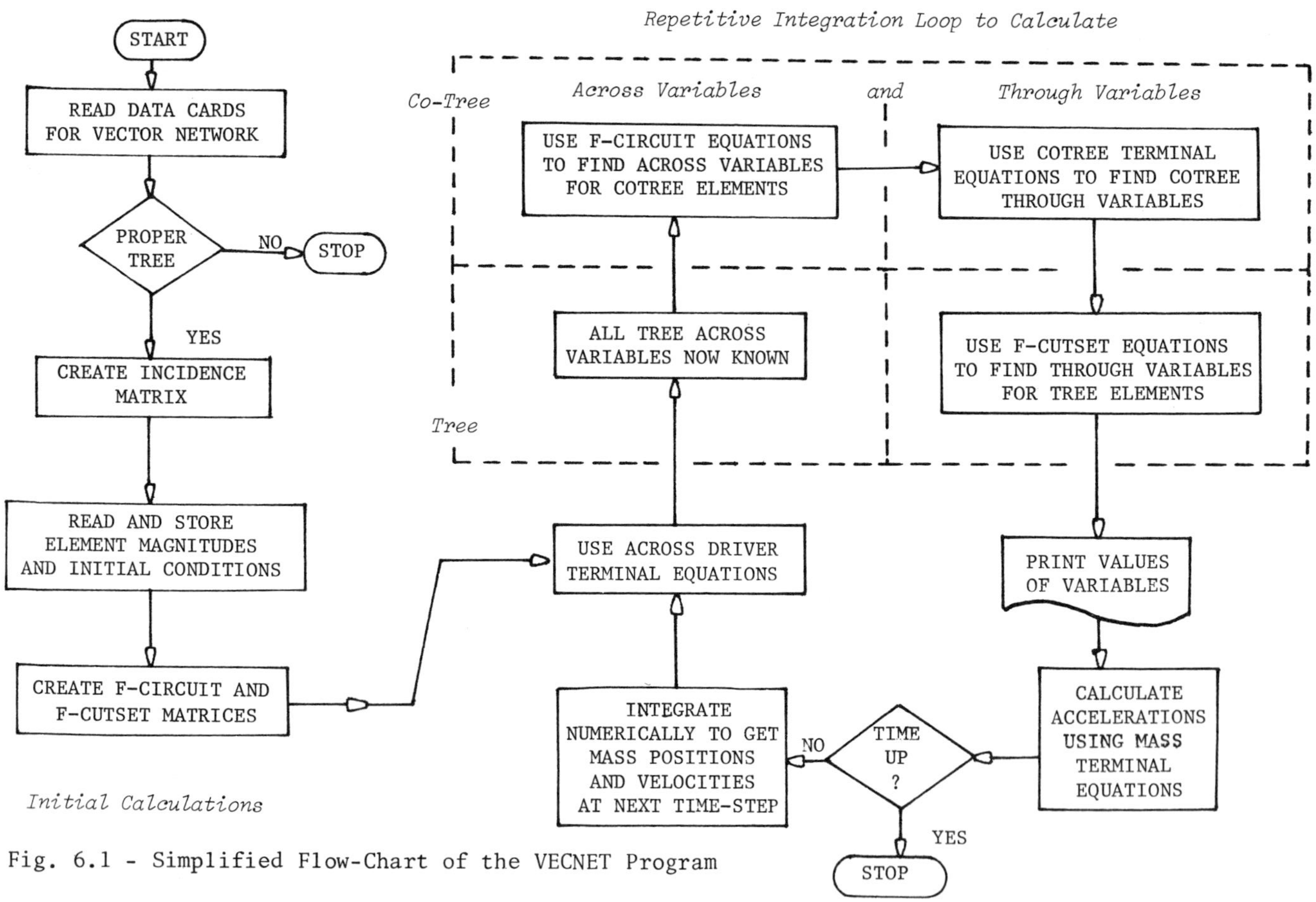

Fig. 6.1 - Simplified Flow-Chart of the VECNET Program

program so that the output of the program could be observed in "real time" on the CRT (cathode-ray-tube) display screen. The final display is a perspective picture in which all three directions of motion can be observed. (Motion perpendicular to the screen results in a change in symbol size.) Using the CRT output, it is possible to photograph the motion of dynamic systems, giving a permanent record of the dynamic response [17, 18]. An example of this type of output is shown in Figure 6.2, which is a photograph of the dynamic simulation of a simple system similar to the one shown in Figure 2.1. It is believed that computer programs based on this algorithm have considerable potential value in the design of dynamic mechanical systems.

7. THE DUAL RELATIONSHIP

7.1 Force and displacement vectors

For many years, polygons of force vectors have been used to illustrate force summations in mechanical systems (in fact, when the force vectors are drawn to scale, the polygon, itself, represents a force summation). However, an interesting new relationship becomes evident when vector-network methods are applied to the study of dynamic systems: the polygons of force vectors are observed to be "dual" vector-network diagrams. The term "dual" is used in the electrical sense, in which two networks are considered to be "duals" if the through-variables in one network are respectively proportional to the across-variables in a second network. In this case, there is a slight difference, since the original vector-network diagram is constructed in (x, y, z) displacement vector-space and the dual is constructed in (F_x, F_y, F_z) force vector-space.

The dual network differs from the original network in that the roles of through- and across-variables are reversed: in the dual

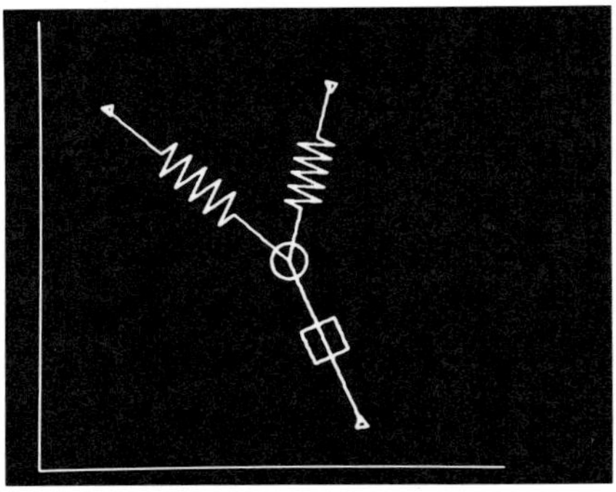

Fig. 6.2 - Photograph of Display Screen Output of
Dynamic Simulation

network, *forces* satisfy the *circuit* postulate and *displacements* satisfy the *vertex* postulate. Because of this reversal, several interesting observations are possible: first, the elements in the tree of the vector-network diagram form the co-tree of the dual; secondly, the f-cutset equations of the vector-network diagram become the f-circuit equations of the dual and, thirdly, the f-circuit equations of the vector-network diagram become the f-cutset equations of the dual.

7.2 Constructing the dual network

The simplest way to illustrate the relationship between the two networks is to use the graphical construction procedure, developed for electrical networks, which turns cutsets into circuits. This method depends directly on manipulating the vector-network diagram and does not require the f-circuit or f-cutset equations to be written [8]; it consists of mentally flattening the vector-network diagram to two dimensions such that no vectors intersect (except at nodes), placing a node in each circuit and joining the nodes with lines, such that one and only one line crosses each vector in the vector-network diagram. The result is a distorted form of the dual network diagram, but it has the correct topology. This procedure is illustrated in Figure 7.1 which shows the vector-network diagram for the simple system in Figure 2.1. Vector $\vec{r}_3$ has been distorted so that it does not intersect vector $\vec{r}_5$; nodes e, f, g, h have been inserted in each closed loop and node k has been added outside the network. The nodes are joined by dashed lines, such that one dashed line crosses each vector in the network. Each dashed line is assigned the same number as the vector that it crosses. The resulting network of dashed lines is obviously distorted in Figure 7.1, but it has the correct order of inter-connection. The dashed lines, when straightened and drawn to scale, would be similar to Figure 7.2 which shows four force polygons. (Three of the polygons contain only two vectors, so vectors $\vec{F}_2$, $\vec{F}_3$ and $\vec{F}_4$ are distorted for clarity.) It is obvious that Figure 7.2 shows all of the force relationships of interest in this simple system. Also, it will be observed that the f-cutset equations (5.15) for the vector-network diagram (Figure 2.2) of this system are the f-circuit equations for the dual network (Figure 7.2), if vectors $\vec{F}_5$, $\vec{F}_6$, $\vec{F}_7$, $\vec{F}_8$ are taken as the tree in the dual network. Similarly, the f-circuit equations (5.16) and (5.17) become the f-cutset equations for the dual.

7.3 Graphical solution procedure

It is evident that the vector-network diagram and its dual could be used to obtain, graphically, the values of all variables in a dynamic mechanical system [12, 14]. This includes the d'Alembert forces associated with mass elements which, in turn, yield the accelerations of the various masses. The graphical solution method would probably be practical only for two-dimen-

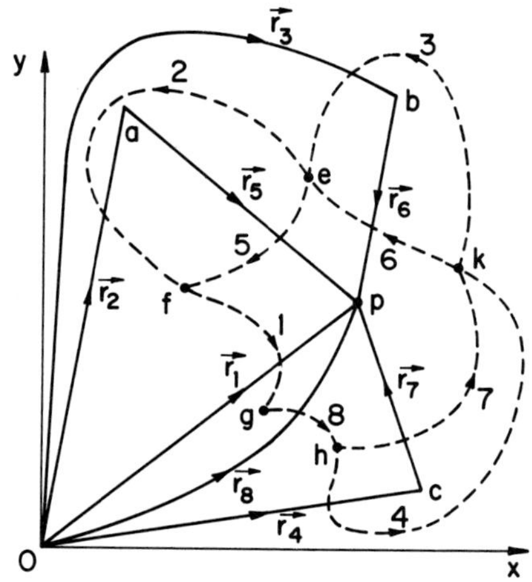

Fig. 7.1 - Graphical Construction of the Dual of the Network in Fig. 2.2

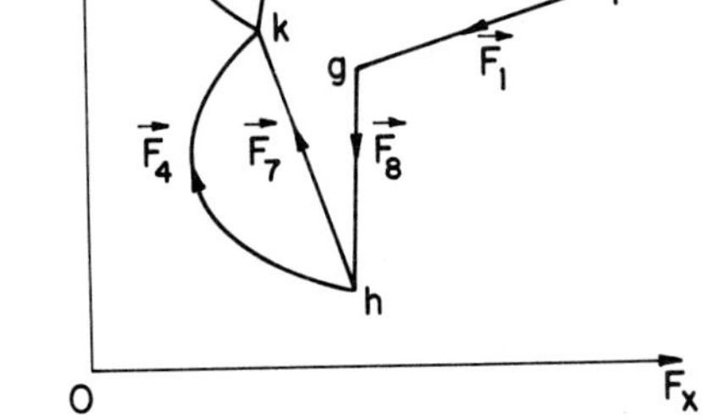

Fig. 7.2 - The Dual Network

sional systems, but the procedure is valid for three-dimensional systems if solid geometry techniques are used; it follows (in a rough fashion) the computer algorithm described earlier, except that the application of the f-circuit equations is done graphically (using the vector-network diagram) and the application of the f-cutset equations is done graphically (using the dual network). However, the graphical solution is tedious and approximate compared to computer techniques and is mentioned here mainly because of its academic interest. A graphical technique (used by engineers for solving static structures), known as Bow's notation, is virtually identical to the procedure described above for constructing the dual network and has been in common use for many years [22].

8. THE PRINCIPLE OF ORTHOGONALITY

In 1952, B.D.H. Tellegen described a new network theorem for electrical networks [19]. The theorem had been used in specific applications prior to that time, but Tellegen was the first to show the versatility and generality of the theorem [20]. Using the vector-network model, it is possible to extend Tellegen's electrical network theorem so that it applies to three-dimensional mechanical systems as well. In this context, the theorem, which is a consequence of the orthogonality of the f-circuit and f-cutset matrices, is recognized as a much more general form of the principle of virtual work [12, 21]. Consequently, the theorem is referred to as the "principle of orthogonality" to emphasize the general nature of its application.

8.1 Statement and derivation of the principle

The principle of orthogonality could be stated most generally as follows: "If the scalar products of the through and across-variables associated with each edge of a system graph are summed over all of the edges in the graph, then the sum will be zero".

More specifically, considering the vector-network model of a three-dimensional mechanical system, if the through and across-variables associated with each and every element are listed in the column matrices $[\vec{Y}]$ and $[\vec{X}]$ respectively (where it is assumed that the entries are in the same order, by element, in each column matrix), then the principle of orthogonality states that:

$$[\vec{Y}]^T \cdot [\vec{X}] = [\vec{X}]^T \cdot [\vec{Y}] = 0 \qquad (8.1)$$

In the equation, the general notation has been used, since $\vec{Y}$ represents force $\vec{F}$ or any of its time-derivatives and $\vec{X}$ represents displacement $\vec{r}$ or any of its time-derivatives, such as velocity $\vec{v}$ and acceleration $\vec{a}$.

The derivation of the principle depends entirely on a basic theorem of graph theory which proves that the f-circuit and f-cutset matrices are orthogonal: assuming that we construct the vector-network diagram for any system, select any tree in it and write the f-cutset and f-circuit equations for the network, we would get equations similar to (3.2) and (3.4) described previously, and, as a consequence of a basic theorem in graph theory, there is a relationship between sub-matrices [A] and [B], as shown earlier in equation (3.11) and repeated below:

$$[A] = -[B^T] \qquad \text{and} \qquad [B] = -[A^T] \qquad (8.2)$$

This relationship is essential in proving the principle of orthogonality. If the through-variables $[\vec{F}]$ and across-variables $[\vec{r}]$ for each element in the vector-network model are multiplied together and their products summed over all of the branches and chords in the graph, and if equations (3.2) and (3.4) are substituted, then we get:

$$[\vec{r}]^T \cdot [\vec{F}] = [\vec{r}_t^T \ \vec{r}_c^T] \cdot \begin{bmatrix} \vec{F}_t \\ \vec{F}_c \end{bmatrix} = [\vec{r}_t]^T \, [U_t \ - \ B^T] \cdot \begin{bmatrix} -A \\ U_c \end{bmatrix} \ [\vec{F}_c] = 0 \qquad (8.3)$$

or alternatively:

$$[\vec{F}]^T \cdot [\vec{r}] = [\vec{F}_t^T \; \vec{F}_c^T] \cdot \begin{bmatrix} \vec{r}_t \\ \vec{r}_c \end{bmatrix} = [\vec{F}_c]^T \; [-A^T \; U_c] \cdot \begin{bmatrix} U_t \\ -B \end{bmatrix} \; [\vec{r}_t] = 0 \tag{8.4}$$

In the above derivations, the last term in each of equations (8.3) and (8.4) can easily be shown to be zero by multiplying out the two central matrices in these terms and noting the results from equation (8.2). Therefore:

$$[\vec{r}]^T \cdot [\vec{F}] = [\vec{F}]^T \cdot [\vec{r}] = 0 \tag{8.5}$$

and the principle is proved, following the steps in the proof for scalar systems in the text by Koenig, Tokad and Kesavan [6].

The above result was known for voltages and currents in electrical networks before Tellegen. However, Tellegen demonstrated the *generality* of the result: since it depends only on topological considerations, then the through and across-variables may come from different networks (with the same topology) or from the same network at different times. Moreover, either the through or across-variables (or both) may be operated upon by a "linear operator" such as multiplication, differentiation, integration, etc., and the principle of orthogonality will still apply.

8.2 Orthogonality and virtual work

The principle of virtual work is the foundation for many of the most significant proofs in statics and dynamics. It appears that many of these derivations can be re-formulated using the principle of orthogonality as the basis, since the principle of virtual work can be shown to be a special case of the principle of orthogonality. To demonstrate, consider the vector-network diagram of any three-dimensional mechanical system and assume that the masses in the system undergo an infinitesimal (virtual) displacement which is consistent with the constraints. These displacements can be assembled into a vector network with the same topology (although not the same size) as the original network. The proof proceeds as follows: The f-circuit equations for any given vector-network diagram may be written as:

$$[B\ U_c] \begin{bmatrix} \vec{r}_t \\ \\ \vec{r}_c \end{bmatrix} = [0] \tag{8.6}$$

If the masses are now subject to a virtual displacement $\delta\vec{r}$, in harmony with the constraints, then the network will retain its order of inter-connection and the circuit equations may be written, after the displacement, as:

$$[B\ U_c] \begin{bmatrix} \vec{r}_t + \delta\vec{r}_t \\ \\ \vec{r}_c + \delta\vec{r}_c \end{bmatrix} = [0] \tag{8.7}$$

Subtracting equations (8.6) from equations (8.7) yields

$$[B\ U_c] \begin{bmatrix} \delta\vec{r}_t \\ \\ \delta\vec{r}_c \end{bmatrix} = [0] \tag{8.8}$$

which indicates that the virtual displacements satisfy the circuit equations. Therefore, these values could be used as the across-variables in a mathematical statement of the principle of orthogonality. Consequently, the principle of virtual work may be considered as a special case of the principle of orthogonality:

$$[\vec{F}_c^T\ \vec{F}_t^T] \cdot \begin{bmatrix} \delta\vec{r}_t \\ \\ \delta\vec{r}_c \end{bmatrix} = 0 \tag{8.9}$$

Although the units in the summation are units of work, it is obvious, in this context, that no work is involved. The principle of virtual work is simply a consequence of the orthogonality of the f-circuit and f-cutset matrices.

8.3 Quasi-work theorem

The principle of virtual work, in almost every application, implies that the forces and virtual displacements exist simultaneously in the same system. However, as the principle of orthogonality illustrates, these variables may exist in different systems or they may come from the same system at different times, so long as the vector-network diagrams of the systems have the same topology. To denote cases where the variables do not exist simultaneously in the same system, the prefix "quasi" is used in electrical networks [20] and seems appropriate for mechanical systems as well. The "quasi-work" theorem could be stated in the following form, as a special case of the principle of orthogonality: "If the vector-network diagrams of two systems have the same topology, then the scalar product of the through-variable associated with each edge in one diagram and the across-variable of the corresponding edge in the other diagram, when summed over all the edges, will total zero". It is assumed that the through and across-variables are chosen such that the scalar products mentioned above have the units of work, although it is possible to have units of power or other time-derivatives.

There are (at least) two well-known mechanical theorems which could be derived using the quasi-work theorem. These are the Method of the Unit Load and Rayleigh's Reciprocal Theorem, both of which apply to stationary systems.

8.4 Orthogonality and conservation of energy

The classical method of deriving the principle of conservation of energy consists of applying Newton's second law to a mass element and integrating the resulting equation with respect to displacements. The characteristics of the mass element are therefore introduced at the beginning of the derivation. However, the conservation of energy is independent of the types of element that are present in a system; it is, in fact, a consequence of the principle of orthogonality and can be derived easily using this principle. If equation (8.5) expressing the principle of orthogonality is written using force $\vec{F}$ as the through-variable and velocity v as the across-variable, then (8.5) becomes:

$$[\vec{F}]^T \cdot [\vec{v}] = [\vec{v}]^T \cdot [\vec{F}] = 0 \tag{8.10}$$

where it is permissible to use velocity as the across-variable, since it satisfies the f-circuit equations, as shown in equation (3.5) earlier.

The above equation (8.10) is immediately recognizable as a statement of the balance of instantaneous power: the power input to the system, added to the rates at which energy is stored or dissipated, must sum to zero. The integration of equation (8.10) over some specified time interval will yield the summation of the energy input, stored and dissipated by the system elements during that time interval. This summation will be zero since the integration of the right-hand side of equation (8.10) over a specified time interval will be zero. In order to convert this conclusion into a more familiar form, it is necessary to substitute the terminal equations for the various elements into equation (8.10) before integrating with respect to time. The result is an indirect statement of the conservation of energy and, significantly, it is obtained without specifying the types of elements that exist in the system.

8.5 A vector formulation similar to Lagrange's equation

The derivation of Lagrange's equation involves integrations that yield kinetic and potential energy terms. In use, the energy terms for a given system are substituted into Lagrange's equation and then differentiated to yield the equations of motion of the system. However, it is possible to obtain equations of motion directly from the vector-network model that are identical to those obtained from Lagrange's equation. The procedure is similar to the Lagrangian techniques: just as the principle of virtual work forms the basis for the development of Lagrange's equations, the principle of orthogonality is the basis for the formulation using the vector-network model. However, the use of the vector-network model eliminates the need for kinetic and potential energy terms. It should, perhaps, be added that this new formulation is simple in concept but practical applications are equal (or greater) in difficulty to Lagrangian formulation and this new procedure is not recommended for practical problems, at present.

Consider any translational system for which the vector-network model has been constructed and a proper tree identified. Select a set of generalized coordinates q_h such that there is one coordinate for each degree of freedom in the system; then, as is assumed in the Lagrangian method, it is possible to express the position vector for each particle mass as a function of the generalized coordinates q_h. Since any linear operation on an across-variable yields another acceptable across-variable, consider the partial derivative of the displacements $\vec{r}$ with respect to q_h. The resulting variable may be substituted into the orthogonality equation (8.5) to yield:

$$[\vec{F}]^T \cdot \left[\frac{\delta \vec{r}}{\delta q_h}\right] = 0 \qquad (8.11)$$

Performing the partial differentiation indicated in the above equation with respect to each of the generalized coordinates, in turn, will yield the complete set of equations of motion for the masses in the system.

To show that equation (8.11) does, indeed, yield equations of motion equivalent to Lagrange's equations, it may be observed that equation (8.11) can be obtained using traditional Lagrangian techniques if all forces (including those associated with the masses and other conservative elements) are treated as generalized forces. If this is done, Lagrange's equation reduces to the (usually trivial) case where the sum of the generalized forces is zero.

8.6 Example problem

The formulation procedure similar to Lagrange's equation, described in the preceding section, can be illustrated using the simple system shown earlier in Figure 2.1. The vector-network diagram is shown in Figure 2.2 and the f-circuit equations (5.16) and terminal equations (5.7) to (5.14) have been previously derived. In this simple problem, there are two degrees of freedom; select x_1, y_1, the coordinates of the mass, to be the generalized coordinates. The mass position vector can then be expressed:

$$\vec{r}_1 = x_1\hat{i} + y_1\hat{j} \tag{8.12}$$

Substituting the f-circuit equations (5.15) into the orthogonality equation (8.10) yields:

$$[\vec{F}_1\ \vec{F}_2\ \vec{F}_3\ \vec{F}_4\ \vec{F}_5\ \vec{F}_6\ \vec{F}_7\ \vec{F}_8]\cdot\frac{\partial}{\partial q_h}\begin{bmatrix}\vec{r}_1\\ \vec{r}_2\\ \vec{r}_3\\ \vec{r}_4\\ \vec{r}_1-\vec{r}_2\\ \vec{r}_1-\vec{r}_3\\ \vec{r}_1-\vec{r}_4\\ \vec{r}_1\end{bmatrix} = 0 \tag{8.13}$$

Then, performing the partial differentiation with respect to x_1 we get:

$$[\vec{F}_1\ \vec{F}_2\ \vec{F}_3\ \vec{F}_4\ \vec{F}_5\ \vec{F}_6\ \vec{F}_7\ \vec{F}_8]\cdot\begin{bmatrix}\hat{i}\\ 0\\ 0\\ 0\\ \hat{i}\\ \hat{i}\\ \hat{i}\\ \hat{i}\end{bmatrix} = 0 \tag{8.13}$$

which simplifies to

$$(\vec{F}_1 + \vec{F}_5 + \vec{F}_6 + \vec{F}_7 + \vec{F}_8)\cdot(\hat{i}) = 0 \tag{8.14}$$

If the terminal equations (5.7) and (5.11) to (5.14) are substituted, we get:

$$\{- m\dot{\vec{v}}_1 - k_5(|\vec{r}_5| - s_5)\hat{r}_5 - k_6(|\vec{r}_6| - s_6)\hat{r}_6 - d(\vec{v}_7\cdot\hat{r}_7)\hat{r}_7 - W\hat{j}\}\cdot(\hat{i})=0 \tag{8.14}$$

A certain similarity can be observed between the above equation (8.14) and the previously-derived equation (5.19). When the scalar multiplication by unit vector $(\hat{i})$ is carried out, the result will be identical to equation (5.21).

Similarly, partial differentiation of equation (8.13) with respect to y_1 will yield:

$$(\vec{F}_1 + \vec{F}_5 + \vec{F}_6 + \vec{F}_7 + \vec{F}_8)\cdot(\hat{j}) = 0 \tag{8.15}$$

When the terminal equations (5.7) and (5.11) to (5.14) are substituted and the scalar multiplication is performed, the result will be identical to previously-derived equation (5.22). Consequently, it can be seen that, for this simple example, the technique yields the same scalar equations of motion (5.21) and (5.22) as derived previously.

9. CONCLUSION

This paper has shown a new way to look at traditional concepts of dynamics. All of the basic concepts established by Newton in 1687 are included in this reorganized format; the result does not detract from traditional concepts, but a few new ideas are added:

(1) It is possible to tabulate all of the equations available for solution of a given system and show that there is a necessary and sufficient set.

(2) It is possible to write a "self-formulating" computer program, using an algorithm based on this technique, which simulates the response of three-dimensional dynamic mechanical systems, given only the system description as input.

(3) The dual nature of displacement vectors used in the vector-network diagram and force vectors used in force polygons is illustrated.

(4) The derivation of a new and basic concept, the principle of orthogonality, is shown. This principle is very general, since many traditional concepts, such as the principle of virtual work and the instantaneous balance of power are seen to be special cases of the principle of orthogonality.

The concepts in this paper which show the greatest potential for future development are, in the opinion of the author, the computer algorithm and the principle of orthogonality.

The paper has omitted any mention of rotational elements (masses with significant rotational inertia) or the development of terminal equations for more complex elements, such as kinematic constraints. These problems are discussed in other publications [12, 13, 15].

ACKNOWLEDGEMENTS

The author should like to acknowledge, with thanks, the contribution of Dr. H.K. Kesavan whose inspiration led to the initiation of this research. The help of G.J. Savage and G.D. Miller in the development of the computer graphics output of the VECNET program is also gratefully acknowledged and the careful typing of Mrs. Linda McIntosh, in the preparation of the manuscript, is very much appreciated. A portion of this research was supported by grant numbers C1283 and A8419 of the National Research Council of Canada.

REFERENCES

1. Seshu, S. and Reed, M.B., *Linear Graphs and Electrical Networks*, Addison-Wesley, 1961.
2. Busacker, R.G. and Saaty, T.L., *Finite Graphs and Networks*, McGraw-Hill, 1965.
3. Trent, H.M., "Isomorphisms between Oriented Linear Graphs and Lumped Physical Systems", *Journal of the Acoustical Society of America*, Vol. 27, No. 3, pp. 500-527, 1955.
4. Koenig, H.E. and Blackwell, W.S., "Linear Graph Theory - a Fundamental Engineering Discipline", *IRE Transaction on Education*, Vol. E3, pp. 42-49, 1960.
5. Koenig, H.E. and Blackwell, W.S., *Electromechanical System Theory*, McGraw-Hill, 1961.
6. Koenig, H.E., Tokad, Y. and Kesavan, H.K., *Analysis of Discrete Physical Systems*, McGraw-Hill, 1967.
7. Roe, P.H., *Networks and Systems*, Addison-Wesley, 1966.
8. Shearer, J.L., Murphy, A.T., and Richardson, H.H., *Introduction to Systems Dynamics*, Addison-Wesley, 1967.
9. Sanford, R.S., *Physical Networks*, Prentice-Hall, 1965.
10. Blackwell, W.A., *Mathematical Modelling of Physical Networks*, MacMillan, 1968.
11. Martens, H.R. and Allen, D.R., *Introduction to System Theory*, Merill, 1969.
12. Andrews, G.C., *The Vector-Network Model - A Topological Approach to Mechanics*, Ph.D. Thesis, University of Waterloo, 1971.
13. Andrews, G.C. and Kesavan, H.K., "The Vector-Network Model: A New Approach to Vector Dynamics", *Mechanism and Machine Theory*, Vol. 10, pp. 57-75, 1975.
14. Andrews, G.C., *Dynamics Using Vector-Network Techniques*, Course Lecture Notes, Dept. of Mech. Eng., University of Waterloo, Waterloo, Ontario, Canada N2L 3G1.

15. Rogers, R.J. and Andrews, G.C., "Simulating Planar Systems Using a Simplified Vector-Network Method", *Mechanism and Machine Theory*, Vol. 10, pp. 509-19, 1975.
16. Andrews, G.C. and Kesavan, H.K., "VECNET: A Simulation Program for 3-D Mechanical Systems", *Proceedings* (abstract only) 3rd Canadian Congress of Applied Mechanics (CANCAM), University of Calgary, Calgary, Canada, 1971.
17. Andrews, G.C. and Savage, G.J., *Simulation of Dynamic Mechanical Systems Using the Vector Network*, 16 mm/colour/sound motion-picture film, University of Waterloo, 1972. Distributed by Canadian Film Institute, Ottawa, Canada.
18. Savage, G.J. and Andrews, G.C., "Visual Simulation of Dynamic Three-Dimensional Systems Using the Vector Network", *Simulation*, Vol. 19, No. 6, pp. 187-191, 1972.
19. Tellegen, B.D.H., "A General Network Theorem, with Applications", *Philips Research Reports*, Vol. 7, No. 4, pp. 259-269, Aug. 1952.
20. Penfield, P., Spence, R., and Duinker, S., *Tellegen's Theorem and Electrical Networks*, Research Monograph 58, M.I.T. Press, 1970.
21. Andrews, G.C. and Kesavan, H.K., "The Principle of Orthogonality: A More General Form of the Principle of Virtual Work" *Proceedings*, (abstract only) 4th Canadian Congress of Applied Mechanics (CANCAM), Ecole Polytechnic, Montreal, 1973.
22. Maxwell, R.L., *Kinematics and Dynamics of Machinery*, Prentice-Hall, 1960.

THE NETWORK CONCEPT AS A UNIFYING PRINCIPLE IN ENGINEERING AND THE PHYSICAL SCIENCES

FRANKLIN H. BRANIN, JR.
System Communications Division Laboratory
IBM Corporation
Kingston, New York

SUMMARY

The concept of a network of interconnected elements pervades many areas of both engineering and the physical sciences. Basic to this concept are the existence of: (1) a pair of functionally related *complementary* or *conjugate* variables that characterize each individual network element and (2) *topological constraints* imposed on these conjugate variables by the very act of interconnecting the network elements. The underlying mathematical description of the network problem belongs to *algebraic topology* which explains clearly why network analogies are so ubiquitous. An understanding of the fundamentals of the network problem and its extension to the *3-network*, in which surface and volume elements are involved, provides a unified basis for analyzing networks of all kinds - whether electrical, mechanical, structural, molecular, acoustic, hydraulic, or thermal, to mention a few.

1. INTRODUCTION

Many physical phenomena can be described by means of a network of interconnected elements, each of which is characterized by a pair of functionally related *complementary* or *conjugate* variables. The properties of each network element are described by *constitutive relations* involving these conjugate variables. In electrical networks, *current* and *voltage* are the appropriate variables whereas *force* and *velocity* are the conjugate variables for mechanical, structural and molecular networks. In hydraulic and acoustic systems, the complementary variables are *fluid flow* and *pressure* while in thermal systems they are *heat flow (or entropy flow)* and *temperature*.

Though physically different, these variables can always be classified in two distinct ways:

(1) *through* variables such as current, force, fluid flow and heat (or entropy) flow -- so-called because they correspond to physical quantities passing or acting *through* a network element;

(2) *across* variables such as voltage, velocity, pressure difference and temperature difference -- so-called because they correspond to physical quantities appearing or acting *across* the network element.

In all of these network problems, the constitutive relations involving through and across variables are mathematically similar in that they are linear or nonlinear algebraic or differential equations. In the case of mechanical, structural, and molecular networks, however, the force and velocity variables may be *vector* quantities while in all the other cases cited they are *scalars*.

Now the very act of *interconnecting* the network elements in all of these instances imposes *topological constraints* of two distinct types on the through and across variables: the through variables, whether vector or scalar quantities, must always sum to zero at every *point* of interconnection whereas the across variables must sum to zero around every *closed path* or *loop* of the network. Accordingly, these topological constraints give rise to strictly linear algebraic equations and so whatever nonlinearities exist must be contributed by the constitutive relations.

The formal mathematical treatment of the constitutive and constraint relations has been most highly developed in connection with electrical networks. But there is nothing inherently or exclusively *electrical* about the network problem as a mathematical entity. Therefore, all phenomena to which the network concept applies properly come within the purview of *network analysis*.

Basically, network analysis is concerned with three major items:

(1) the characterization of individual network elements by means of appropriate constitutive relations;

(2) the determination of all the pertinent topological constraint equations imposed by the network connections;

(3) the solution of the resulting system of equations by analytic and/or numerical methods.

The first of these three items is primarily discipline-dependent, requiring some detailed understanding of the physical variables involved. The second item is also somewhat discipline-dependent but can be dealt with in terms of topological matrices that are obtainable from a *linear graph* depicting the network connections. These matrices may be used in several different (but equivalent) ways to express all of the necessary topological constraints.

Now the *formulation* aspect of network analysis consists solely of deriving all of the necessary constitutive and constraint equations; once this has been accomplished, the formulation process is complete and the *solution* task must be undertaken. Thus the formulation process in analyzing <u>any</u> network problem is trivially simple in concept, although it may involve a sizable bookkeeping chore. Unfortunately, but for historically justifiable reasons, the line of demarcation between the formulation and solution phases of network analysis has become somewhat blurred.

Part of our objective, therefore, will be to put these two aspects of network analysis into better perspective. This is particularly important in view of the great impact on the subject that has resulted from the development of computer programs for solving network problems automatically.

Before the advent of computers, when numerical solutions had to be done manually or with digital calculators, it was quite natural to attack the solution of any network problem by first applying whatever algebraic or other *symbolic* manipulations were available to minimize the number of equations. This understandable tendency led to the development of several different classical methods of formulation that actually involve the solution process of symbolic Gauss elimination to reduce the number of equations describing the network. Those formulations resulting in the fewest equations for a given problem were considered as being in some sense the "best."

However, with the recent emphasis on exploiting the sparseness properties of the network equations, it is now widely recognized that symbolic Gauss elimination, if carried out without due regard to its influence on sparsity, can lead to a serious loss of efficiency in solving the network equations on a computer. This is just as relevant in treating mechanical, structural, or other such networks as it is for electrical. Indeed, the paper by Orlandea and Calahan in this book particularly demonstrates this point.

Taking the viewpoint that formulation consists solely of deriving all the constitutive and constraint equations while solution includes all subsequent manipulations, symbolic and/or numeric, we will show the basic simplicity as well as the interrelations of several different methods of formulation. This treatment will show that the formulation process is well understood. It will also help to emphasize the fact that the solution process is where we collectively need the most help. For unfortunately, the great facility that we achieve in formulation by using topological matrices does not appear to be of any real significance in providing leverage for obtaining *numerical* solutions.

Basically, our solution processes for linear and nonlinear algebraic/differential equations rely very heavily on clever programming techniques (e.g.sparse matrix computations), high speed "number crunching" computers, extensive arithmetic and algebra, and considerable calculus. To be sure, very commendable progress is being made so that we can now handle routinely networks involving more than 1000 equations. But we really need to face up to the challenging question, "What shall we do for an encore?"

In other words, must we continue to rely on relatively rudimentary mathematics or is there some really powerful mathematical principle that we can envision, adapt, rediscover, or invent to streamline the solution process? Can we somehow take advantage of of the underlying *structure* of the network problem to improve the efficiency of our solution technique?

One possible ray of hope regarding this last question comes from the recognition that the equations for highly repetitive network problems, such as those resulting from the numerical solution of partial differential equations, can be formulated in such a way that the corresponding coefficient matrix will also display this repetitivity. The coefficient matrix takes the form of a *bordered block diagonal* matrix in which the diagonal blocks may be identical submatrices. [1,2,3] This obvious repetitivity can, of course, be exploited during LU factorization [4,5] but whether or not any overall computational advantage can be gained remains to be shown.

Before we can expect to make real progress in applying higher mathematics to the solution of network problems, we naturally must understand their underlying mathematical properties quite clearly. For this purpose, Roth's algebraic-topological characterization of the network problem [6] is very helpful because it shows precisely why and where the network concept applies. Roth treats only the *1-network* problem, however, which is based on a *linear graph* consisting of interconnected points and lines.

But a much broader class of network problems exists in which are involved not only points and lines (0- and 1-dimensional objects) but also *surface* and *volume* elements (2- and 3-dimensional objects) all topologically interconnected and comprising what we may call a *3-network*. The author [7,8] has extended Roth's method of characterization to the 3-network problem with rather interesting results. In particular, by introducing the concept of a *dual 3-network*, the author was able to show the direct and intimate relationship that exists between network theory -- in this more expanded sense -- and the *vector calculus*. Moreover, this relationship justifies the use of network models for two large classes of partial differential equations that include, among many others, *Maxwell's equations* for the electromagnetic field. In a very significant paper elsewhere in this book, Tonti has greatly expanded on this idea and shown its remarkable compass.

In this paper, we shall begin by reviewing the fundamental laws of *1-networks* in matrix form and showing the similarity of their structure in several different physical situations. We will explain the many basic ways of formulating the 1-network equations in *tableau* form [9,10] and how they relate to the classical formulations; we will also discuss some of the important computational considerations to be taken into account in solving these network equations. We will then present Roth's abstract characterization of the 1-network problem. Finally, we will describe the 3-network problem, its formal solution, and its relation to the vector calculus and Maxwell's equations. In this way, we seek to display in some detail the tremendous scope of the network concept and its application as a unifying principle in both theory and practice to the analysis of problems in several fields of engineering and the physical sciences. Unfortunately, this approach does not result in any breakthroughs in solution technique. But it does clearly define the many areas of application affected by whatever computational and/or theoretical advances are made in the future.

2. THE 1-NETWORK LAWS IN MATRIX FORM

Mathematically interpreted, the network problem has two major ingredients: (1) an underlying *topological* structure and (2) an associated *algebraic* structure. These ingredients are intimately (often unrecognizedly) blended with the *physics* of the network problem in which the interrelations between the physical variables current and voltage, force and velocity, fluid flow and pressure, heat flow (or entropy flow) and temperature, etc., are usually of primary interest. Important as these physical quantities are, however, it is only the underlying mathematical structure of the network problem that can provide any real unity of approach amidst a variety of applications. This structure we must identify.

The expression of the network laws in matrix-vector form is a powerful aid in clarifying the mathematical character of the network problem as well as in making possible the development of computer programs that automatically compile and solve the network equations. Indeed, there has been a very fruitful symbiosis between the computer and the matrix approach to network theory in the last two decades, the results of which are strongly reflected in the presentation to follow.

We will deal first with the constitutive laws relating the through and across variables in electrical, mechanical, hydraulic, and thermal systems, treating all four of these cases in an analgous manner so as to show their essential identity of structure. Similarly, we will show on the basis of linear graph theory how the topological constraints in all of these cases can be handled on a common basis using certain elementary topological matrices.

Many other instances of the 1-network problem can be cited, as shown by the work of Smerage and especially by that of Tonti reported elsewhere in this book. But the four examples we have chosen suffice to show that the basic pattern of the 1-network laws and the methods of analysis are largely independent of the field of application.

2.1 *THE CONSTITUTIVE LAWS*

In electrical, mechanical, hydraulic, and thermal networks, the *fundamental* across variables are voltage, velocity, pressure (difference) and temperature (difference) while the *fundamental* through variables are current, force, fluid flow and heat flow (or entropy flow). In all these cases, with but one exception (*i.e.* thermal networks) there exist physically meaningful *integral* and *differential* cognates of these fundamental through and across variables.

For example, in electrical networks, the integral of voltage is *flux* while the integral of current is *charge*. In mechanical systems, the integral of velocity is *displacement* while its differential form is *acceleration;* the integral of force, of course, is *impulse*. In hydraulic systems, the integral of pressure is *pressure impulse* while the integral of fluid flow is *fluid mass* or *fluid volume*. Finally, in thermal systems, the integral of heat flow (or entropy flow) is *heat content* (or *entropy*); but apparently, *there exists no recognized counterpart of the integral of temperature*. (More about this later.)

To exhibit the similarity of the constitutive laws in these four different fields, we make use of the diagrams shown in Figs. 1 to 4. In each case, the (integral, fundamental, and differentilal) through variables appear across the top of the diagram while the (differential, fundamental, and integral) across variables appear across the bottom.

In electrical systems, the current and voltage are related by the physical quantities *resistance* R and *conductance* G according to the equations

$$V = RI \tag{1}$$

and

$$I = GV \tag{2}$$

If we regard Eqs. (1) and (2) as matrix-vector expressions, then they may be used to describe in this very concise form *all* the constitutive relations for any resistive network of arbitrary size and configuration. [11,12] Thus, the symbols I and V are *vectors* whose individual components represent the currents through and voltages across each of the resistive elements in the network.

Usually, the R (or G) *matrix* will be diagonal, as dictated by the physics of the network. But mathematically, off-diagonal elements are quite permissible and may be used effectively in modeling physical systems where couplings exist between branches of the network whether or not they are topologically connected. Indeed, the constitutive laws are always taken to be valid quite *independently* of whatever constraints may be imposed on the network variables by the connections.

Now the integral through variable *charge* Q is related to the voltage by the physical quantity *capacitance* C as follows

$$Q = CV \tag{3}$$

Similarly, the integral across variable *flux* Φ is related to the current by the physical quantity *inductance* L thus,

$$\Phi = LI \tag{4}$$

In the case of constant capacitance and/or inductance, differentiation of Eqs. (3) and (4) yields the relations

$$I = \dot{Q} = C\dot{V} \tag{5}$$

and

$$V = \dot{\Phi} = L\dot{I} \tag{6}$$

Once again, the symbols C and L may represent matrices with C being diagonal but with L frequently having off-diagonal terms to represent mutual inductances.

All of the relationships in Eqs. (1) to (6) can be summarized by the *transformation diagram* shown in Fig. 1. Here each of the arrows represents a transformation of one variable into another by the action of a *premultiplying* algebraic, integral, or differential operator. (The transformations expressed by Eqs. (5) and

(6) are represented by broken lines to indicate that they are restricted to the case of constant C and constant L.

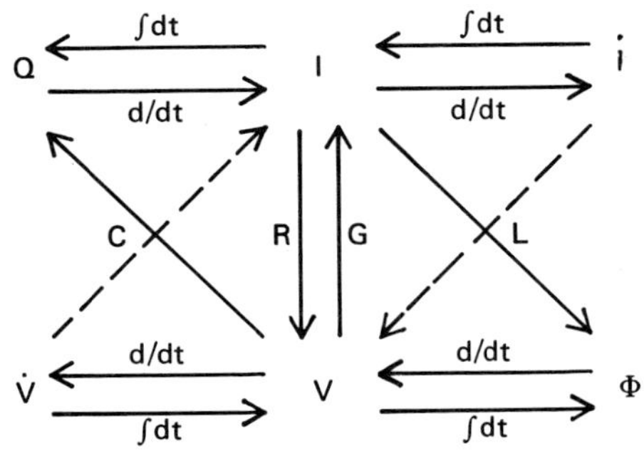

Fig. 1 - Constitutive Relations for Electrical Network Elements

For nonlinear resistive elements, Eqs. (1) and (2) are replaced by the expressions

$$V = R(I)I \tag{7}$$

and

$$I = G(V)V \tag{8}$$

or, in most general form by

$$f(V,I) = 0 \tag{9}$$

Similarly, nonlinear capacitances and inductances require equations of the form

$$Q = C(V)V \tag{10}$$

and

$$\Phi = L(I)I \tag{11}$$

Finally, voltage and current *sources*, which may be either constant or time-dependent, are defined thus:

$$V_E = E(t) \tag{12}$$

and

$$I_J = J(t) \tag{13}$$

where the functions E(t) and J(t) are known beforehand.

In the case of mechanical systems, the through and across variables force and velocity are interrelated by the equations

$$F = RV \tag{14}$$

and

$$V = R^{-1}F \tag{15}$$

where R is the *mechanical resistance* and R^{-1} is its inverse. Here we note that mechanical resistance is analagous to electrical *conductance*. This curious discrepancy appears to be the result of the historical habit of regarding both force and voltage as *causes* while velocity and current are viewed as *effects*. Mathematically, of course, we can have both voltage and current sources in electrical systems with currents and voltages -- elsewhere in the system -- being the responses, or effects. Similarly, in mechanical systems, we can have both force and velocity *sources*, although the latter are perhaps as unusual today as current sources used to be.

In any event, it is quite clear that irrespective of the opposite characters of mechanical and electrical resistance, force and current must be regarded as cognates since they are both through variables while velocity and voltage are cognates since they are both across variables. It follows, therefore, that the proper analog of Eq. (3) is

$$I = MV \tag{16}$$

where I is the *impulse* and M is the mass. By the same token, the proper analog of Eq. (4) is

$$D = SF \tag{17}$$

where D is the displacement and S the compliance (reciprocal stiffness) of a spring. Differentiating Eqs. (16) and (17) and assuming constant mass and constant spring stiffness, we have

$$F = M\dot{V} \tag{18}$$

which is familar as Newton's second law, and

$$V = S\dot{F} \tag{19}$$

which is seldom if ever used in mechanics.

These relations are all summarized in Fig. 2 which corresponds directly to Fig. 1.

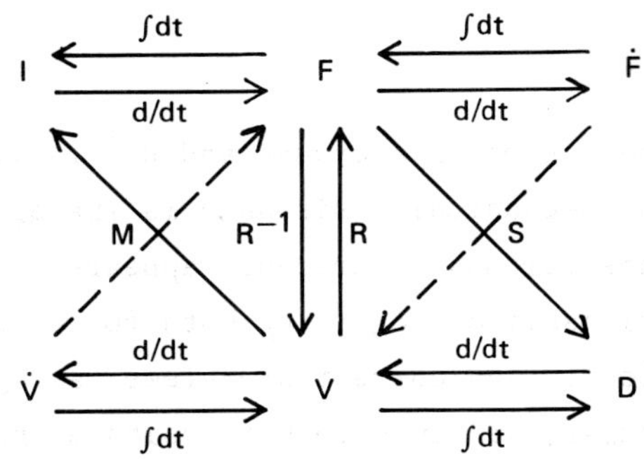

Fig. 2 - Constitutive Relations for Mechanical Network Elements

One important distinction between electrical and mechanical systems is that in the former, the through and across variables are *scalar* quantities whereas in the latter they may be *vector* quantities as well. Indeed, each force and velocity variable (and their related integral and differential variables) may be a vector with up to six components -- three for translational and three for rotational degrees of freedom. In this case, R, M, and S are properly regarded as *tensors*.

Another distinction is that electrical systems seldom if ever require explicit solutions for the flux variable Φ. But mechanical systems frequently require that the displacement be computed explicitly because the *geometry* of the interconnected parts may have a significant effect in determining the system equations. Indeed, the role of geometry in mechanics has no counterpart in electrical systems, even though both systems obey the same *topological* laws.

From the topological nature of through and across variables, it is evident that the only acceptable analogy between mechanical and electrical systems is the *mass-capacitance* analogy indicated by Figs. 1 and 2. This analogy, first proposed by Firestone [13], was preceded by the *mass-inductance* analogy which even today is widely used. But the latter analogy is fundamentally erroneous because it (unwittingly, no doubt) interchanged the roles of through and across variables. This interchange is permissible if and only if the graph of the system is planar or can be mapped on a sphere -- in which case a *dual* graph exists wherein the roles of through and across variables are actually interchanged. But when the graph does not have a dual -- and there are mechanical systems for which this is indeed the case [14] -- the mass-inductance analogy breaks down completely. The mass-capacitance analogy, on the other hand, always works. Accordingly, we regard the mass-inductance analogy as an unfortunate historical mistake and fully concur with Firestone's concluding remark: "The author nominates for oblivion the conventional left-handed mechanical-electrical analogy."

One final point which the correspondence between Figs. 1 and 2 makes clear: Eq. (16) shows that the equality of impulse I and *momentum* MV is a consequence of a *constitutive* law -- no more, no less. Furthermore, since impulse rather than momentum is a through variable, it is *impulse* that is "conserved" -- as a consequence of topology -- rather than *momentum*. Thus, the familiar law of conservation of momentum is really the result of a topological constraint (on impulse) *and* a constitutive law (relating impulse to momentum).

By a straightforward extension of the foregoing treatment, we arrive at the diagrammatic representation of the constitutive relations for hydraulic and/or acoustic systems as shown in Fig. 3. Here, the fluid flowrate F (in units of either mass/unit time or volume/unit time) is regarded as the through variable and pressure (difference) P as the across variable. Accordingly, the flowrate and pressure are related by the equations

$$P = RF \tag{20}$$

and

$$F = GP \tag{21}$$

where R is the fluid resistance and G its reciprocal. The integrated through variable Q, (representing either fluid mass or fluid volume) is related to the pressure difference by

$$Q = CP \quad (22)$$

where C is the capacitance, measured in appropriate units. Finally, the integrated across variable, *pressure impulse* Γ, is related to the flowrate by

$$\Gamma = IF \quad (23)$$

where I is the *fluid inertance*. [15]

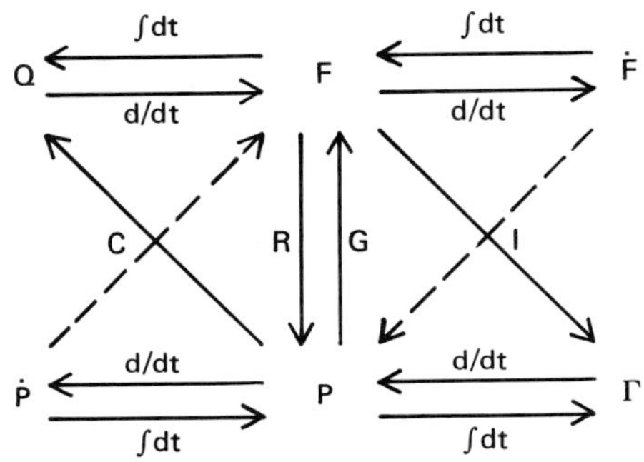

Fig. 3 - Constitutive Relations for Hydraulic or Acoustic Systems

The constitutive diagram for thermal systems, shown in Fig. 4, is easily obtained. Here, heat flowrate $\dot{Q}$ is the primary through variable and temperature (difference) T the primary across variable related by

$$\dot{Q} = GT \quad (24)$$

and

$$T = R\dot{Q} \quad (25)$$

where G is the thermal conductance and R the thermal resistance. The integrated through variable Q is the *heat content* related to temperature by

$$Q = CT \quad (26)$$

where C is the thermal capacity.

Strangely enough, there appears to be no recognizable *integrated* across variable in thermal systems (what is the meaning of the time integral of temperature?) and so no analog of thermal inductance. Even so, as pointed out by Shearer, Murphy and Richardson [16], the phenomenon of "second sound" in liquid helium at very low temperature, where heat flows with wavelike motion, suggests strongly that there must exist some form of thermal inductance. For in all other systems, -- electrical, mechanical, hydraulic, acoustic -- wave motion or oscillation can not occur without both capacitance-like *and* inductance-like elements. This observation provides food for thought.

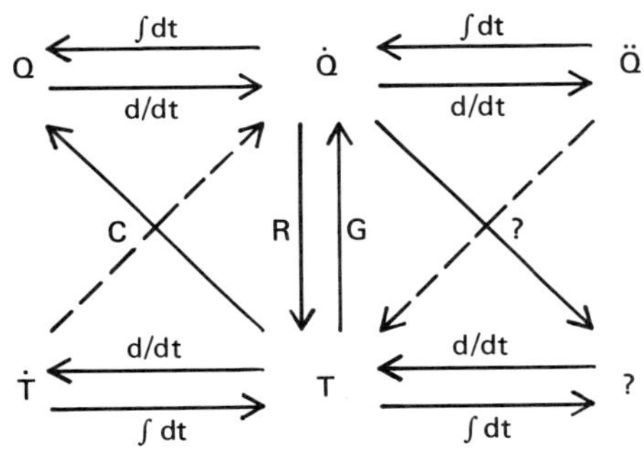

Fig. 4 - Constitutive Relations for Thermal Systems

Before concluding this section, it is worth mentioning that Chua [17] has recently discovered what he calls the *memristor*, a missing circuit element. From Fig. 1 it is obvious that although there are recognized electrical network elements relating Q and V, I and V, and I and Φ, there is none relating Q and Φ directly. Chua postulated the existence of such an element, defined in general by the relation

$$M(Q,\Phi) = 0 \tag{27}$$

and called it the memristor because it involves both charge and flux, both of which -- being *integrated* variables -- imply a dependence on history or memory.

Chua has actually found uses for the concept of the memristor in device modeling [18] and Oster [19] has discussed three physical devices that exemplify the properties of a memristor: two of them are electrolytic cells and one a tapered dashpot. At present, however, the memristor remains largely a mathematical curiosity.

2.2 LINEAR GRAPH THEORY AND TOPOLOGICAL MATRICES

The most efficient way of describing the *topological constraints* on the network variables is by means of four topological matrices: A, B_T, C, and D. [12] These matrices, in turn, are readily defined in terms of linear graph theory which treats the connectivity properties of systems of interconnected points and lines. Actually, only a few rudimentary concepts from graph theory suffice to provide all the formal apparatus needed to define and characterize the topological matrices used in network analysis. We may summarize the gist of this apparatus in the following way.

One important graph theoretical concept is that of a *tree* which is any graph that contains no closed paths or *loops*. Generally, the tree of a graph means a tree that contains all the nodes of the graph; at least one such tree can always be found. Relative to any given tree, each of the remaining branches of the graph is called a *link* and the collection of all links comprises the *co-tree* or *tree-complement*. Fig. 5a shows a typical linear graph -- with one possible choice of tree depicted by solid lines in Fig. 5b, the links being drawn as broken lines.

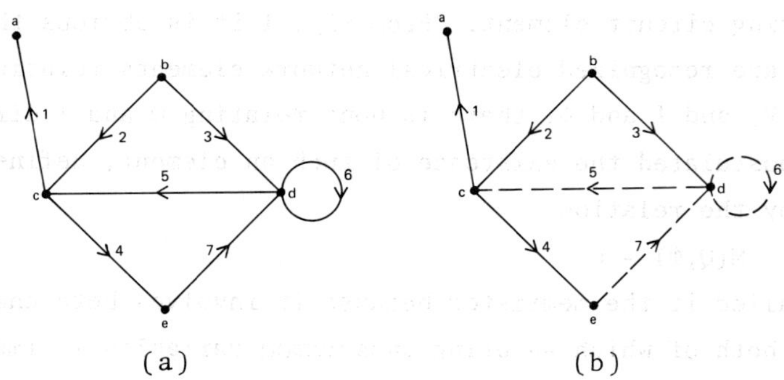

Fig. 5 - Linear Graph with Typical Tree-Link Assignment

Another useful concept is that of a *cutset*; this is any set of branches which, when removed from the graph (by cutting) separates the graph into two or more disconnected parts. A simple way to define any cutset is by means of a closed surface that passes through the graph so as to cut any number of branches (just *once*) without cutting any of the nodes. The set of branches cut by this *dissecting surface* constitutes a cutset. When these branches are removed, the remaining parts of the graph inside and outside the surface are obviously disconnected. Fig. 6 shows two such dissecting surfaces defining the cutsets [2,5,7] and [3,5,7].

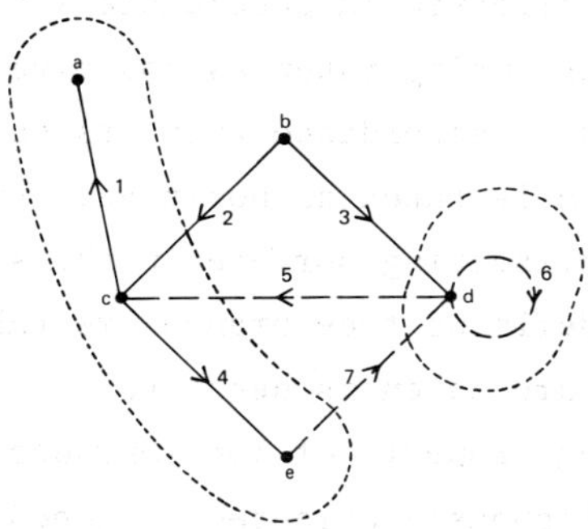

Fig. 6 - Linear Graph Showing Two F-Cutsets

Now once a particular tree is chosen in a graph, the closed path or loop formed by adding any single link to the tree is called a *fundamental* loop or f-loop. Similarly, if any dissecting surface passes through the graph so that it cuts only *one* tree branch (and any number of links), the corresponding cutset is called a *fundamental* cutset or f-cutset. Clearly, since the number of the f-cutsets in a graph is equal to the number of tree branches while the number of f-loops is equal to the number of links, the total of these two quantities must be equal to the number of branches in the graph.

One other concept that plays a central role in the graph theory relating to networks is that of the *loop-cutset*; this is any set of branches that belong *simultaneously* to a loop and to a cutset. For instance, in Fig. 6 the loop-cutset corresponding to the f-loop [2,3,5] and the f-cutset [2,5,7] consists of the two branches [2,5]. The number of branches in a loop-cutset is always *even*; this is because a closed curve and a closed surface can intersect eachother only in an even number of points.

In any graph, the nodes, branches, loops, and cutsets may all be *oriented*. For example, branch 4 is oriented *from* node c *to* node e; thus node c is taken to be *positively* oriented with respect to branch 4 while node e is *negatively* oriented with respect to this branch.

Each f-loop is oriented in accordance with its corresponding link, this link always being taken as positive. Similarly, each f-cutset is oriented in accordance with its corresponding tree branch -- which again is taken as positive. Branches in a cutset that penetrate the dissecting surface in the same (inward or outward) sense as the defining tree branch are taken as positive while those opposed are taken as negative.

A distinguishing property of *loop-cutsets*, however, is that they cannot be unambiguously oriented. Indeed, this lack of orientability is of fundamental significance. To illustrate this property, we need to define one final concept: the *neighboring pair*. This is defined as any two branches in a loop-cutset that are encountered *in sequence* when the loop is traversed in either sense.

What we may properly call *the fundamental lemma* of graph theory relating to network analysis may now be stated as follows:

> If both the members of a neighboring pair of branches in a loop-cutset are oriented *alike* with respect to the loop, they must orient *oppositely* with respect to the cutset -- and conversely.

This lemma is central to the proof of certain basic properties relating the topological matrices described below.

The four topological matrices A, B_T, C, and D used to characterize the connectivity properties of any linear graph show how the branches are connected to the nodes (A matrix), included in the node-to-datum paths of a tree (B_T matrix), included in the loops

(C matrix) and cutsets (D matrix). Formally, the elements of these matrices may be defined thus:

$$\begin{Bmatrix} a_{ij} \\ b_{ij} \\ c_{ij} \\ d_{ij} \end{Bmatrix} = [+1,-1,0] \text{ if the i-th branch is [pos,neg,not]} \begin{Bmatrix} \text{incident on} \\ \text{included in} \\ \text{included in} \\ \text{included in} \end{Bmatrix} \text{the j-th} \begin{Bmatrix} \text{node} \\ \text{node-datum path} \\ \text{loop} \\ \text{cutset} \end{Bmatrix}$$

For the graph of Fig. 5 the branch-node incidence matrix A is

	a	b	c	d	e
1	-1	0	1	0	0
2	0	1	-1	0	0
3	0	1	0	-1	0
4	0	0	1	0	-1
5	0	0	-1	1	0
6	0	0	0	0	0
7	0	0	0	-1	1

A matrix

(Here, each row shows which nodes are connected to -- incident on -- each branch while each column shows which branches are incident on each node. Row 6 is null because branch 6 is both positively and negatively incident on node d.)

Since the sum of elements in each row of the incidence matrix is zero, the columns are linearly dependent -- that is, any column is the negative sum of all the others and so may be deleted without loss of information. The column selected for deletion, say e in the present case, corresponds to the *datum* or reference (or ground) node.

After deletion of the datum column, the number of columns remaining is always equal to the number of tree branches. If, then, we arrange the rows so that those corresponding to the branches of the tree precede those corresponding to the links, the upper partition of the matrix, A_T, is always square and invertible. Indeed, the inverse of A_T turns out to be directly related to the node-to-datum path matrix, B_T. [20]

For the tree shown in Fig. 5b, the B_T matrix is

	a	b	c	d
1	-1	0	0	0
2	0	1	0	1
3	0	0	0	-1
4	1	1	1	1

B_T matrix

(Here, the rows show which node-to-datum paths include each branch while the columns show which branches constitute each node-to-datum path.)

For any tree, it can be shown that

$$A_T^{-1} = B_T^t \tag{28}$$

The proof, which is most easily demonstrated by showing that $B_T^t A_T = U_T$ where U_T is a unit matrix, depends on two basic ideas. First is the fact that only the first branch of any node-to-datum path can be incident on the initial node of the path -- and if positively [negatively] incident thereon, then positively [negatively] included in this path. (This accounts for the diagonal terms of U_T, as the reader can easily prove for himself.) Second is the fundamental lemma of the neighboring pair which accounts for the off-diagonal terms.

Although the branch-loop (or branch-mesh) matrix C may be defined no matter what loops (or meshes) are chosen, it is usual to choose the f-loops corresponding to a given tree. This choice is simplest because it allows the C matrix to be derived explicitly from the A and B_T matrices. [20]

Using the tree-link structure of Fig. 5b, the f-loop matrix C turns out to be

	5	6	7
1	0	0	0
2	-1	0	1
3	1	0	-1
4	0	0	1
5	1	0	0
6	0	1	0
7	0	0	1

C matrix

Using the same tree-link structure, the f-cutset matrix D, which can be derived directly from the foregoing C matrix, becomes

	1	2	3	4
1	1	0	0	0
2	0	1	0	0
3	0	0	1	0
4	0	0	0	1
5	0	1	-1	0
6	0	0	0	0
7	0	-1	1	-1

D matrix

In the C and D matrices, the rows show which loops or cutsets contain each branch while the columns show which branches constitute each loop or cutset. Moreover, because of the convention that each f-loop contains its corresponding link with positive orientation, the link-portion of the C matrix is a unit matrix. For analogous reasons, the tree-portion of the D matrix is also a unit matrix. Accordingly, the *tree*-portion C_T of the f-loop matrix and the *link*-portion D_L of the f-cutset matrix contain the real essence of these two matrices.

Now for *any* loop matrix -- whether referring to f-loops or not -- and for *any* cutset matrix -- whether referring to f-cutsets or not -- the following remarkable relation holds true:

$$C^t D = 0 \tag{29}$$

The reason for this property is easy to identify since when any row of C^t (corresponding to a column of C) is multiplied by any column of D, the *matching nonzero terms* correspond to just those branches that belong to a loop-cutset. What is more, the number of resulting nonzero products is always *even*, with equal numbers of them being positive and negative -- this by virtue of the fundamental lemma of the neighboring pairs. Hence, the sum of these products is guaranteed to vanish and Eq. (29) follows directly.

Since the branches incident on a node correspond to what we may call a *nodal-cutset*, the A matrix can be regarded as a special case of the D matrix. Accordingly, analogous to Eq. (29) we have the additional relation

$$C^t A = 0 \tag{30}$$

If we apply this to the case of the f-loop matrix and partition the A matrix according to the same tree-link structure, it is easy to show, using Eq. (28), that

$$C_T = -B_T A_L^t \tag{31}$$

Finally, for the case of the f-loop and f-cutset matrices corresponding to any choice of tree, Eq. (29) implies that

$$D_L = -C_T^t \tag{32}$$

These five relations between A, B_T, C, and D comprise the essence of the topological apparatus need for network analysis.

The principal use -- indeed, the *only* use -- of these topological matrices in network analysis is to express the constraints imposed on the through and across variables by the very act of interconnecting the elements of any network. To illustrate how succinctly these matrices may used for this purpose, we will confine our discussion to the electrical case with the understanding, of course, that analagous procedures can be used for mechanical, structural, hydraulic, acoustic, thermal, and other such network problems.

To simplify our discussion, we will adopt the *composite branch* convention wherein each branch of the network is generalized by allowing it to contain both a *voltage source* E and a *current source* J as well as a resistance, conductance, capacitance, or inductance element. The composite branch is depicted in Fig. 7.

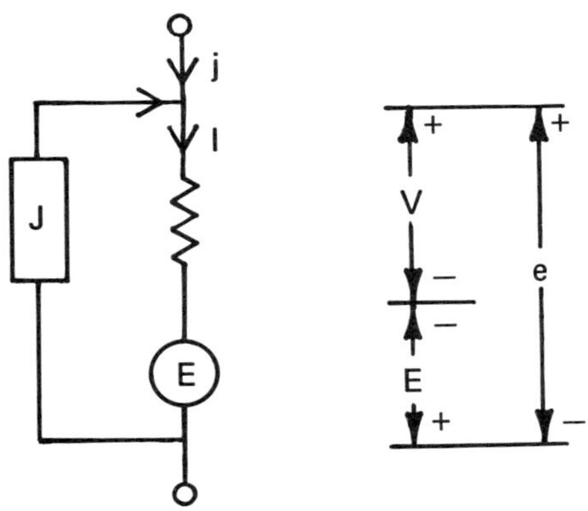

Fig. 7 - Composite Branch Convention

Here, the *element voltage* V and the *element current* I are taken across and through the R, G, C, or L element itself while the *branch voltage* e and the *branch current* j are taken across and through the entire branch. Thus, the variables e and j are what the rest of the network sees at the terminals of the branch.

With the polarities (orientations) as shown the relations

$$V = E + e \tag{33}$$

and

$$I = J + j \tag{34}$$

follow as consequences of the Kirchhoff voltage and current laws applied within the composite branch.

Now in any (electrical) network, the branch voltages can be expressed as a linear combination of the node-to-datum voltages according to Kirchhoff's voltage law. If we let e represent the vector of all branch voltages and e' the vector of all node (to datum) voltages, then the voltage law can be written with the aid of the branch-node incidence matrix A as follows:

$$e = A\ e' \tag{35}$$

Thus, the A matrix provides precisely the correct combination of node voltages. Another way of expressing the branch voltages is in terms of a linear combination of tree-branch voltages in which case the cutset matrix D may be used thus:

$$e = D\ e_T \tag{36}$$

Here, the vector e_T represents the tree-branch voltages.

Finally, an alternative (and mathematically equivalent) way of describing the voltage law is to state that the sum of the branch voltages around every loop must be zero. This can be expressed in terms of the loop matrix C in the following way:

$$C^t\ e = 0 \tag{37}$$

The C matrix here may refer to any set of loops or meshes.

The cognate forms of Kirchhoff's current law may easily be written using these matrices in a somewhat different manner. For example, the statement that the branch currents are a linear combination of the loop currents may be expressed thus:

$$j = C\, j_L \tag{38}$$

where the C matrix provides the appropriate combination of link currents j_L. Alternatively, if we choose any independent set of meshes and represent the mesh currents by the vector j', we may also write

$$j = \overline{C}\, j' \tag{39}$$

where $\overline{C}$ is the corresponding branch-mesh incidence matrix.

Finally, since the sum of currents in any cutset must be zero, we have

$$D^t\, j = 0 \tag{40}$$

or, in terms of the *nodal cutsets*

$$A^t\, j = 0 \tag{41}$$

In summary, Eqs. (35) to (37) are *equivalent* forms of the voltage law while Eqs. (38) to (41) are *equivalent* forms of the current law. These topological constraint relations have their direct counterparts in all network problems where across and through variables exist and in all such cases, the matrices A, C, and D provide the simplest way of defining these contraints precisely.

Just as there are three (or four) different but equivalent ways of stating each of the two main types of topological constraint, so there are three general ways of formulating the constitutive relations. First, using the vectors V and I to represent all the element voltages and element currents in a network and the matrix Z to represent the R, G, C, and/or L elements in *impedance* form (that is as R, G^{-1}, dL/dt, or $\int C^{-1}dt$), we can write

$$V = Z\ I \tag{42}$$

Alternatively, we may use the matrix Y to represent the R, C, G, and L elements in *admittance* form (that is as R^{-1}, G, $\int L^{-1}dt$, or dC/dt) and write

$$I = Y\ V \tag{43}$$

Finally, if we regard some of the elements as admittances (with subscript y) and the others as impedances (with subscript z), we may use the *hybrid* form of the constitutive relations, namely

$$\begin{bmatrix} I_y \\ V_z \end{bmatrix} = \begin{bmatrix} Y_y & \\ & Z_z \end{bmatrix} \begin{bmatrix} V_y \\ I_z \end{bmatrix} \tag{44}$$

Again, Eqs. (42) to (44) are equivalent forms of the constitutive relations.

In formulating the *network equations*, it is necessary to include any one of the three forms of the constitutive law, any one of the three forms of the across variable constraint law and any one of the four forms of the through variable constraint law -- and this is true for every kind of network problem involving these three basic laws. Accordingly, there are 3 x 3 x 4 = 36 *different* ways of writing the network equations, all of them equivalent. In each case, moreover, the number of variables always matches the number of equations exactly. In substance, then, the *formulation* task in network analysis is trivially simple in concept since it consists solely of writing the three basic network laws as described above. However, much effort has been expended in the development of numerous methods of "formulation" which actually involve part of the *solution* process as well. This will be discussed more fully in the next section.

3. THE 1-NETWORK PROBLEM

In this section, we will refer specifically to the *1-network problem* (instead of the "network problem") since we will later be discussing its higher dimensional analog, the *3-network problem*. Basically, the 1-network problem can be stated formally as follows in electrical terminology -- with appropriate changes for other applications:

> Given a network of known topology, known elements C, G, R, and/or L, and known voltage and current sources E and J, find all the response voltages and currents e and j such that the constitutive laws and topological constraint laws are satisfied simultaneously.

A somewhat different version of this 1-network problem has been formulated and solved by Roth [6,7], but this need not be discussed here.

3.1 *FORMULATION OF THE NETWORK EQUATIONS*

As specified in the formal statement of the 1-network problem just given, the constitutive and topological constraint laws must be satisfied *simultaneously*. Accordingly, any of the many equivalent ways of stating these laws in matrix form gives rise to a set of *simultaneous equations* that "describe the network." The task of formulating the network equations, then, is quite simple in principle. However, there are so many variations on the central theme, that it may be desirable to explain some of the more important methods of formulation. In all cases, we will use the composite branch convention described earlier, but with the understanding that our treatment can be readily extended to the simple branch convention as well. [9,10]

Suppose first that the *impedance* form of the constitutive law

$$V = Z\ I \tag{42}$$

is used to describe all the elements of a network. By combining Eqs. (33) and (34) with Eq. (42), we can write

$$e - Z\, j = -(E - Z\, J) \tag{45}$$

where the unknown variables e and j are collected on the left. Then, using Eqs. (37) and (39), we can express the topological constraints in terms of the C matrix alone as follows:

$$j - C\, j' = 0 \tag{46}$$

and

$$C^t e = 0 \tag{47}$$

Collecting Eqs. (45) -(47) together, we can write these equations in "tableau" form as follows:

$$\begin{bmatrix} U & -Z & 0 \\ 0 & U & -C \\ C^t & 0 & 0 \end{bmatrix} \begin{bmatrix} e \\ j \\ j' \end{bmatrix} = \begin{bmatrix} -(E - ZJ) \\ 0 \\ 0 \end{bmatrix} \tag{48}$$

and our formulation task in complete. Alternatively, we may use the *admittance* form of the constitutive law and express the topological constraints in terms of the A matrix alone. In this case, Eq. (48) would be replaced by

$$\begin{bmatrix} U & -Y & 0 \\ 0 & U & -A \\ A^t & 0 & 0 \end{bmatrix} \begin{bmatrix} j \\ e \\ e' \end{bmatrix} = \begin{bmatrix} -(J - YE) \\ 0 \\ 0 \end{bmatrix} \tag{49}$$

In both of these cases, we have introduced extra variables -- j' or e' -- in stating the topological constraints. But we can avoid this by defining a specific tree and cotree so that Eq. (38) is used in place of Eq. (39) and Eq. (36) in place of Eq. (35).

Then, if we partition the Z and Y matrices and the E and J vectors according to the tree-branches and links, we can write the network equations either in the form

$$\begin{bmatrix} U_T & 0 & -Z_T & 0 \\ 0 & U_L & 0 & -Z_L \\ 0 & 0 & U_T & -C_T \\ C_T^t & U_L & 0 & 0 \end{bmatrix} \begin{bmatrix} e_T \\ e_L \\ j_T \\ j_L \end{bmatrix} = \begin{bmatrix} -(E_T - Z_T J_T) \\ -(E_L - Z_L J_L) \\ 0 \\ 0 \end{bmatrix} \tag{50}$$

or in the cognate form

$$\begin{bmatrix} U_L & 0 & -Y_L & 0 \\ 0 & U_T & 0 & -Y_T \\ 0 & 0 & U_L & -D_L \\ D_L^t & U_T & 0 & 0 \end{bmatrix} \begin{bmatrix} j_L \\ j_T \\ e_L \\ e_T \end{bmatrix} = \begin{bmatrix} -(J_L - Y_L E_L) \\ -(J_T - Y_T E_T) \\ 0 \\ 0 \end{bmatrix} \tag{51}$$

Continuing in this manner, we can choose any other combination of Eqs. (35) through (44) that completely specifies all three network laws. Three combinations are of special interest because they use the *hybrid* form of the constitutive law. Since the hybrid form permits some branches to be viewed as admittances and the rest as impedances, it is the most general (hence the *preferred*) form of the constitutive law. In particular, for networks containing both C's and L's, the hybrid form must be used if *differential* rather than integral or integro-differential equations are desired.

The simplest of these three formulations is the *hybrid nodal* approach [21] in which Eq.(44) is used for the constitutive law while the A matrix alone is used to express the topological constraints. If we partition the vectors in Eqs. (33) and (34) appropriately, we can combine these equations with Eq. (44) to obtain a hybrid analogy to Eq. (45) having the form

$$\begin{bmatrix} j_y - Y_y e_y \\ e_z - Z_z j_z \end{bmatrix} = \begin{bmatrix} -(J_y - Y_y E_y) \\ -(E_z - Z_z J_z) \end{bmatrix} \tag{52}$$

Eqs. (35) and (41), correspondingly partitioned, become

$$\begin{bmatrix} e_y \\ e_z \end{bmatrix} - \begin{bmatrix} A_y \\ A_z \end{bmatrix} e' = 0 \tag{53}$$

and

$$\begin{bmatrix} A_y^t & A_z^t \end{bmatrix} \begin{bmatrix} j_y \\ j_z \end{bmatrix} = 0 \tag{54}$$

where, as in Eq. (49), we have introduced the node voltages e' as auxiliary variables. The tableau form of Eqs. (52)-(54) can now be written as follows [2]

$$\begin{bmatrix} U_y & 0 & -Y_y & 0 & 0 \\ 0 & U_z & 0 & 0 & -Z_z \\ 0 & 0 & U_y & -A_y & 0 \\ A_y^t & 0 & 0 & 0 & A_z^t \\ 0 & -U_z & 0 & A_z & 0 \end{bmatrix} \begin{bmatrix} j_y \\ e_z \\ e_y \\ e' \\ j_z \end{bmatrix} = \begin{bmatrix} -(J_y - Y_y E_y) \\ -(E_z - Z_z J_z) \\ 0 \\ 0 \\ 0 \end{bmatrix} \tag{55}$$

Again, to avoid using node voltages, we can choose a tree and cotree -- with admittance branches taken first and impedance branches taken last in the tree-link sorting process. As a result of this ordering, the path-in-tree corresponding to each y-link will contain only y-tree branches whereas the path-in-tree corresponding to each z-link may contain both y- and z-tree branches. Consequently, the tree-portion of the f-loop matrix, namely C_T, takes the special form

$$C_T = \begin{bmatrix} C_{Tyy} & C_{Tyz} \\ 0 & C_{Tzz} \end{bmatrix} \tag{56}$$

where the null submatrix indicates the absence of any paths-in-z-tree corresponding to y-links.

Using this special form of C_T and partitioning the tree-branches and links according to y- and z-types, we can write the topological constraint law $j_T = C_T j_L$ as follows [9,10]

$$\begin{bmatrix} j_{Ty} \\ j_{Tz} \end{bmatrix} = \begin{bmatrix} C_{Tyy} & C_{Tyz} \\ 0 & C_{Tzz} \end{bmatrix} \begin{bmatrix} j_{Ly} \\ j_{Lz} \end{bmatrix} \tag{57}$$

while the cognate relation $e_L = D_L e_T = -C_T^t e_T$ -- using Eq. (32) -- becomes

$$\begin{bmatrix} e_{Ly} \\ e_{Lz} \end{bmatrix} = \begin{bmatrix} -C_{Tyy}^t & 0 \\ -C_{Tyz}^t & -C_{Tzz}^t \end{bmatrix} \begin{bmatrix} e_{Ty} \\ e_{Ty} \end{bmatrix} \tag{58}$$

Finally, by partitioning Eq. (44) according to y- and z-type tree-branches and links and combining the result with Eqs. (57) and (58), we can write the *hybrid tableau* equations in the form

$$\begin{bmatrix} U_{Lz} & 0 & 0 & 0 & 0 & 0 & 0 & -Z_L \\ 0 & U_{Ty} & 0 & 0 & 0 & 0 & -Y_T & 0 \\ 0 & 0 & U_{Tz} & 0 & 0 & -Z_T & 0 & 0 \\ 0 & 0 & 0 & U_{Ly} & -Y_L & 0 & 0 & 0 \\ 0 & 0 & 0 & 0 & U_{Ly} & 0 & C_{Tyy}^t & 0 \\ 0 & 0 & 0 & 0 & 0 & U_{Tz} & 0 & -C_{Tzz} \\ 0 & U_{Ty} & 0 & -C_{Tyy} & 0 & 0 & 0 & -C_{Tyz} \\ U_{Lz} & 0 & C_{Tzz}^t & 0 & 0 & 0 & C_{Tyz}^t & 0 \end{bmatrix} \begin{bmatrix} e_{Lz} \\ j_{Ty} \\ e_{Tz} \\ j_{Ly} \\ e_{Ly} \\ j_{Tz} \\ e_{Ty} \\ j_{Lz} \end{bmatrix} = \begin{bmatrix} -(E-ZJ)_{Lz} \\ -(J-YE)_{Ty} \\ -(E-ZJ)_{Tz} \\ -(J-YE)_{Ly} \\ 0 \\ 0 \\ 0 \\ 0 \end{bmatrix} \tag{59}$$

where the right hand side has been abbreviated in an obvious way.

An alternative form of Eq. (59) may be obtained by rearranging rows and columns so as to produce the expression

$$\begin{bmatrix} U_{Ty} & 0 & 0 & 0 & 0 & 0 & -C_{Tyy} & -C_{Tyz} \\ 0 & U_{Tz} & 0 & 0 & 0 & 0 & 0 & -C_{Tzz} \\ 0 & 0 & U_{Ly} & 0 & C^t_{Tyy} & 0 & 0 & 0 \\ 0 & 0 & 0 & U_{Lz} & C^t_{Tyz} & C^t_{Tzz} & 0 & 0 \\ U_{Ty} & 0 & 0 & 0 & -Y_T & 0 & 0 & 0 \\ 0 & -Z_T & 0 & 0 & 0 & U_{Tz} & 0 & 0 \\ 0 & 0 & -Y_L & 0 & 0 & 0 & U_{Ly} & 0 \\ 0 & 0 & 0 & U_{Lz} & 0 & 0 & 0 & -Z_L \end{bmatrix} \begin{bmatrix} j_{Ty} \\ j_{Tz} \\ e_{Ly} \\ e_{Lz} \\ e_{Ty} \\ e_{Tz} \\ j_{Ly} \\ j_{Lz} \end{bmatrix} = \begin{bmatrix} 0 \\ 0 \\ 0 \\ 0 \\ -(J-YE)_{Ty} \\ -(E-ZJ)_{Tz} \\ -(J-YE)_{Ly} \\ -(E-ZJ)_{Lz} \end{bmatrix} \quad (60)$$

Both of these forms of the hybrid equations will be referred to in the next section.

The third combination that uses the hybrid form of the constitutive law is the so-called *state variable* formulation in which the objective is to derive a minimal set of first order differential equations in the form $\dot{x} = A\,x + f(t)$ to describe the network. [22] Although we will not describe the method in detail, it is based on an elaboration of the hybrid equations just given. [See Appendix I of Ref. 20]

In particular, the state variable approach uses the simple branch convention in which each voltage and current source is regarded as an individual branch. What is more, the y-type and z-type branches are further subdivided into C- and G-types and R- and L-types, respectively. Then, before a tree-link sorting process is carried out, the branch types are placed in the order E, C, G, R, L, J; this insures that the earlier branches are given priority in the tree and the later ones priority in the cotree.

In this way, any *E-links* and *J-tree branches* are identified so that they can be discarded as inadmissible. [23] Moreover, the maximum number of C-type tree branches and L-type links will also be identified. [24] The state variables themselves, then, are the *C-type tree branch voltages* and the *L-type link currents* -- and these comprise the *differential* variables symbolized below by x_1. In addition, the voltages across and currents through all the G- and R-type branches constitute the set of *algebraic* variables symbolized by x_2. The net result of this phase of the state variable formulation is a set of algebraic and differential equations of the general form

$$P \dot{x}_1 + Q_{11}x_1 + Q_{12}x_2 = f_1(t) \tag{61a}$$

$$Q_{21}x_1 + Q_{22}x_2 = f_2(t) \tag{61b}$$

These equations can be reduced to pure differential equations by solving Eq. (61b) for x_2, the algebraic variables, and substituting the result in Eq. (61a), thus:

$$P \dot{x}_1 + [Q_{11} - Q_{12} Q_{22}^{-1} Q_{21}] x_1 = f_1(t) - Q_{12} Q_{22}^{-1} f_2(t) \tag{62}$$

Finally, the canonical form $\dot{x} = A x + f(t)$ can be obtained by inverting the P matrix and noting that

$$A = -P^{-1}[Q_{11} - Q_{12} Q_{22}^{-1} Q_{21}] \tag{63}$$

and

$$f = P^{-1}[f_1 - Q_{12} Q_{22}^{-1} f_2] \tag{64}$$

Once again, it should be remarked that although all of the preceding discussion on formulation has been couched in electrical terminology, it applies with equal force to all 1-network problems, no matter what their physical significance may be. This point will be more fully appreciated in light of the discussion in the next section on solution of the network equations, particularly by sparse matrix techniques.

3.2 *SOLUTION OF THE NETWORK EQUATIONS*

Traditionally, network analysis methods have been motivated by the effort to minimize the number of equations that "describe the network." This motivation is quite in keeping with the necessity to reduce the burden of numerical solution by essentially *manual* methods. But with the advent of digital computers, coupled with sparse matrix techniques, [4,5] this motivation requires careful reexamination; it may be disastrous when applied indiscriminately to large network problems where matrix sparsity is of primary concern.

In order to emphasize this point, we have purposely chosen to regard the *solution phase* of network analysis as commencing immediately after the equations expressing the three fundamental network laws have been written, using the methods just described. On this basis, all the classical "formulations" are seen to include a sizable component of the solution process in the form of *symbolic Gauss elimination*.

For example, the *mesh method* of analysis, although starting with Eqs. (45)-(47), continues its "formulation" process by eliminating the variables e and j, yielding the *mesh equations* for the network in the form

$$C^tZC\,j' = C^t(E - ZJ) \tag{65}$$

To show that this process is formally equivalent to symbolic Gauss elimination, we may refer to the tableau in Eq. (48). Here, as in subsequent examples, the coefficient matrix has been ordered so as to make it as nearly upper triangular as possible. Hence, to complete the triangularization process symbolically, we first zero out the C^t submatrix by taking the appropriate linear combination of the first and third rows. The new third row produced by this operation is:

$$\begin{bmatrix} \vdots & & \vdots \\ 0 & C^tZ & 0 \end{bmatrix}\begin{bmatrix} \vdots \\ j' \end{bmatrix} = \begin{bmatrix} \vdots \\ C^t(E - ZJ) \end{bmatrix} \tag{66}$$

Combining this with the second row, we can eliminate C^tZ to yield

$$\begin{bmatrix} \vdots & & \vdots \\ 0 & 0 & C^tZC \end{bmatrix}\begin{bmatrix} \vdots \\ j' \end{bmatrix} = \begin{bmatrix} \vdots \\ C^t(E - ZJ) \end{bmatrix} \tag{67}$$

which is identical to Eq. (65).

In the same manner, Eq. (49) can be symbolically reduced to the *node equations*,

$$A^tYA\ e' = A^t(J - YE) \tag{68}$$

Eq. (50) to the *f-loop equations*,

$$(C_T^tZ_TC_T + Z_L)j_L = C^tZC\ j_L = C^t(E - ZJ) \tag{69}$$

and Eq. (51) to the *f-cutset equations*,

$$(D_L^tY_LD_L + Y_T)e_T = D^tYD\ e_T = D^t(J - YE) \tag{70}$$

In the case of Eq. (55), the symbolic elimination of j_y, e_z, and e_y can be carried out yielding the *hybrid nodal equations*,

$$\begin{bmatrix} A_y^tY_yA_y & A_z^t \\ A_z & -Z_z \end{bmatrix}\begin{bmatrix} e' \\ j_z \end{bmatrix} = \begin{bmatrix} A_y^t(J_y-Y_yE_y) \\ (E_z-Z_zJ_z) \end{bmatrix} \tag{71}$$

For Eq. (59), symbolic reduction leads to the *hybrid equations*,

$$\begin{bmatrix} D_y^tY_yD_y & -C_{Tyz} \\ C_{Tyz}^t & C_z^tZ_zC_z \end{bmatrix}\begin{bmatrix} e_{Ty} \\ j_{Lz} \end{bmatrix} = \begin{bmatrix} D_y^t(J_y-Y_yE_y) \\ C_z^t(E_z-Z_zJ_z) \end{bmatrix} \tag{72}$$

where we have made the abbreviations

$$D_y = \begin{bmatrix} U_{Ty} \\ -C^t_{Tyy} \end{bmatrix} \tag{73}$$

and

$$C_z = \begin{bmatrix} C_{Tzz} \\ U_{Lz} \end{bmatrix} \tag{74}$$

Finally, if we abbreviate Eq. (60) as follows:

$$\begin{bmatrix} U & -T \\ H_1 & H_2 \end{bmatrix} \begin{bmatrix} w \\ x \end{bmatrix} = \begin{bmatrix} 0 \\ f \end{bmatrix} \tag{75}$$

we can perform a symbolic reduction to obtain the result

$$\begin{bmatrix} U & -T \\ 0 & H_1T+H_2 \end{bmatrix} \begin{bmatrix} w \\ x \end{bmatrix} = \begin{bmatrix} 0 \\ f \end{bmatrix} \tag{76}$$

which is an alternative form of the hybrid equations, but not reduced as far as Eq. (72).

Now whenever we solve a set of (linear algebraic) network equations, we *always* carry out Gauss elimination -- symbolically and/or numerically -- until only one equation in one unknown variable remains. We can properly say, then, that this single equation "describes the network" *in terms of the remaining variable*. Indeed, we can also say that at each stage of the elimination process, whether symbolic or numeric, the set of equations at hand "describe the network" in terms of the retained variables. It is fatuous, therefore, to argue about which of these various sets of equations are THE network equations; they all are. We may, however, refer to any of the various sets of tableau equations as the *fundamental* network equations.

The real point of our discussion relates to the question of

matrix sparsity. In order to take full advantage of the inherent sparsity of the coefficient matrices in the majority of network problems, the *order* in which the variables and equations are processed during Gauss elimination has a major effect on the computational cost. [4,5] To allow optimal ordering, then, full freedom to interchange rows and columns of the coefficient matrix must be permitted. But this freedom is constrained by any form of symbolic elimination. Accordingly, if sparse matrix techniques are to be applied to any network problem, it is better to apply them to the *tableau* equations directly rather than to a reduced set such as the mesh, node, f-loop, f-cutset, or either of the hybrid sets of equations.

To illustrate this point, we may cite the example of two recent network analysis programs, ECAP II [23] and ASTAP, [25] both of which use sparse matrix techniques. The formulation used in ECAP II is the first of the hybrid techniques, given by Eq. (72) whereas ASTAP uses the second, given by Eq. (76). Tests on the computational speed of these two programs showed that the second formulation, Eq. (76), gave significantly better performance due to the fact that the symbolic reduction had not been carried so far as in the case of Eq. (72). In other words, Eq. (76) retains more freedom to attain optimal ordering than does Eq. (72). The logical consequence of this observation, of course, is that Eq. (59) -- or Eq. (60) -- provides *complete* freedom to interchange rows and columns and so is a better choice than either Eq. (72) or Eq. (76).

But even Eq. (59) is not optimum because it expresses the topological constraints in terms of the C matrix -- which is almost always considerably more dense than the A matrix. Therefore, if the hybrid form of the constitutive law is required, the hybrid nodal tableau, Eq. (55), is likely to be better than Eq. (59) in terms of sparsity, even though it requires the use of node voltages e' as auxiliary variables. If the hybrid form of the constitutive law is not needed, then the nodal tableau, Eq. (49) is likely to be superior to the mesh, f-loop, and f-cutset tableau Eqs. (48), (50), and (51) for the same reason.

In the field of structural analysis [26] the flexibility method and the stiffness method are most widely used, the first being the counterpart of the mesh method, using Eq. (65), and the second being the counterpart of the node method, using Eq. (68). In view of our previous observations, structural analysis can take most effective advantage of matrix sparsity by using the stiffness method at the level of the *tableau* or fundamental network equations, analogous to Eq. (49). There is some qualitative evidence that not much sacrifice in optimal ordering results from reducing the nodal tableau to the node equations. But to the author's knowledge, no definitive comparison on this point has yet been made. No matter what the outcome of such a comparison, the results should be equally applicable to all 1-network problems.

We will not deal with the question of the existence of a solution to the network problem. [6] And the only observation that we will make about the numerical solution of differential equations is that *implicit* methods, which are commonly used today, rely heavily on the efficient solution of the basic algebraic problem Ax = b -- by sparse matrix techniques. The same is true for the numerical solution of nonlinear algebraic equations. Therefore, we conclude this section with two recommendations:

1. If *either* the y- or z-form of the constitutive law is acceptable, then the *nodal tableau*, Eq. (49), is probably the most efficient formulation for any 1-network problem from the sparsity point of view; if the hybrid form of the constitutive law is required, then the *hybrid nodal tableau*, Eq. (55) is the preferred formulation.
2. If matrix sparsity is of no concern, then whatever formulation and symbolic reduction process satisfies the particular requirements of the problem at hand and leads to the smallest set of equations, will obviously be preferred.

3.3 *ABSTRACT CHARACTERIZATION OF THE 1-NETWORK PROBLEM*

In a very significant contribution to network theory, Roth [6] developed a completely abstract characterization of the 1-network problem in terms of algebraic topology. He also showed conditions for the existence and proved the uniqueness of a solution to this problem. Roth explained the underlying character of the 1-network problem as having two ingredients: (1) a *topological structure*, called a linear graph (or 1-network) and (2) an associated *algebraic structure* with certain unique properties. By identifying the admissible class of mathematical entities that can inhabit this algebraic structure, Roth provided a foundation for establishing network analogies on a rigorous rather than an intuitive basis.

The essence of Roth's approach is the following. Each node, branch, and mesh (loop) of a linear graph may have associated with it *two* different mathematical entities (of the same category) -- which we have heretofore called the *through* and *across* variables. From a theoretical point of view, these entities are completely abstract, although they may later be assigned physically meaningful interpretations.

We begin by assigning to each *node* (except the datum node) an *arbitrary* mathematical quantity (an element of any Abelian or additive group) and denote the collection of these nodal quantities by the symbol e'. The existence of the A matrix, then, "induces the assignment" of a *derived* set of quantities e associated with the *branches* according to the relation e = A e'. Next, we can assign to the branches an additional set of *arbitrary* quantities E. The existence of the C matrix, then, "induces the assignment" of a *derived* set of quantities associated with the *meshes*: <u>zeroes</u>, according to the relation $C^t e = 0$, and <u>nonzeroes</u> E' according to the relation $E' = C^t E$. The interrelationships between these *Type-1* variables e', e, E, and E' are shown diagrammatically in Fig. 8 as a sequence of transformations or mappings known in algebraic topology as a *cohomology sequence*.

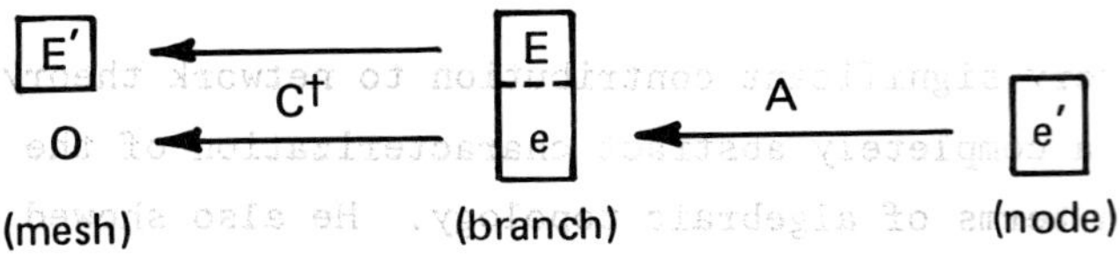

Fig. 8 - Interrelations Between Network Variables of Type 1

In similar fashion, we may assign *arbitrary* quantities to the *meshes* of the graph and symbolize these by j'. The existence of the C matrix, then, "induces the assignment" of *derived* quantities j on the *branches* according to the relation j = C j'. Once again, we can also assign the *arbitrary* quantites J to the branches and the A matrix "induces the assignment" of a *derived* set of quantities associated with the *nodes*: zeroes, according to the relation $A^t j = 0$, and nonzeroes J' according to the relation $J' = A^t J$. The interrelations between these *Type-2* variables j', j, J, and J' are shown diagrammatically in Fig. 9 as a *homology sequence*.

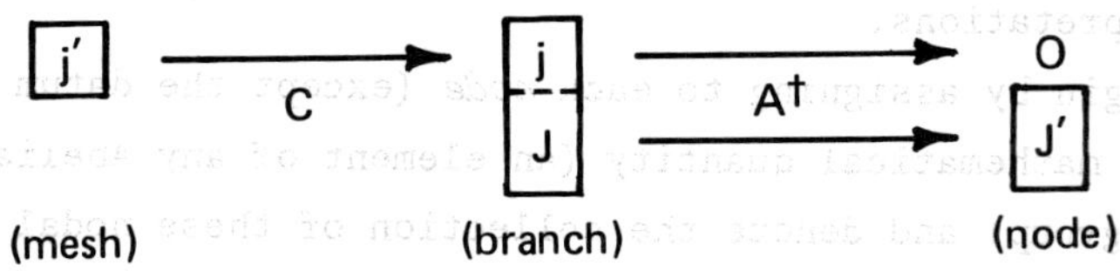

Fig. 9 - Interrelations Between Network Variables of Type 2

It is worth noting that no matter what set of quantities e' we assign to the nodes, the derived branch quantities e (= Ae') always sum to zero on the meshes. This follows from the basic relation $C^t A = 0$ since $C^t e = C^t(Ae') = (C^t A)e' = 0\ e' = 0$. Also, since $A^t C = 0$, it follows that no matter what set of quantities j' we assign to the meshes, the corresponding derived branch quantities j (= Cj') must always sum to zero on the nodes. The *arbitrary* branch quantities E and J need not sum to zero, however.

The algebraic structure depicted by the homology and cohomology sequences of Figs. 9 and 8 is rather loose in that the Type-1 variables are completely independent of the Type-2 variables, and *vice versa*. As a consequence of the looseness of this structure, a very broad class of mathematical objects -- elements of any Abelian group -- can be associated therewith. We recognize, moreover, that this structure corresponds to the Kirchhoff voltage and current laws in electrical networks and hence to the topological constraint relations in any 1-network problem. But the counterpart of the constitutive laws is missing.

Roth took the additional step, therefore, of assuming the existence of a 1-1 transformation (isomorphism) between the Type-1 and Type-2 variables associated with the branches, as shown in Fig. 10. This transformation, involving the Z operator (matrix) and its inverse Y, we recognize as archetypical of the constitutive laws discussed above. Hence the diagram in Fig. 10 completely epitomizes the 1-network problem. Even the composite transformations C^tZC and A^tYA we recognize as corresponding to the *mesh impedance* and *nodal admittance* matrices of Eqs. (65) and (68).

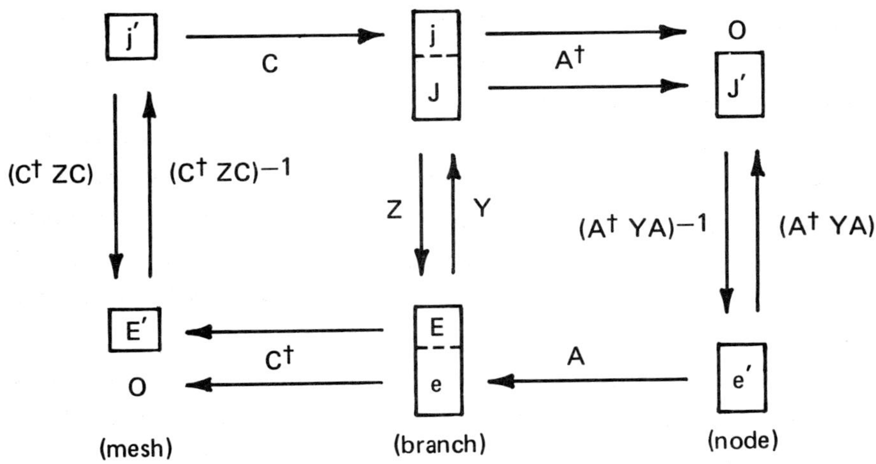

Fig. 10 - Roth's Transformation Diagram Depicting the Algebraic Structure of the 1-Network Problem

If the elements of the Z (and/or Y) matrix are real or complex numbers, then the admissible mathematical entities that can be associated with the nodes, branches, and meshes of a 1-network are *elements of any vector space*. This is a much more restrictive class of objects than the Abelian groups, but the algebraic structure is correspondingly richer because the Type-1 and Type-2 variables are now intimately related to eachother. Even so, vector spaces encompass an enormous range of mathematical objects.

One immediate consequence of the inclusion of vector space elements in the algebraic structure of Fig. 10 is the fact that mechanical and structural analysis both properly come within the purview of the 1-network problem. To be sure, the forces and displacements (velocities, accelerations) associated with each node, branch, and mesh of a structural or mechanical (even molecular) system can be 6-vectors with three translational and three rotational components. But this is no impediment since elements of any vector space are admissible.

There is, however, one additional complication which mechanical, structural, and molecular systems involve and that is the necessity to take *geometry* explicitly into account along with the constitutive and topological constraint laws. Indeed, geometry comes into play simply because the topological constraint equations require that each of the vector variables involved be expressed in the same coordinate system. Fortunately, the geometric transformations required to accomplish this can be treated as a separate operation [26] leaving the constitutive and constraint laws intact. And in the presence of large displacements, *geometric nonlinearities* will arise in what may otherwise be a strictly linear system. Nevertheless, mechanical, structural, and molecular systems can still be regarded as legitimate 1-network problems.

Now within the algebraic structure of Figs. 8, 9, and 10 are certain substructural details of some interest. To show these, let us simplify our discussion by assuming that only real or complex numbers are associated with the nodes, branches, and loops.

Then, if there are n nodes, the e' vector will be (n-1)-dimensional. If b is the number of branches, the e vector will have b elements -- but it will always be confined to an *(n-1)-dimensional subspace*. This is because it is derived from the (n-1)-dimensional vector e' and must additionally satisfy the linear dependency constraint, $C^t e = 0$. The E vector, on the other hand, being wholly arbitrary, occupies the entire b-dimensional space. Finally, the E' vector is m-dimensional, where m = b - (n-1), even though it is derived from the b-dimensional E vector. For the C^t operator itself reduces the dimension to m. Incidentally, the E' vector has no constraints to satisfy.

Quite similarly, the j' vector is m-dimensional and gives rise to a b-element j vector that is confined to an *m-dimensional subspace* in which every vector must satisfy the constraint $A^t j = 0$. The J vector, again being arbitrary, occupies the entire b-dimensional space while the J' vector is (n-1)-dimensional.

The foregoing discussion on the algebraic structure of the 1-network problem has not touched on the *statement* of the problem. Clearly, we do not actually set up the problem by assigning values to the vectors e' and j' since these may be used as auxiliary variables in *solving* the problem as described earlier. We may, of course, assign the vectors E and J arbitrarily but then we must find the corresponding values of e and j.

Roth has proposed an alternative 1-network problem that is best described in terms of the diagram of Fig. 11 where the substitutions V = E + e and I = J + j, from Eqs. (33) and (34), are implied.

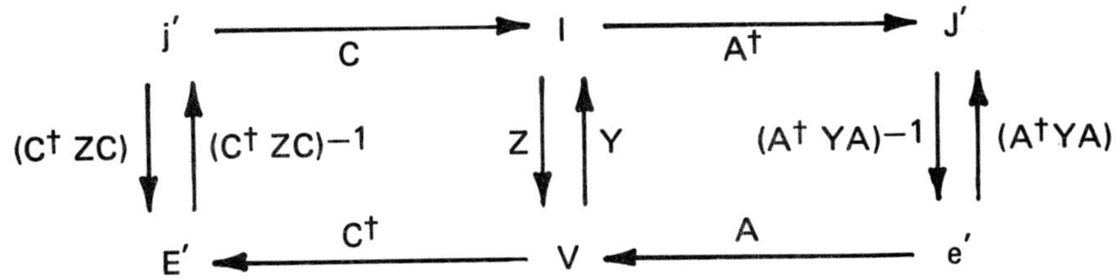

Fig. 11 - Roth's Transformation Diagram, Simplified

Roth's problem may be stated as follows:

Given a 1-network of known topology (which defines the A and C matrices), known element values (which define the Z and/or Y matrices), and known vectors E' and J', find the vectors V and I such that: $V = ZI$ (or $I = YV$), $C^tV = E'$, and $A^tI = J'$.

Roth established a condition for the existence of a solution to this problem and showed that the following solution (which can be literally read directly from the diagram of Fig. 11) is unique.

$$V = A(A^tYA)^{-1}J' + ZC(C^tZC)^{-1}E' \tag{77a}$$

$$I = C(C^tZC)^{-1}E' + YA(A^tYA)^{-1}J' \tag{77b}$$

This version of the 1-network problem -- though not its solution -- is actually closer to the structural analysis problem [26] than the "electrical network problem" described earlier. For in structural analysis, the *joint forces* (counterparts of J') and the *ring displacements* (counterparts of E') are generally given rather than the counterparts of J and E which have no recognized meaning.

In summary, then, the 1-network problem is characterized by:

1. An underlying topological structure, the linear graph, with each of whose nodes, branches, and loops we can associate the elements of any vector space.
2. An associated algebraic structure, depicted by Roth's transformation diagrams, in which the algebraic-topological archetypes of the constitutive and constraint laws appear.

Therefore, any physical or mathematical system which possesses *all* of these characteristics can be treated as a 1-network problem. Admittedly, the special structure of the 1-network problem does not presently give us any significant computational leverage in *solving* the equations that describe the network. But it does provide us with a unified basis for *formulating* these equations. What is more, it clearly identifies the areas of common interest in science and engineering wherein theoretical and/or computational advances in any one field can be shared by all.

4. THE 3-NETWORK PROBLEM

Having arrived at a definitive mathematical characterization of the 1-network problem, we now turn our attention to its higher dimensional homolog, the 3-network problem. This extension to higher dimension is necessary because many physical systems require the use of surface and volume elements to characterize them properly. In other words, such systems imply the existence of an underlying topological structure consisting of a set of interconnected points, lines, surface, and volume elements -- which we call a *3-network* -- and an associated algebraic structure similar to what we found in the case of the 1-network. Without a clear understanding of these topological and algebraic structures, then, we cannot fully appreciate the import of "network analogies" that relate to this class of physical phenomena.

The original motivation for the author's study of the 3-network problem was the publication by Kron [27] of an attempted explanation of Maxwell's equations for the electromagnetic field using a modification of Roth's transformation diagram. Kron's approach, however, did not appear to be consistent and it differed markedly from what the author eventually developed. Even so, it did serve to stimulate interest in a network-topological treatment of electromagnetic theory using an extension of Roth's ideas.

4.1 *EXTENSION OF LINEAR GRAPH CONCEPTS TO THE 3-NETWORK*

In algebraic topology "networks" of interconnected 0-, 1-, 2-, 3-,...,n-dimensional objects, called *n-cells* or *n-simplexes*, are studied to find those properties which depend solely on the connectivity relations between the various objects. These connectivity or *topological* properties are those which are invariant under all possible geometric distortions -- bendings, twistings, stretchings, etc. -- that do not break any of the connections.

The topological matrices and the algebraic structure that we have used to characterize the 1-network problem can be generalized to handle oriented n-networks, or *n-complexes*, in a straightforward manner. We shall consider only the 3-network in any detail, however, and to do so we will need some new terminology.

Instead of referring to points, lines, surface, and volume elements, we will refer to 0-, 1-, 2-, and 3-cells. The *faces* of any p-cell are the (p-1)-cells incident thereon and the *boundary* of a p-cell is determined by the set of all its faces -- with due regard to orientation. Similarly, the *cofaces* of any p-cell in an n-complex where $0<p\leq n$ are the (p+1)-cells -- if any -- that are incident thereon. Moreover, the *coboundary* of a p-cell is determined by the set of all its cofaces, again with due regard to orientation. When applied to an entire n-complex, the boundary and coboundary *operators* can be expressed as matrices whose elements, like those of the A and C matrices of the 1-network, consist only of +1, -1, and 0. Indeed, the A^t matrix represents the boundary operator for the 1-cells and A the coboundary operator for the 0-cells of a 1-network. Since there are no 2-cells, however, the C matrix does *not* correspond to a boundary nor C^t to a coboundary operator. [6,7]

In the 3-network, we will use the symbols C_{ij} to represent the boundary operators (or connection matrices) showing the incidence relations between all the i-cells and j-cells. For example, C_{01} represents the incidence matrix between 0-cells and 1-cells; hence it is identical with A^t. Similarly, we will let C^{ji} (or C^t_{ij}) symbolize the coboundary operators. Thus, both C^{10} and C^t_{01} are identical with the A matrix.

To decide the relative orientations of 1- and 2-cells in setting up C_{12}, we orient each 2-cell as if it were a loop and then use the 1-network convention. But in setting up the C_{23} boundary operator, we will need the following rule: consider as *positively incident* on the 3-cell any (bounding) 2-cell that appears oriented *clockwise when viewed from outside the 3-cell.*

To simplify our treatment of the 3-network problem, we will consider only *simply connected cubical complexes* such as the one shown in Fig. 12. By "simply connected" we mean that all of the outermost 1-, 2-, and 3-cells have faces that belong to the complex and that all interior points belong to some 0-, 1-, 2-, or 3-cell of the complex. Thus, there are no "holes".

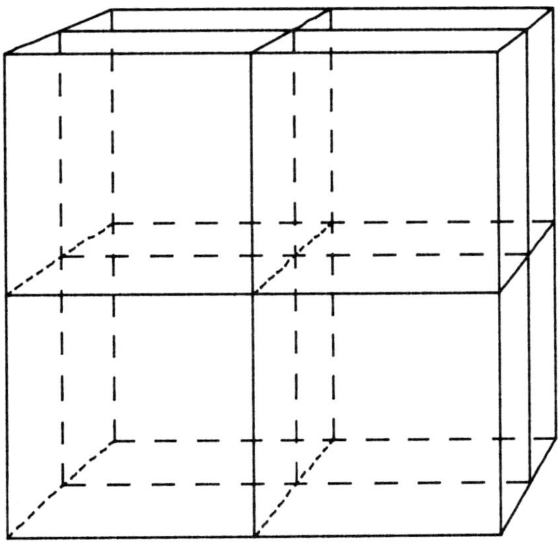

Fig. 12 - Simply Connected Cubical Complex, or 3-Network

The algebraic structure that we can associate with this 3-network is similar to that of the 1-network in that two types of variable may be associated with each p-cell, p = 0, 1, 2, 3. Since the terms "through" and "across" are no longer appropriate to describe the variables associated with 2- and 3-cells, these terms will be dropped. We will use the term *p-chain* for the set of all mathematical quantities (elements of some vector space) assigned to the p-cells of a complex if these quantities can be operated upon by a *boundary* operator. The cognate set of quantities, operated upon by a *coboundary* operator, we call a *p-cochain*. [28]

When a boundary operates on a p-chain, it produces ("induces the assignment of") a related (p-1)-chain on the (p-1)-cells; similarly, when a coboundary operates on a p-cochain, it produces a related (p+1)-cochain on the (p+1)-cells. In a simply connected 3-network, then, we have a *chain sequence* involving the successive boundary operators C_{23}, C_{12}, and C_{01}; we also have the cognate *cochain sequence* involving the successive coboundary operators C^{10} (or C^t_{01}), C^{21} (or C^t_{12}), and C^{32} (or C^t_{23}).

Now any p-chain whose boundary is zero is called a *p-cycle* and if this p-cycle also happens to be the boundary of a (p+1)-chain, then it is called a *bounding p-cycle*. Again, any p-cochain whose coboundary is zero is called a *p-cocycle* and if it is also the coboundary of a (p-1)-cochain, then it is called a *cobounding p-cocycle*. In a 3-complex that is not simply connected, there will be some *nonbounding* p-cycles and possibly also some *noncobounding* p-cocyles. This complicates the situation considerably by requiring *homology* and *cohomology* sequences to be used in place of the *chain* and *cochain* sequences -- which are simpler to describe. That is why we limit our discussion to simply connected 3-networks.

In linear graphs, we stressed the fundamental importance of the relation $A^tC = 0$. The counterpart of this in the higher dimensional homologs is the basic statement: *the boundary of the boundary is zero*. The cognate statement, *the coboundary of the coboundary is zero*, is also true. [28] Thus, for any set of adjacent dimensions such that $i + 1 = j$ and $j + 1 = k$, we have

$$C_{ij}C_{jk} = 0 \tag{78}$$

and

$$C^{kj}C^{ji} = C^t_{jk}C^t_{ij} = 0 \tag{79}$$

To set up the algebraic structure that characterizes the 3-network, we will let the symbol Q_p (for p = 0, 1, 2, 3) represent the *space of all p-chains*; similarly, we will let Q^p represent the *space of all p-cochains*. Next, we assume that the resulting chain and cochain sequences are interrelated by an isomorphism (or 1-1 transformation) S_1/T_1 between the 1-chains and 1-cochains which is identical, except for symbols, to the Z/Y isomorphism used in the 1-network problem. Finally, we also assume the existence of a second interrelation in the form of the isomorphism S_2/T_2 between the 2-chains and 2-cochains. The resulting algebraic structure is depicted in Fig. 13.

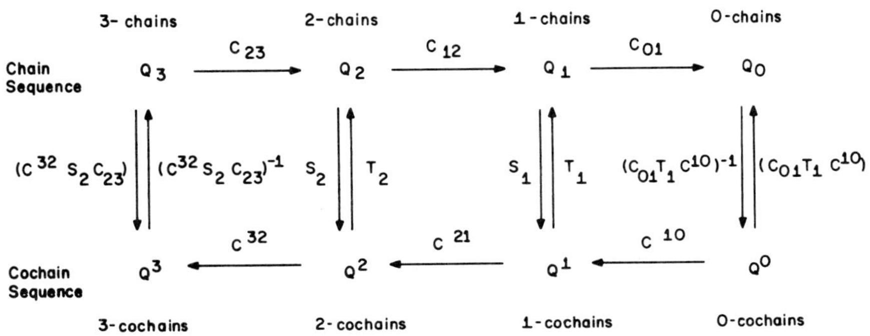

Fig. 13 - Algebraic Structure of a Simply Connected 3-Network

Here, the existence of the isomorphism S_1/T_1, together with the boundary and coboundary C_{01} and C^{10}, *induces* the isomorphism $C_{01}T_1C^{10}$ (or, alternatively, $C_{01}T_1C_{01}^t$) between the 0-cochains and 0-chains, this isomorphism being identical -- except for symbols -- to A^tYA in the 1-network. But there is no counterpart of C^tZC; instead, there is the induced isomorphism $C^{32}S_2C_{23}$ (or $C_{23}^tS_2C_{23}$) between the 3-chains and 3-cochains. (A solution to the 3-network problem exists if and only if the inverses of both of these induced isomorphisms exist. Roth's existence theorem for the 1-network problem [6] can be shown [7] to apply here also.)

As in the case of the 1-network problem, we can have both *arbitrary* and *derived* quantities within the chain and cochain sequences of Fig. 13. The derived quantities arise from applying boundary or coboundary operators to arbitrarily assigned p-chains or p-cochains. What is more, these derived quantities are always *p-cycles* or *p-cocycles* -- in a simply connected complex -- because *the boundary of the boundary is zero* and *the coboundary of the coboundary is zero*. Accordingly, the p-cycles and p-cocycles are confined to *subspaces* for which we use the symbols M_p and M^p. Arbitrarily defined p-chains and p-cochains, on the other hand, are free to occupy the entire space and for these we use the symbols P_p and P^p. Fig. 14 shows the details of this infrastructure.

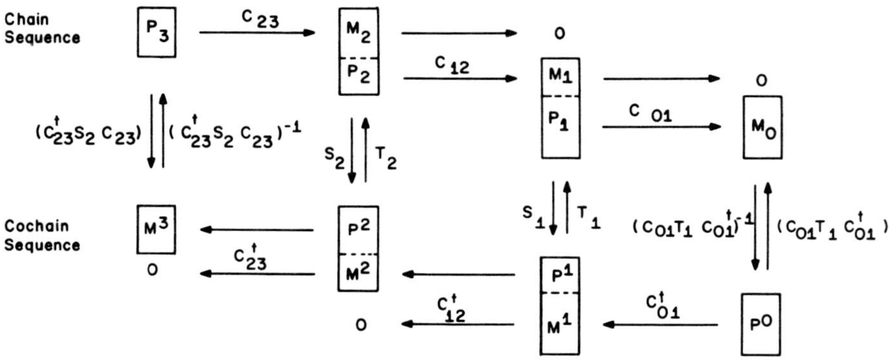

Fig. 14 - Detailed Algebraic Structure of the 3-Network

The dimensions of these various spaces and subspaces are topological invariants determined by the ranks r_1, r_2, and r_3 of the connection matrices (boundary operators) C_{01}, C_{12}, and C_{23}, respectively. If we let n_0, n_1, n_2, and n_3 be the numbers of 0-, 1-, 2-, and 3-cells in a simply connected 3-complex, then it turns out that [29]

$$r_1 = n_0 - 1 \tag{80}$$

$$r_2 = n_1 - r_1 \tag{81}$$

$$r_3 = n_2 - r_2 = n_3 \tag{82}$$

Thus, C_{01} is an r_1 x n_1 matrix with rank r_1, C_{12} is an n_1 x n_2 matrix with rank r_2, and C_{23} is an n_2 x n_3 matrix with rank r_3.

Table I summarizes the dimensions of the various spaces and subspaces shown in Fig. 14.

Table I - Space and Subspace Dimensions for a Simply Connected 3-Network

Space/Subspace		Dimension
M_0	P^0	n_0-1 $(=r_1)$
P_1	P^1	n_1
P_2	P^2	n_2
P_3	M^3	n_3 $(=r_3)$
	M^1	r_1
M_1	M^2	r_2
M_2		r_3

4.2 *FORMULATION AND SOLUTION OF THE 3-NETWORK PROBLEM*

The topological constraint equations implicit in the algebraic structure of the 3-network are as follows:

$$C_{23}P_3 = M_2 \quad \Longleftrightarrow \quad C_{12}M_2 = 0 \tag{83}$$

$$C_{12}P_2 = M_1 \quad \Longleftrightarrow \quad C_{01}M_1 = 0 \tag{84}$$

$$C_{01}P_1 = M_0 \tag{85}$$

$$C^t_{01}P^0 = M^1 \quad \Longleftrightarrow \quad C^t_{12}M^1 = 0 \tag{86}$$

$$C^t_{12}P^1 = M^2 \quad \Longleftrightarrow \quad C^t_{23}M^2 = 0 \tag{87}$$

$$C^t_{23}P^2 = M^3 \tag{88}$$

where the symbol <===> means *implies and is implied by* -- in other words, *mathematical equivalence*. The two "constitutive laws" are:

$$P_1 + M_1 = T_1(P^1 + M^1) \quad \Longleftrightarrow \quad P^1 + M^1 = S_1(P_1 + M_1) \tag{89}$$

$$P_2 + M_2 = T_2(P^2 + M^2) \quad \Longleftrightarrow \quad P^2 + M^2 = S_2(P_2 + M_2) \tag{90}$$

With the foregoing equations as a basis, the author originally expected to find two versions of the 3-network problem analogous to those associated with the 1-network. However, every attempt to identify a Roth-type problem with M_0 and M^3 prescribed failed utterly -- apparently because such a problem is ill-posed. The only 3-network problem that could be established turned out to be a homolog of the electrical network problem, but with notable differences. Specifically, it is not necessary to prescribe all four of the vectors P_1, P^1, P_2, and P^2 because P^1 and P_2 *are determined by* P_1 and P^2. Accordingly, the 3-network problem may be stated as follows:

> Given: a simply connected 3-network; its connection matrices C_{01}, C_{12}, C_{23}; the isomorphisms S_1 and S_2 (or their inverses T_1 and T_2); and the vectors P_1 and P^2; find the vectors M_0, M_1, M_2, P_2, P_3 and P^0, P^1, M^1, M^2, M^3 such that Eqs. (83) - (90) are satisfied.

Using Eqs. (85) and (88), the vectors M_0 and M^3 can be evaluated immediately from the given values of P_1 and P^2. To show that P^1 and P_2 can also be calculated, we can use algebraic manipulations much like those used to solve the 1-network problem. In particular, we first write the counterpart of Eq. (68) and three cognate relations thus:

$$C_{01}T_1C_{01}^t\ P^0 = C_{01}(P_1 - T_1P^1) \tag{91}$$

$$C_{12}^tS_1C_{12}\ P_2 = C_{12}^t(P^1 - S_1P_1) \tag{92}$$

$$C_{12}T_2C_{12}^t\ P^1 = C_{12}(P_2 - T_2P^2) \tag{93}$$

$$C_{23}^tS_2C_{23}\ P_3 = C_{23}^t(P^2 - S_2P_2) \tag{94}$$

Next, we solve Eq. (92) for P_2 and Eq. (93) for P^1 by inverting the matrices $C_{12}^t S_1 C_{12}$ and $C_{12} T_2 C_{12}^t$. (Roth's existence theorem applies here.) Then we substitute the resulting relations into Eqs. (93) and (92), respectively to obtain the expressions

$$C_{12}[(C_{12}^t S_1 C_{12})^{-1} - T_2] C_{12}^t P^1 = C_{12}(C_{12}^t S_1 C_{12})^{-1} C_{12}^t S_1 P_1 + C_{12} T_2 P^2 \quad (95)$$

$$C_{12}^t[(C_{12} T_2 C_{12}^t)^{-1} - S_1] C_{12} P_2 = C_{12}^t (C_{12} T_2 C_{12}^t)^{-1} C_{12} T_2 P^2 + C_{12}^t S_1 P_1 \quad (96)$$

If the coefficient matrices of P^1 and P_2 in Eqs. (95) and (96) are nonsingular -- and this places a restriction on the isomorphisms S_1/T_1 and S_2/T_2 -- then we can solve for both P^1 and P_2 directly.

With these vectors determined, M^2 and M_1 can easily be evaluated from Eqs. (87) and (84). Finally, to solve for M^1 and M_2, we must first compute the "auxiliary variables" P^0 and P_3 using Eqs. (91) and (94); then Eqs. (86) and (83) give the desired values of M^1 and M_2, completing the solution.

As an alternative to the foregoing procedure, we can write a set of tableau equations to describe the problem. We will ignore Eqs. (85) and (88) as trivial and omit the specific evaluation of P^0 and P_3 as unnecessary. The resulting tableau equations are

$$\begin{bmatrix} U & 0 & -S_1 & 0 & U & 0 \\ 0 & U & 0 & -T_2 & 0 & U \\ 0 & 0 & U & 0 & 0 & -C_{12} \\ 0 & 0 & 0 & U & -C_{12}^t & 0 \\ C_{12}^t & 0 & 0 & 0 & 0 & 0 \\ 0 & C_{12} & 0 & 0 & 0 & 0 \end{bmatrix} \begin{bmatrix} M^1 \\ M_2 \\ M_1 \\ M^2 \\ P^1 \\ P_2 \end{bmatrix} = \begin{bmatrix} S_1 P_1 \\ T_2 P^2 \\ 0 \\ 0 \\ 0 \\ 0 \end{bmatrix} \quad (97)$$

Since this tableau is upper triangular except for the last two rows, we can use symbolic Gauss elimination to produce the following result:

$$\begin{bmatrix} -C_{12}^t & C_{12}^t S_1 C_{12} \\ C_{12} T_2 C_{12}^t & -C_{12} \end{bmatrix} \begin{bmatrix} P^1 \\ P_2 \end{bmatrix} = \begin{bmatrix} -C_{12}^t S_1 P_1 \\ -C_{12} T_2 P^2 \end{bmatrix} \tag{98}$$

But this is equivalent to Eqs. (92) and (93) and so has the same solution. The other variables, M^1, M_2, M_1, and M^2, are obtained by back substitution. For numerical solutions, of course, Eq. (97) should be used directly with sparse matrix techniques.

4.3 *THE DUAL 3-NETWORK -- A LINK TO THE VECTOR CALCULUS*

There is a strong link between algebraic topology and the vector calculus which some mathematicians have long since recognized; for the *coboundary operators* are intimately related to the vector calculus operators *grad, curl, and div*. [30] Actually, the algebraic structure of the 3-network problem reflects -- or rather is a *characterization of* -- the *nonmetric* properties of Euclidean 3-space which the vector calculus seeks to describe as a continuum. Therefore, it is not too surprising to find the vector calculus emerging from the 3-network problem as we let the 1-, 2-, and 3-cells approach a limit of zero size.

However, in order to effect the linkage between the 3-network algebra and the vector calculus, we need first to introduce the concept of a *dual 3-network*. This necessity devolves from the curious fact that there is no role for the *chain sequence* to play. What we require is an algebraic structure with *two cochain sequences* and a set of transformations that are isomorphic to those of Fig. 14.

Now any linear graph that can be mapped on a sphere has a *dual graph* (also mappable on the sphere) in which each node is in 1-1 correspondence with the meshes of the primal graph, and vice versa. Moreover, the branches of both the primal and dual graphs are also in 1-1 correspondence. For planar graphs, however, the dual requires an extra node to be placed outside the primal graph. [31]

This concept of duality may be extended to higher dimensional complexes whenever it is possible to set up a 1-1 correspondence between the p-cells of the *primal* n-complex and the (n-p)-cells of the *dual* n-complex. [29] The extension to dual 2-complexes that map onto a plane or sphere is immediate on the basis of dual linear graphs having the same property: simply identify each 2-cell of the former with a mesh of the latter.

In the case of simply connected 3-complexes, a dual 3-complex can be established as shown in Fig. 15. Here, each 0-cell of the dual lies at the center of each 3-cell of the primal 3-network on a body-centered cubic lattice, so to speak, and obviously in 1-1 correspondence. Moreover, each 1-cell of the dual passes through the center of a 2-cell of the primal, again forming a 1-1 correspondence. To complete this process, we must join together at a

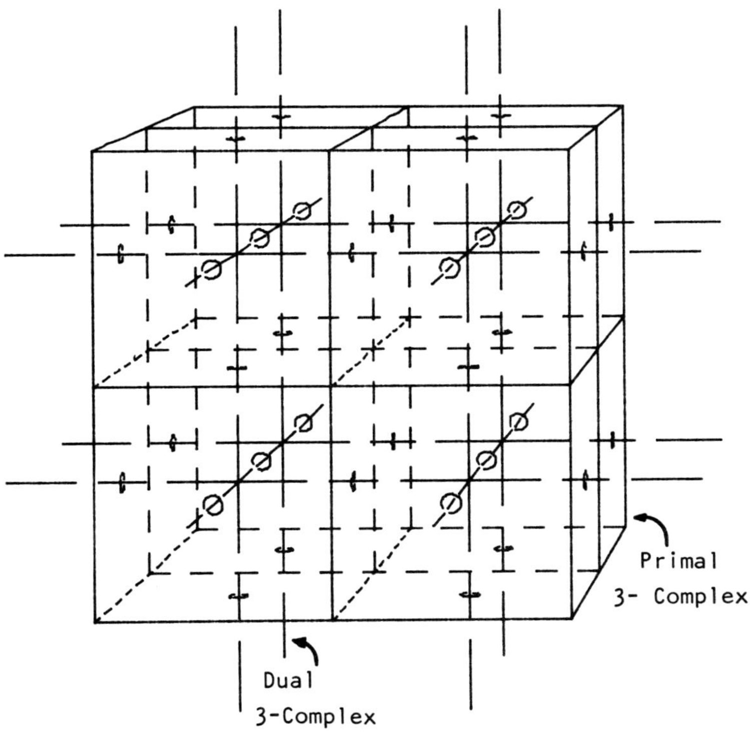

Fig. 15 - Primal and Dual Cubical Complexes

point all the exterior 0-cells of the dual 3-network and then "fill in" the 2- and 3-cells implied by this additional connection. Thus, the 2-cells (3-cells) of the dual complex can be made to correspond 1-1 with the 1-cells (0-cells) of the primal.

As a direct consequence of these 1-1 correspondences between the primal p-cells and the dual (3-p)-cells, there is an *induced identity* between the primal and dual boundary and coboundary operators. What is more, there is also an induced identity between the primal and dual chains and cochains. These identities are summarized in Table II with underscores being used to designate the *dual* boundary and coboundary operators.

Table II - Primal/Dual 3-Network Correspondences

Primal 3-Network		Dual 3-Network
p-cells	$\simeq$	(3-p)-cells
p-chains	$\simeq$	(3-p)-cochains
p-cochains	$\simeq$	(3-p)-chains
C_{01}	$\simeq$	$\underline{C}^{32} = \underline{C}^t_{23}$
C_{12}	$\simeq$	$\underline{C}^{21} = \underline{C}^t_{12}$
C_{23}	$\simeq$	$\underline{C}^{10} = \underline{C}^t_{01}$
C^t_{01}	$\simeq$	$\underline{C}_{23}$
C^t_{12}	$\simeq$	$\underline{C}_{12}$
C^t_{23}	$\simeq$	$\underline{C}_{01}$

The net result, then, is that the *dual cochain sequence* is isomorphic with the *primal chain sequence* and therefore can replace it in the algebraic structure shown in Fig. 14. Moreover, the *dual coboundary operators* being identical with the inversely corresponding *primal boundary operators*, we can retain the latter in writing the dual cochain sequence, as shown in Fig. 16. Thus we have the needed algebraic structure with two *cochain* sequences.

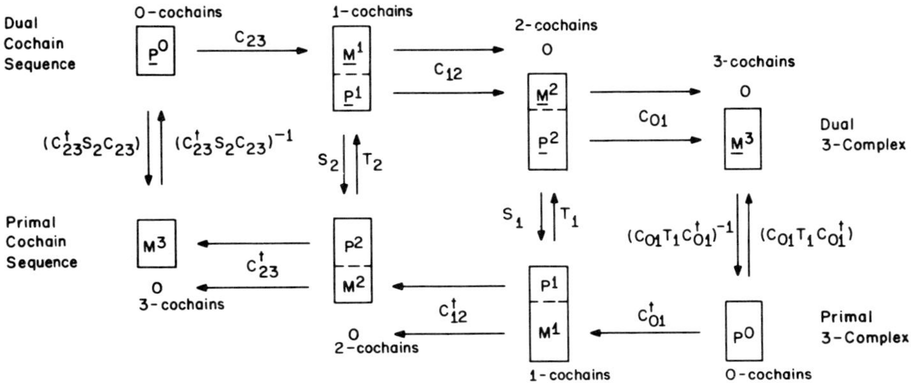

Fig. 16 - Algebraic Structure of the Primal/Dual 3-Network

4.4 *THE OPERATIONAL STRUCTURE OF THE VECTOR CALCULUS*

The vector calculus operators *grad, curl, div* follow the pattern of a *cochain* sequence rather than a chain sequence. To show this, let us assume that an arbitrary *scalar* field $\phi(x,y,z)$ is defined throughout a simply connected region of 3-space which is subdivided by a finite cubical complex or 3-network. If we evaluate this function at each 0-cell of the 3-network, we thereby define a 0-cochain belonging to P^0. The line integral of $\overline{grad}\ \phi$ between two adjacent 0-cells a and b is

$$\int_a^b \overline{grad}\ \phi \cdot \overline{dl} = \phi(b) - \phi(a) \tag{99}$$

This defines a number assignable to the 1-cell connecting a and b in precisely the same way that is implied by the relation $e = Ae'$ discussed in Section 3.3.

If we apply the line integral of Eq. (99) to every 1-cell, we derive a 1-cochain belonging to M^1. For infinitesimal 1-cells, with vector respresentation $\overline{dl}$, the elements of this 1-cochain become $-\overline{grad}\ \phi \cdot \overline{dl}$. Thus, the line integral of the gradient acts like the coboundary operator C_{01}^t for 1-cells of finite size whereas the (negative) gradient operator itself acts like this coboundary operator in the limiting case of infinitesimal 1-cell size.

In similar fashion, we can define an arbitrary 1-cochain in P^1 by assuming the existence of a vector field $\overline{A}(x,y,z)$ and evaluating the line integral of $\overline{A}$ along each 1-cell of the 3-network. The number thus assigned to each 1-cell contributes to a *derived* quantity assignable to every 2-cell that is a coface thereof according to the relation $M^2 = C_{12}^t P^1$. In other words, the coboundary operating on these line integrals defines a *derived* 2-cochain in M^2. Looked at another way, the element of this 2-cochain assigned to any particular 2-cell is computed by evaluating the line integral around the *boundary* of the 2-cell in question.

Referring to Stokes theorem,

$$\oint \overline{A}\cdot\overline{dl} = \int_S curl\ \overline{A}\cdot\overline{ds} \tag{100}$$

we see that for infinitesimal 2-cells, the elements of the derived 2-cochain under discussion are given by $curl\ \overline{A}\cdot\overline{ds}$ where $\overline{ds}$ is the vector representation of the 2-cell.

Finally, we can define an arbitrary 2-cochain in P^2 by assuming the existence of another vector field $\overline{B}(x,y,z)$ and evaluating the surface integral of $\overline{B}$ over each 2-cell. The number thus computed for each 2-cell contributes to a derived quantity assignable to every 3-cell that is a coface thereof according to the equation $M^3 = C_{23}^t P^2$. Thus, the coboundary operating on these surface integrals defines a derived 3-cochain in M^3. Again, the element of this 3-cochain assigned to any particular 3-cell is found by evaluating the surface integral over the boundary of the 3-cell. From Gauss' theorem

$$\int_S \overline{B}\cdot\overline{ds} = \int_V div\ \overline{B}\ dv \tag{101}$$

we recognize that for infinitesimal 3-cells, the elements of the derived 3-cochain are given by $div\ \overline{B}\ dv$ where dv represents the (volume of) the 3-cell.

Now an essential point to appreciate is that in all three of the cases discussed the *topological character* of the coboundary operation remains *unchanged* in passing from the discrete (finite) structure to the continuum. In other words, the *grad, curl,* and *div* operators are still topologically *coboundary operators* even though they represent limits as the 1-, 2-, and 3-cells approach zero dimensions. Another point to recognize is that the 1- and 2-cochains obtained by applying the coboundary operators are actually *1- and 2-cocycles* since their coboundaries, in turn, are zero. This leads, of course, to the fact that the fundamental relations $curl\ \overline{grad} = 0$ and $div\ curl = 0$ are simply reflections of the basic rule that *the coboundary of the coboundary is zero.*

We can summarize the foregoing discussion by means of Fig. 17 where ϕ is an arbitrary scalar function; $\overline{V} = \overline{A} - \overline{grad}\ \phi$, with $\overline{A}$ an arbitrary vector function; $\overline{W} = \overline{B} + curl\ \overline{A}$, with $\overline{B}$ another arbitrary vector function; and $\omega = div\ \overline{B}$ a derived scalar function.

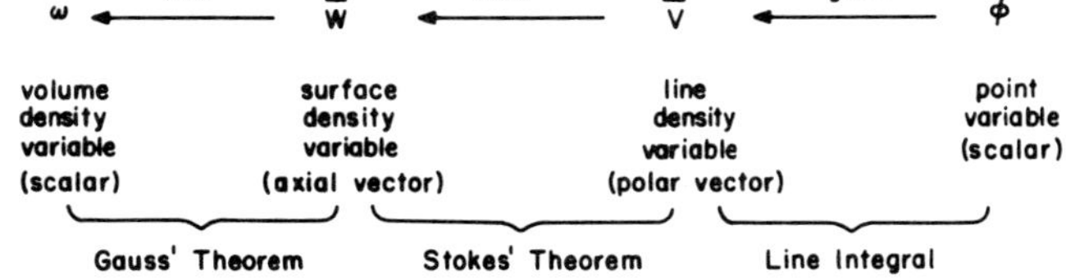

Fig. 17 - Cochain Sequence for the Vector Calculus

It is worth pointing out that there are *two topologically distinct types of vector* implied by Fig. 17: a *line density* vector, such as $\overline{V}$, and a *surface density* vector, such as $\overline{W}$. According to the algebraic structure we have developed, it is possible to take the *curl* of a line density vector but NOT its *divergence* and to take the *divergence* of a surface density vector, but NOT its *curl*. Yet this fundamental rule is frequently violated in the literature because the distinction between these two types of vector is not appreciated. We might conjecture, incidentally, that there may be a relation between these two vector types and *polar* and *axial* vectors.

Since we have shown that a cochain sequence involving the vector calculus operators can be associated with an infinitesimal 3-network, we can also set up a dual cochain sequence with its infinitesimal *dual* 3-network, thus leading to the vector calculus counterpart of Fig. 13, as shown in Fig. 18.

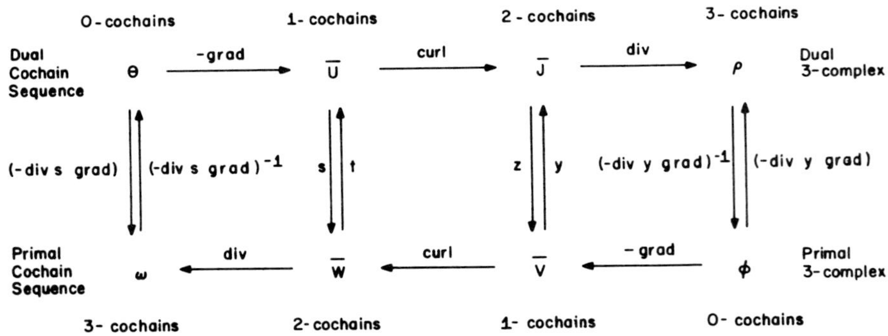

Fig. 18 - Operational Structure of the Vector Calculus

In this diagram, the isomorphisms z/y and s/t that interrelate the primal and dual cochain sequences play the role of *constitutive parameters* that characterize the medium or space in which some phenomenon, describable by the vector calculus, takes place. This will become clear in the next section when we develop an interpretation of Maxwell's equations in terms of this diagram.

4.5 *A 3-NETWORK INTERPRETATION OF MAXWELL'S EQUATIONS*

We have chosen Maxwell's equations as a vehicle for interpreting the operational structure of the vector calculus because they illustrate some particularly interesting features of this structure. Our approach will be to deal first with the electrostatic and magnetostatic fields and then with the electrodynamic and magnetodynamic fields. This will lead directly to a full treatment of the electromagnetic field equations. The resulting interpretation is completely consistent dimensionally and operationally with all aspects of the vector calculus and Maxwell's equations and it provides an elegant framework for understanding both.

Fig. 19 depicts the operational structure of the electrostatic field, occupying only a portion of the entire structure shown in Fig. 18. Here, the electrostatic variables and their dimensions are: ϕ, the potential in *volts;* $\overline{E}$, the electric field in *volts/meter;* $\overline{D} + \overline{d}$, the displacement in *coulombs/meter*^{2}*;* ρ, the charge density in *coulombs/meter*^{3}.

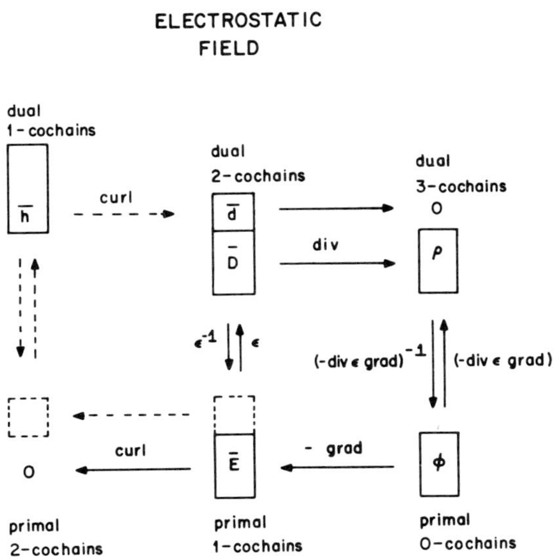

Fig. 19 - Operational Structure of the Electrostatic Field

To be precise, we have specified the *displacement vector* as the sum of two components, $\overline{D}$ and $\overline{d}$, where the component $\overline{d}$ corresponds to the *subspace of 2-cocycles*. In more familiar terms, $\overline{d}$ is the *solenoidal component of displacement*, having zero divergence. The fundamental relations: $div\ \overline{d} = 0$, $div\ \overline{D} = \rho$; $-\overline{grad}\ \phi = \overline{E}$ and $\overline{D} + \overline{d} = \varepsilon\overline{E}$, where ε is the dielectric constant, are consistent with their algebraic counterparts in the primal/dual 3-network problem. Incidentally, we recognize the existence of a vector potential $\overline{h}$, having the dimensions of *coulombs/meter*, such that $curl\ \overline{h} = \overline{d}$; but this relation may not be particularly useful.

By making the appropriate substitutions of the symbols in Fig. 19 for those in Figs. (16) or (14), we can immediately transcribe all of the pertinent equations for the primal/dual 3-network problem into their electrostatic counterparts. For example, the electrostatic equivalent of the solution of Eq. (91) is

$$\phi = -(div\ \varepsilon\ \overline{grad})^{-1}\ div\ \overline{D} \tag{102}$$

Here we could replace $div\ \overline{D}$ by ρ, in which case we would have the formal solution to the electrostatic field problem, nanely: Given the charge density $\rho(x,y,z)$ (and the appropriate boundary conditions), find the electric potential $\phi(x,y,z)$.

Fig. 20 shows the operational structure of the magnetostatic field, occupying a different portion of the diagram in Fig. 18. In this case, the magnetostatic variables and their dimensions are: $\overline{A}$, the vector potential in *webers/meter*; $\overline{B}$, the magnetic induction in *webers/meter*^{2}; $\overline{H}$, the magnetic field in *amperes/meter*; and $\overline{i}$, the current density in *amperes/meter*^{2}. Again, the fundamental relations: $curl\ \overline{H} = \overline{i}$; $curl\ \overline{A} = \overline{B}$; and $\overline{B} = \mu\overline{H}$, where μ is the magnetic permeability, are fully consistent with the algebra of the 3-network problem. The solution to Eq. (93), when properly transcribed, yields the basic magnetostatic equation

$$\overline{A} = (curl\ \mu^{-1}\ curl)^{-1}\ \overline{i} \tag{103}$$

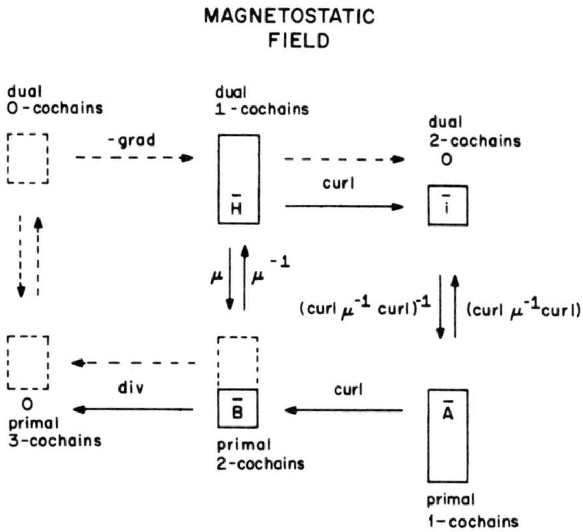

Fig. 20 - Operational Structure of the Magnetostatic Field

Figs. 19 and 20 are readily modified to describe the electro-*dynamic* and magneto*dynamic* fields. We simply introduce the time derivative operator $\partial/\partial t$, insert the derivatives $\dot{\rho}$, $\dot{\bar{D}}$, $\dot{\bar{d}}$, $\dot{\bar{A}}$, $\dot{\bar{B}}$, and change the sign of μ, $\dot{\bar{A}}$, and $\dot{\bar{B}}$ as shown below in Fig. 21.

It is now evident that both $\bar{i}$ and $\dot{\bar{d}}$, the solenoidal component of the displacement *current*, are topologically and dimensionally identical since they both belong to the subspace of *dual* 2-cocycles, $\underline{M}^2$. Accordingly, these quantities may be merged --even *identified*, as Maxwell correctly discerned -- thereby forming a link between the electrodynamic and magnetodynamic fields.

But it is also clear that $-\dot{\bar{A}}$ and $\bar{E}$ are dimensionally identical and that they belong to the space of *primal* 1-cochains, P^1, although $\bar{E}$ is confined to the subspace M^1. Hence, these two vectors may be merged -- this time by addition -- forming yet another link between the electro and magnetodynamic fields.

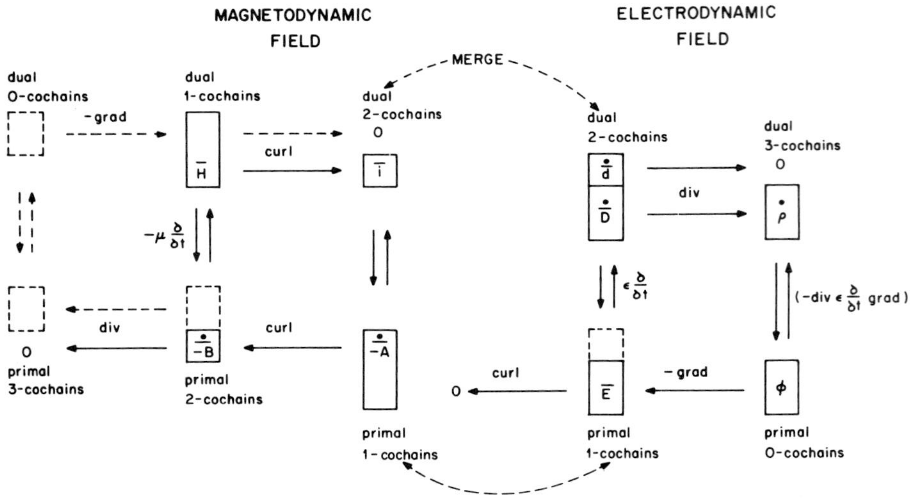

Fig. 21 - Operational Structure of Electro and Magnetodynamic Fields

The final step in arriving at an algebraic-topological description of the electromagnetic field is to introduce the electrical conductivity σ and a *prescribed current density vector* $\overline{I}$. As indicated in Fig. 22, we can combine the conductive current with the displacement current in the constitutive relation

$$\overline{I} + \overline{i} = (\sigma + \varepsilon\partial/\partial t)(\overline{E} - \dot{\overline{A}}) \tag{104}$$

analogous to Eq. (89). We also have its companion

$$-\dot{\overline{B}} = (-\mu\partial/\partial t)\ \overline{H} \tag{105}$$

By using the algebraic structure of the 3-network problem as a guide, we conclude that the <u>only</u> source of the electromagnetic field is the *current density vector* $\overline{I}$. This follows from Eqs. (95) and (96), which show that only the vectors P_1 and P^2 need be prescribed, and from the fact that while $\overline{I}$ is the counterpart of P_1, no counterpart of P^2 exists in the electromagnetic field problem.

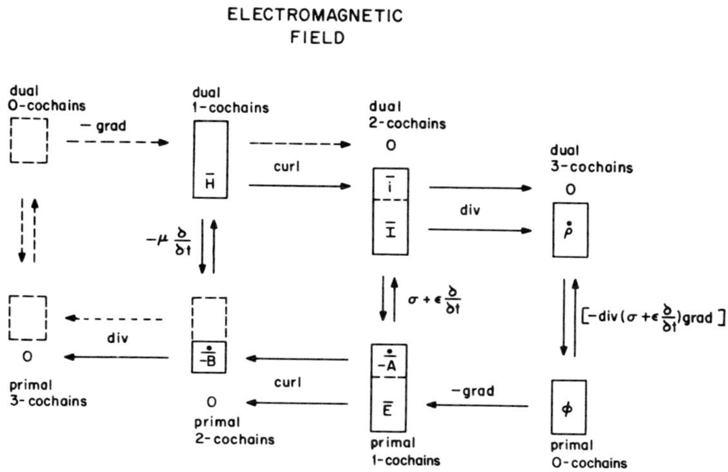

Fig. 22 - Operational Structure of Maxwell's Equations

To demonstrate that the current density vector $\overline{I}$ is the only source of the electromagnetic field, we can write Maxwell's equations in *tableau* form using Eq. (97) as a model. However, since there is no counterpart of M_2, we will delete the second column and last row of the coefficient matrix and then move the second row to last place. Then, by transcribing symbols, we have:

$$\begin{bmatrix} 1 & -S_1 & 0 & 1 & 0 \\ 0 & 1 & 0 & 0 & -curl \\ 0 & 0 & 1 & -curl & 0 \\ curl & 0 & 0 & 0 & 0 \\ 0 & 0 & -T_2 & 0 & 1 \end{bmatrix} \begin{bmatrix} \overline{E} \\ \overline{i} \\ -\dot{\overline{B}} \\ -\dot{\overline{A}} \\ \overline{H} \end{bmatrix} = \begin{bmatrix} S_1\overline{I} \\ 0 \\ 0 \\ 0 \\ 0 \end{bmatrix} \qquad (106)$$

Here, we abbreviate thus: $S_1 = (\sigma + \varepsilon\partial/\partial t)^{-1}$ and $T_2 = (-\mu\partial/\partial t)^{-1}$ and we underscore the *dual* curl operators. By symbolic elimination, we can reduce the last two rows of Eq. (106) to the counterpart of Eq. (98), namely

$$\begin{bmatrix} -curl & curl\ S_1\underline{curl} \\ \underline{curl}\ T_2 curl & -\underline{curl} \end{bmatrix} \begin{bmatrix} -\dot{\bar{A}} \\ \bar{H} \end{bmatrix} = \begin{bmatrix} -curl\ S_1 \bar{I} \\ 0 \end{bmatrix} \tag{107}$$

This shows explicitly that the vectors $\bar{E}$, $\bar{i}$, $\dot{\bar{B}}$, $\dot{\bar{A}}$, and $\bar{H}$ can all be determined from $\bar{I}$. To show that ϕ can also be found from $\bar{I}$, we can write the electromagnetic counterpart of Eq. (49) thus:

$$\begin{bmatrix} 1 & -T_1 & 0 \\ 0 & 1 & \overline{grad} \\ \underline{div} & 0 & 0 \end{bmatrix} \begin{bmatrix} i \\ \bar{E} \\ \phi \end{bmatrix} = \begin{bmatrix} T_1(\dot{\bar{A}} + \bar{I}) \\ 0 \\ 0 \end{bmatrix} \tag{108}$$

Here, we have used the abbreviation $T_1 = (\sigma + \varepsilon\partial/\partial t)$ and have underscored the *dual* divergence operator. Symbolic elimination then leads to the result,

$$\phi = -(\underline{div}\ T_1\ \overline{grad})^{-1}\ \underline{div}\ T_1(\dot{\bar{A}} + \bar{I}) \tag{109}$$

Now there is a significant difference between this equation and Eq. (102). In the electro*static* case, the charge density ρ is the source of the potential field; in the electro*magnetic* case, the current density $\bar{I}$ is the source. The reason for this distinction is easy to discern. The electrostatic field is analogous to the 1-network problem whereas the electromagnetic field is analogous to the 3-network problem. Since the sources differ in the two algebraic structures, they must also differ in the vector calculus analogs.

One final point of interest: the quantity analogous to M_3 is missing from Fig. 22. If present, it would correspond to a *magnetic* charge density. But Fig. 22 is symmetric with respect to the center of the diagram and if we rotate it 180° around this center, we obtain another structure that describes a companion set of "magnetoelectric" field equations in which the *magnetic* charge, but NOT the electric charge, is fundamental. Page and Adams have shown not only that this possibility exists but also that any *linear combination* of electric and magnetic charges leads to a self-consistent theory of electromagnetism. [32] Under such an assumption, the entire diagram of Fig. 22 would be "occupied."

4.6 *3-NETWORK MODELS FOR PARTIAL DIFFERENTIAL EQUATIONS*

The intimate relation between the vector calculus and the 3-network problem provides a sound theoretical basis for setting up network models for a large class of partial differential equations, Maxwell's equations being but one example. Sometimes, a 1-network model will suffice, but in other cases the fuller structure associated with the 3-network is required.

For instance, the general class of partial differential equations having the form

$$div\ \varepsilon\ \overline{grad}\ \phi = k + a_0 \quad + a_1 \frac{\partial\phi}{\partial t} + a_2 \frac{\partial^2\phi}{\partial t^2} \tag{110}$$

requires only a 1-network model -- with RC elements if the first time derivative term is present and with RLC elements of the second time derivative term is present. The 1-network model will suffice because the *dual 2- and 3-cochains* in Fig. 19 are isomorphic to the *primal 1- and 0-chains*.

In the case of the class

$$curl\ \mu^{-1}\ curl\ \overline{A} = \overline{K} + b_0\overline{A} + b_1 \frac{\partial\overline{A}}{\partial t} + b_2 \frac{\partial^2\overline{A}}{\partial t^2} \tag{111}$$

however, it is necessary to invoke the structure associated with a 2-network since the curl operator produces 2-cochains. Maxwell's equations require even more structure since they involve simultaneous equations of both types.

Tonti [33] has greatly extended the idea of characterizing physical phenomena in algebraic-topological terms. As his chapter in this book illustrates, this approach encompasses a very significant portion of theoretical physics -- a fact that should not be too surprising in view of the fundamental position of algebraic topology in modern mathematics. It remains to be seen, of course, what practical (i.e. *computational*) advantages will accrue from such theorizing. But the clarification of thought that inevitably follows from efforts of this kind is sure to bring about a wider appreciation of the underlying *unity* of a great diversity of physical phenomena.

5. SUMMARY AND CONCLUSIONS

1. The main thrust of this exposition has been to show the great breadth of application of the concept of a *network of interconnected elements*. At present, this concept has been applied almost exclusively to physical systems in terms of *linear graph* models, consisting of interconnected points and lines. But higher dimensional homologs exist which involve interconnected points, lines, surface, and volume elements. Their use greatly expands the area of application of network theory.
2. For 1-networks, the process of formulation consists merely of writing equations that express the three fundamental network laws (the constitutive law and two topological constraint relations) in any of a large number of equivalent forms or *tableau* equations. For 3-networks, the same principles apply except that there are two constitutive laws and six topological constraints to consider. In either case, numerical solution by sparse matrix techniques, with optimal ordering, is recommended as being most efficient for large problems. But when matrix sparsity is not important, symbolic reduction of the basic network equations to the greatest possible extent is preferrable.

3. We have described Roth's abstract characterization of the 1-network problem and shown how it can be extended to the 3-network problem. We have also shown that the vector calculus operators *grad, curl, div* are essentially identical in character to the *coboundary* operators in the 3-network problem. Finally, by introducing the concept of a *dual 3-network*, we have derived an *operational structure* for the vector calculus which is isomorphic to the *algebraic structure* of the 3-network problem. On the basis of this isomorphism, we have explained the algebraic-topological character of Maxwell's equations for the electromagnetic field.

4. In summary, by examining and exploiting the algebraic-topological foundations of network theory, we have shown that the network concept is indeed a powerful unifying principle in engineering and the physical sciences.

6. REFERENCES

1. Wu, F. F., "Diakoptic Network Analysis", *Proc. IEEE Power Ind. Comp. Appl.*, pp. 364-37], June 1975.
2. Branin, F. H., "A Sparse Matrix Modification of Kron's Method of Piecewise Analysis", *Proc. IEEE Int'l Symp. on Circuits and Systems*, pp. 383-386, April 1975.
3. Bhat, M. V. and Kesavan, H. K., "Sparse Matrix Techniques in the Theory of Decomposition", *IEEE Power Eng. Soc. Summer Mtg*, Paper A75 450-7, July 1975.
4. Tinney, W. F. and Walker, J. W., "Direct Solutions of Sparse Network Equations by Optimally Ordered Triangular Factorization", *Proc. IEEE*, vol. 55, pp. 1801-1809, Nov. 1967.
5. Hachtel, G. D., Brayton, R. K., and Gustavson, F. G., "The Sparse Tableau Approach to Network Analysis and Design", *IEEE Trans. on Circuit Theory*, vol. CT-18, pp. 101-113, Jan. 1971.
6. Roth, J. P., "An Application of Algebraic Topology to Numerical Analysis: On the Existence of a Solution to the Network Problem", *Proc. Nat. Acad. Sci.*, vol. 41, pp. 518-521, 1955.

7. Branin, F. H., "The Algebraic-Topological Basis for Network Analogies and the Vector Calculus", *Proc. Symp. Generalized Networks*, vol. 16, Microwave Research Inst. Symp. Series, Polytechnic Institute of Brooklyn, pp.453-491, Apr. 1966.
8. Branin, F. H., "The Network Problem: Its Mathematical and Practical Significance", *Digest Rec. 1969 Joint Conf. on Math. and Comp. Aided Design*, (ACM/SIAM/IEEE), pp. 92-109, Oct. 1969.
9. Branin, F. H., "A Unifying Approach to the Classical Methods of Formulating Network Equations", *Proc. IEEE Symp. on Circuits and Systems*, pp. 750-754, Apr. 1974.
10. Branin, F. H., "A Unifying Approach to the Classical Methods of Formulating Network Equations", *Tech. Rep. UWEE 73.9*, Dept. of Elec. Eng., University of Waterloo, Ontario, Dec. 1973.
11. Kron, G., *Tensor Analysis of Networks*, John Wiley, New York, 1939.
12. Branin, F. H., "Computer Analysis of Networks", *Proc. IEEE*, vol. 55, pp. 1787-1801, Nov. 1967.
13. Firestone, F. A., "A New Analogy Between Mechanical and Electrical Systems", *J. Acoust. Soc. Am.*, vol. 4, pp. 249-267, Jan. 1933.
14. Soroka, W. W., *Analog Methods in Computation and Simulation*, McGraw-Hill, New York, 1954, p. 260.
15. Olson, H. F., *Dynamic Analogies*, VanNostrand, New York, 1943.
16. Shearer, J. L., Murphy, A. T., and Richardson, H. H., *Introduction to System Dynamics*, Addison-Wesley, Reading, 1971.
17. Chua, L. O., "Memristor -- The Missing Circuit Element", *IEEE Trans. on Circuit Theory*, vol. CT-18, pp. 507-519, Sept. 1971.
18. Chua, L. O. and Tseng, C. W., "A Memristive Circuit Model for P-N Junction Diodes", *Conf. Rec. 7th Asilomar Conf. on Circuits, Systems, and Computers*, pp. 637-643, Nov. 1973.
19. Oster, G. F., "A Note on Memristors", *IEEE Trans. on Circuits and Systems*, vol. CAS-21, p. 152, Jan. 1974.
20. Branin, F. H., "DC and Transient Analysis of Networks Using a

Digital Computer", *IRE Int'l. Conv. Rec. Part 2*, pp. 236-256, Mar. 1962. (See Appendix III.)

21. Ho, C. W., Ruehli, A. E., and Brennan, P. A., "The Modified Nodal Approach to Network Analysis", *Proc. IEEE Int'l Symp. on Circuits and Systems*, pp.505-509, April 1974.
22. Bashkow, T. R., "The A-Matrix, New Network Concept", *IRE Trans. on Circuit Theory*, vol. CT-4, pp. 117-119, Sept. 1957.
23. Branin, F. H., Hogsett, G. R., Lunde, R. L., and Kugel, L. E., "ECAP II -- A New Electronic Circuit Analysis Program", *IEEE Journ. Solid-State Circuits*, vol. SC-6, pp. 146-166, Aug. 1971.
24. Bryant, P. R., "The Explicit Form of Bashkow's A-Matrix", *IRE Trans. on Circuit Theory*, vol. CT-9, pp. 303-306, Sept. 1962.
25. Weeks, W. T., Jimenez, A. J., Mahoney, G. W., Mehta, D., Qassemzadeh, H., and Scott, T. R., "Algorithms for ASTAP -- A Network Analysis Program", *IEEE Trans. on Circuit Theory*, vol. CT-20, pp. 628-634, Nov. 1973.
26. Fenves, S. J. and Branin, F. H., "A Network-Topological Formulation of Structural Analysis", *J. Struct. Div.*, Amer. Soc. Civil Engr., vol. 89, pp. 483-514, Aug. 1963.
27. Kron, G., *Diakoptics: The Piecewise Solution of Large-Scale Systems*, McDonald, London, 1963.
28. Hurewicz, W., and Wallman, H., *Dimension Theory*, Princeton University Press, Princeton, N. J., 1948. (See Chap. VIII.)
29. Veblen, O., *Analysis Situs*, 2nd Ed., vol. 5, part 2, Am. Math. Soc. Colloquium Publications, New York, 1931.
30. Lefschetz, S., *Introduction to Topology*, Princeton University Press, Princeton, N. J., 1949. (See pp. 23-25.)
31. Seshu, S., and Reed, M. B., *Linear Graphs and Electrical Networks*, Addison-Wesley, Reading, 1961.
32. Page, L. and Adams, N. I., *Electrodynamics*, VanNostrand, New York, 1940.
33. Tonti, E.,"On the Formal Structure of Physical Theories", *Quad. di Gruppi di Ricerca Matematica*, Istituto di Matematica del Politecnico Milano, 1975.

SYSTEMS STRUCTURE AND STABILITY

F. J. EVANS & A. ABD-ALI
Dept. of Electrical & Electronic Eng.
Queen Mary College
University of London
London, England

The future belongs to those who can manipulate entropy. Those who only understand but energy will be only accountants.

KEFFNER

1. INTRODUCTION

The recognition and identification of the *physical* structure of any complex system can be the source of more profound insight into dynamic behaviour. Mathematical structure has provided the means for explicit or analytic solution in linear analysis, but the physics is the basis of the behaviour mechanism, and the nature of this is the same in all systems regardless of linearity. Ideally the mathematics and the physics in any study should be brought into the closest union, and it is in electrical network theory that this occurs most naturally. Such systems possess the unique property that their physical structure is isomorphic to their abstract topology, which can be represented by a linear graph, from which the descriptive equations can be methodically derived. It is this fact that allows distinctive physical function to be associated with a particular element of the system matrix. Thus if the structure of some arbitrary set of equations can be related to, or transformed into, a canonical network form, considerable understanding can result of the physical mechanism which governs the transient motion of the system described. It is also possible to show how the natural time evolution of the system is related to the general principles of irreversible thermodynamics. This approach transcends all linear/nonlinear classifications and so provides a general global view of dynamic systems behaviour, and, further, it is often possible to show how the techniques of linear analysis are embedded as special cases within this general view. Indeed, a qualitative assessment can be provided in cases where the nonlinearity is so severe that even numerical methods produce no conclusive results.

The conventional 'state space' description invokes a rather superficial geometric interpretation, but this can be greatly strengthened by the imposition of a more fundamental geometric framework which 'governs' the whole transient motion. This can be achieved without explicit solution by exploiting the structure exhibited by the equations.

The well ordered symmetrical or skew-symmetrical form of network equations is not, unfortunately, present in most models presented to, or even derived by, the systems analyst. It could be claimed in such cases that this implies that the equations are physically invalid (i.e. invalid as descriptions of physical processes), but it would be impossibly idealistic to demand well structured equations as a necessary condition for analysis. Experience has shown that there is some taxonomy of structural forms that bear both physical and mathematical significance, and that in many cases a wide variation can be accommodated, within certain broad limits, without the loss of characteristic properties.

The basis of this conceptual approach to qualitative analysis will be described and its application to certain grossly non-linear problems, such as those occurring in protein synthesis in living cells, will be demonstrated. It can also be shown that the existence and nature of limit cycles can be effectively studied together with problems of global and local stability, bifurcation and coupled non-linear oscillations. In systems for which there is no accepted 'physics' (e.g. economics) and in those for which identification is hindered by the fact that a correlation between the nature of the input and the internal processes is completely unknown (biological circaidian rhythms), the ability to impose some equivalent physical mechanism on a minimal mathematical model, that provides a qualitative fit to observed behaviour, can provide a useful foundation for analysis.

2. SOME ASPECTS OF SYSTEMS STRUCTURE

The central position occupied by network theory in discussions of physical analogies suggests that it is networks, per se, that form the archetypal class of system. It would be nearer the truth, however, to say that it is in network theory that the underlying algebraic structure of physical theories is most explicitly demonstrated. The work of Branin [1], which developed from that of Kron [2], can now be seen in the much broader frame of generalised algebraic topology that Tonti proposes, and describes in another chapter of this volume.

Here we will discuss the problems that face the systems analyst if either he is presented with some set of equations which, it is purported, describe some physical (or non-physical) system, or he is faced himself with the task of identifying, or introducing, the physics into the problem. It is suggested that an understanding of how dynamic systems 'work' in general, that is of the 'mechanism' by means of which they operate, can provide a useful tool in design and analysis. Also such an understanding can be employed in mapping the given problem back into some more recognizable network form, the structure of which can then be exploited.

It has been shown elsewhere [3] [4] that the significant aspects of physical theory that impinge on this problem can be found in thermodynamics, and variational mechanics in addition to electrical network theory. It is a naturally attractive concept that the physical world is governed by 'variational laws'; that is, that events occur in the most economical manner. There still remains much work to be done in establishing links between formally established potential functions, and those which arise naturally from the pre-identified physical structure of the system.

Normally the definition of potential functions, a priori, rests heavily on an adequate definition of physical variables [5], and it was classically in thermodynamics that such questions were originally posed [6]. The need to deal with non-equilibrium situations has been the motivation, more recently, for the development of modern irreversible thermodynamics [7]. In thermodynamics the definition of significant 'potentials' arose naturally from the need for an aggregation of information, and also for some 'qualitative' method of system description.

It is an interesting historical fact that it was Poincare who first coined the phrase 'Qualitative Dynamics', and also who first conceived the nature of algebraic topology on which Tonti has based a formal structure for physical theories.

The most usual form of descriptive equations adopted in modern dynamic systems analysis is the state space form of sets of first order differential equations

$$\dot{x} = A(x)\ x + \text{(independent input terms)} \tag{1}$$

These express a relationship between two vectors; a displacement vector $\underset{\sim}{x}$ and its associated velocity vector $\dot{\underset{\sim}{x}}$, and the nature of this relationship is contained in the operator, linear or nonlinear, which has a matrix form. By considering the simple second order oscillator, various associated interpretations can be given to the time behaviour of the system, and some of these can be generalized to more complex systems.

Any matrix A can be expressed as a sum of two components, A^+ a symmetric matrix, and A^- a skew-symmetric matrix, such that

$$A = A^+ + A^- \tag{2}$$

where $$A = \frac{1}{2}(A'+A) + \frac{1}{2}(A'-A)$$

and since it is always true that

$$x'.A^-x = 0 \tag{3}$$

we see that the velocity vector $\dot{x}$ is the resultant of two vectors one of which is orthogonal to x as shown in figure 1.

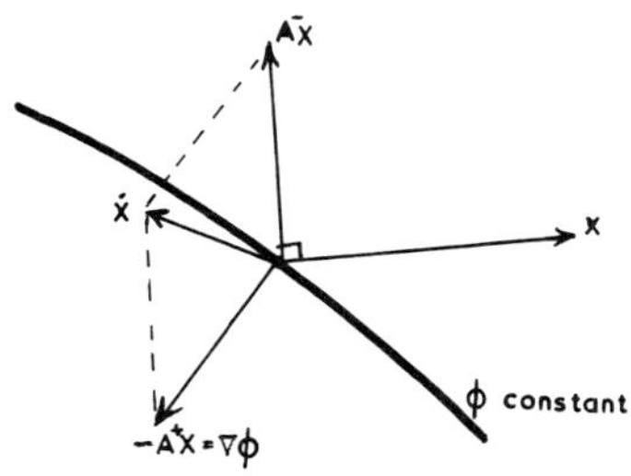

Fig.1

The stability of the system will be governed by the phase relationship of the x vector and the $\dot{x}$ vector, for if it can be shown, for example, that this is *always* greater than 90° the system must be globally stable. If, however, it is *always* less than 90° the system will globally unstable. This will be reflected, in the linear case, by the sign properties of the real part of the eigenvalues. The problem is more difficult in the case where the operator A(x) is nonlinear, for then the phase relationship may vary across both of these conditions, and no net stability properties may be deducible. It is interesting to see how this simple interpretation can also be related to another of a functional form, even in n-space.

The angle θ between the two n-vectors is given by

$$\cos \theta = \frac{\langle x.Ax \rangle}{|x|\,|Ax|} \tag{4}$$

and since, by definition, the denominator of this form is positive, it is a necessary condition that the quadratic form $\langle x.\ Ax \rangle$ be negative definite for the system to have global stability. The relation of this condition to more sophisticated Liapunov analysis is obvious, and it is worth noting at this point that it is the difficulty of being able to recognize naturally occurring potential functions, associated with arbitrary sets of equations, that has been the motivation for all such functional analysis.

These concepts can be further supported by considering the simple second order differential equation.

$$\ddot{x} + a\dot{x} + bx = 0$$

For a spring, mass, damper system such an equation can be considered as a force equation

$$M\ddot{x} + B\dot{x} + Kx = 0$$

in which M is the mass, B the damper coefficient and K the spring constant. This equation then states another vectorial relationship, and if $B = 0$ then the other two force vectors must be equal and opposite, and the implied velocity vector must be orthogonal to both of these. If, however, $B \neq 0$ then the equilibrium required can only be achieved by the three vectors as shown in figure 2, and we see immediately that this demands a non-orthogonal relationship depending on the sign of B. If the velocity always has an in-phase component with the x vector ($B<0$), the

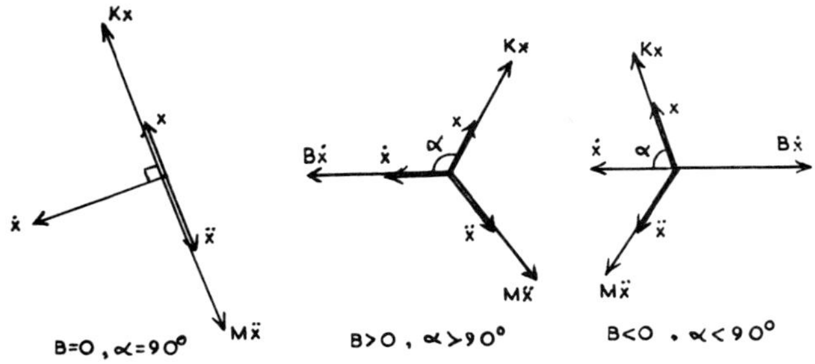

Fig. 2

system will be unstable, and if B>0 the system will be stable because the velocity will have a counter-phase component with the x vector.

If the question is now asked 'Can we extract from the structure of any operator A the necessary information to define the field structure in x-space that must generate any time trajectory?' the analysis can be further extended.

Associated with any symmetric matrix there can exist a quadratic form ϕ such that

$$-\text{grad}\ \phi = A^{+}x \tag{5}$$

providing that,

$$\text{curl}\ A^{+}x = 0 \tag{6}$$

Furthermore, associated with any skew-symmetric matrix there can exist a vector potential function Ψ such that

$$\text{curl}\ \Psi = A^{-}x \tag{7}$$

providing that,

$$\text{div}\ A^{-}x = 0 \tag{8}$$

Assuming that both of these secondary conditions are satisfied, it is now possible to construct a complete representation of the underlying field structure, as shown in a later section of this chapter. But firstly we will attempt to relate these two potential functions - one scalar and the other vectorial - to the possible physical structure of the equations.

It is well known that in the case of electrical networks it is possible to correlate, systematically, the values of the actual physical components with specific elements in the matrix, A, of the network state space equations. As stated in the introduction, this results from the isomorphism between the actual physical structure of the network (i.e. its connectivity) and the abstract linear graph from which the equations can be systematically derived. The following simple circuit will serve as an example.

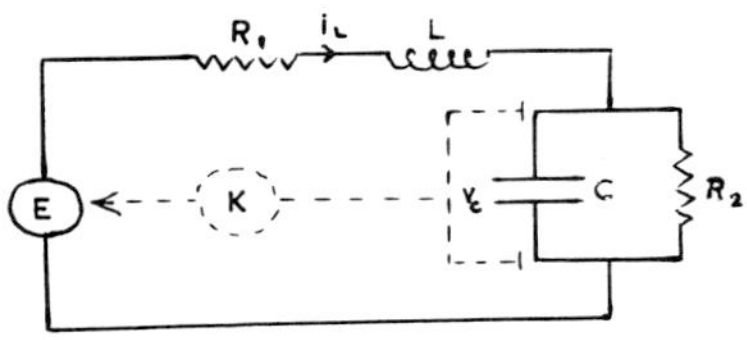

Fig.3

There is an added advantage to the network type problem, namely that the separation of the network into tree and co-tree provides immediately a set of independent variables. The choice of which set to adopt still remains, of course, with the analyst himself, and cannot be completely dissociated from further considerations of the nature of physical variables, as will be shown later. Whatever tree is chosen it is still possible to state *always*

$$\begin{bmatrix} i_t \\ \\ v_1 \end{bmatrix} = \begin{bmatrix} 0 & D \\ \\ -D' & 0 \end{bmatrix} \begin{bmatrix} v_t \\ \\ i_1 \end{bmatrix} \tag{9}$$

Embedded in this simple relationship are the origins of Tellegen's fundamental theorem that, over the whole network

$$\langle i, v \rangle = 0 \tag{10}$$

and also the existence of a certain class of self-adjoint operators defining significant bi-linear forms [3].

From equation (9) and the substitution of the constitutive relationships for the network elements we can obtain a set of state space equations of the form

$$\begin{bmatrix} \dot{i}_L \\ \dot{v}_C \end{bmatrix} = \begin{bmatrix} L^{-1} & 0 \\ 0 & C^{-1} \end{bmatrix} \begin{bmatrix} -R_1 & -1 \\ 1 & -R_2^{-1} \end{bmatrix} \begin{bmatrix} i_L \\ v_C \end{bmatrix} + \begin{bmatrix} L^{-1} \\ 0 \end{bmatrix} E$$

Partitioning of the matrix gives

$$\begin{bmatrix} L^{-1} & 0 \\ 0 & C^{-1} \end{bmatrix} \left[\begin{bmatrix} -R_1 & 0 \\ 0 & -R_2^{-1} \end{bmatrix} + \begin{bmatrix} 0 & -1 \\ 1 & 0 \end{bmatrix} \right]$$

and it can be seen immediately that the resistive elements only affect the symmetrical component, so that the ϕ function will be of a dissipative nature. Also the skew-symmetric component is governed only by the storage elements, and therefore, the rotational properties of the resultant field (i.e. the periodicity) are, as one would expect, dependent only on the presence of the inductance and capacitance. These properties are generalisable to networks of any order.

To generate the state space equations above, the choice of tree is restricted to that which contains all the capacitors and the voltage sources, with all inductors and current sources (if any) in the co-tree and resistors allocated as required. Although this form gives the distinct advantage, from some points of view, of differential equations, it is, from others, mathematically more complex. It is well-known that the so-called Brayton-Moser Mixed Potential Function [8] is related to the state space form in the following way

$$\begin{bmatrix} v_L \\ i_C \end{bmatrix} = \begin{bmatrix} L\dot{i}_L \\ C\dot{v}_C \end{bmatrix} = \begin{bmatrix} -R_1 & -1 \\ 1 & -R_2^{-1} \end{bmatrix} \begin{bmatrix} i_L \\ v_C \end{bmatrix} + \begin{bmatrix} E \\ 0 \end{bmatrix} = \begin{bmatrix} \dfrac{\partial P}{\partial i_L} \\ \dfrac{-\partial P}{\partial v_C} \end{bmatrix} \tag{13}$$

which is not a 'pure gradient' form, unlike equations obtained from other tree selections.

This point will be given greater significance in a later section.

The simple foregoing example can also be used to illustrate certain significant aspects of the introduction of feedback to system models. If it is assumed that the source term E is controlled by the relationship $E = kv_C$ the equations now become

$$\begin{bmatrix} v_L \\ i_C \end{bmatrix} = \begin{bmatrix} L\dot{i}_L \\ C\dot{v}_C \end{bmatrix} = \begin{bmatrix} -R_1 & -1 + k \\ 1 & -R_2^{-1} \end{bmatrix} \begin{bmatrix} i_L \\ v_C \end{bmatrix} \tag{14}$$

and we see that the skew-symmetry has been destroyed. It can, however, be restored by a simple variable transformation

$$v_C^* = (1 - k)v_C$$

We then obtain

$$\begin{bmatrix} L\dot{i}_L \\ \frac{C}{(1-k)}\dot{v}_C^* \end{bmatrix} = \begin{bmatrix} -R_1 & -1 \\ 1 & -\frac{1}{R_2(1-k)} \end{bmatrix} \begin{bmatrix} i_L \\ v_C^* \end{bmatrix} \tag{15}$$

This can be considered as a 'passive' network equivalent of the previously 'active' feedback system, and it can be seen that the numerical values of some components have been modified in such a way that they could become negative, depending on the value of k. Such conditions would clearly relate to the conditions for instability given by normal eigenvalue analysis. It should be noted that in equation (13) the ability to express network equations in terms of a potential function depends on the existence, from an 'integrability' point of view, of the potential itself, and that this is governed by the skew-symmetry of the equations, since it is necessary that

$$\frac{\partial}{\partial v_C}\left(\frac{\partial P}{\partial i_L}\right) = \frac{\partial}{\partial i_L}\left(\frac{\partial P}{\partial v_C}\right) \tag{16}$$

In the simple example above the restoration of the skew-symmetry is possible because all functions of two variables possess a convenient integrating factor, but in higher order systems this most probably will not be the case. Therefore by the addition of 'feedback' certain very important physical properties may have been destroyed, and this is due to the implicit assumption that it is possible to conceive of feedback of this class, which is 'perfect' in the sense that the measurement of the controlling variable is possible without losses. In reality this is not possible, and we can see that approximations, in a very general way, during the modelling process can introduce physically 'unacceptable' properties in spite of the model itself being a reasonable representation of the overall system.

Thermodynamics, probably more than any other discipline, is intimately concerned with the nature of physical variables, as stated earlier. Furthermore, from a thermodynamic point of view, stability is dependent on a simple inequality between the rate at which energy is stored in a system and the rate at which it is being dissipated. If the second is always greater than the first the system must, by necessity, be globally stable. However, this apparently simple criterion is often not decidable for reasons similar to those illustrated above, in the network problem. One can only expect a rigorous thermodynamic interpretation of system behaviour to be possible if the given description is thermodynamically 'acceptable'. As, in principle, it is not possible to infer complexity from simplicity, and to reintroduce detail lost by earlier approximations in the model, one can only resort to exploiting that structure which *does* remain in other ways.

Such possibilities will be the subject of the remaining part of this chapter, although reference to earlier work concerning the classification of physical variables and its relation to thermodynamic, variational and stability properties of system dynamics can certainly provide added insight [6] [3] [4] [9].

3. A CLASSIFICATION OF SYSTEMS

The following is a brief re-statement of work reported elsewhere [10].

Considering the basic autonomous form of state space equations

$$\dot{x} = A(x)x$$
$$= A^{+}(x).x + A^{-}(x).x \tag{17}$$

and we will assume in

case (i) that $\text{curl}\ (A^{+}(x).x) = 0$ (18)

and $\text{div}\ (A^{-}(x).x) = 0$ (19)

Then $\dot{x} = -\text{grad}\phi + \text{curl}\ \Psi\ .x$ (20)

If we now define F_σ to be the resolved part of $\dot{x}$ along $A^{+}(x).x$ and also F_τ to be the resolved part of $\dot{x}$ along $A^{-}(x).x$ then the following relations exist

$$F_\sigma = \left\langle \dot{x}, \frac{-\nabla\emptyset}{|\nabla\emptyset|} \right\rangle \frac{-\nabla\ \emptyset}{|\nabla\ \emptyset|} = -\sigma \left|\frac{\nabla\ \emptyset}{|\nabla\ \emptyset|}\right|^2 \tag{21}$$

in which $\sigma = \langle \dot{x}, -\nabla\emptyset \rangle = -F_\sigma'\ \nabla\emptyset = -\frac{\partial x}{\partial t} \cdot \frac{\partial\ \emptyset}{\partial\ x} = -\frac{\partial\emptyset}{\partial t}$ (22)

Also $F_\tau = \left\langle \dot{x}. \frac{A^{-}x}{A^{-}x} \right\rangle \frac{A^{-}x}{|A^{-}x|} = \tau \left|\frac{A^{-}x}{|A^{-}x|}\right|^2$ (23)

and $\tau = \langle \dot{x}.A^{-}x \rangle = F_\tau'.A^{-}x$ (24)

$$\sigma = \langle \dot{x}.A^{+}x \rangle = F_\sigma'.A^{+}x \tag{25}$$

In some cases it is also useful to define a function

$$G = \langle \nabla\emptyset, \nabla\emptyset \rangle \tag{26}$$

The inter-relationship of these terms is shown diagrammatically in figure 4.

If one refers back to the earlier network example it is seen that the function σ above, defining as it does the rate of change

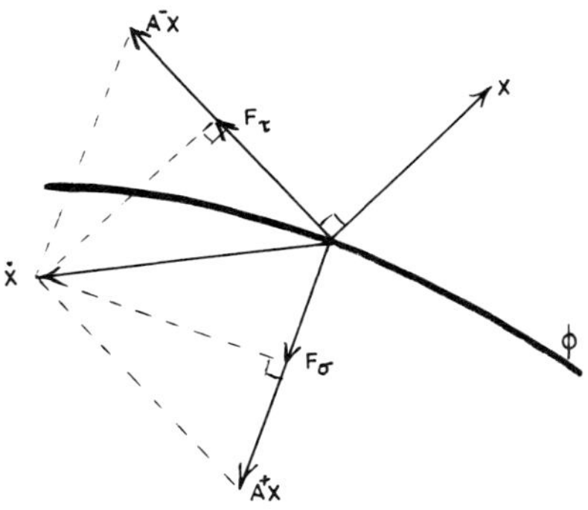

Fig.4

of $\emptyset$, can be interpreted as a measure of dissipation in the equivalent network, and must therefore have considerable bearing on the general stability properties of the system.

If $\sigma = 0$, the x vector must be tangential to the $\emptyset$ contours. Clearly the system can never remain within a closed $\sigma = 0$ contour, although a stationary point can exist *on* a $\sigma = 0$ contour, and stationary points must occur at *all* intersections of $\tau = 0$, and $\sigma = 0$ contours. The $\emptyset$ contours will only define stable or unstable regions if, and only if they enclose a stationary point.

If $G = \langle \nabla\emptyset, \nabla\emptyset \rangle = 0$, then $\nabla\emptyset = 0$ and it follows that

$$\dot{x} = A^{-}(x).x \text{ everywhere on this contour.}$$

Even if, within this class, the skew-symmetric component does not possess zero divergence, and hence a solenoidal field cannot be defined, it will always be possible to exploit the fact that $x.\ A^{-}x = 0$. Under this condition the skew-symmetric component can always be considered as forming a set of hyperspheres in state space.

Case (ii) The state space equations are partitioned into the form

$$x_i = f_i(x_i) + g_i(x_1 \ldots x_{i-1}, x_{i+1} \ldots x_n) \qquad (27)$$

in which $f_i(x_i)$ is a 'diagonal' function (i.e. *only* a function of x_i, and g_i has a zero divergence, due to the absence of x_i.

Case (iii) If the state equations are reformulated in 'phase variable' form, especially with case (ii), then the $\emptyset$ function is much more easily defined.

These cases are clearly not exhaustive, and associated with these it is often helpful in the analysis to identify one of the following forms of A matrix.

a) A 'constant' structure A matrix in which any nonlinearity is a function only of the state vector element for which it is a coefficient.

b) An A matrix in which the non-linearity is a function of a state vector element other than that for which it is a coefficient.

c) A variable structure A matrix in which more than one choice exists for the disposition of any nonlinear term within the matrix. (These forms are often less tractable).

It has been shown [10] that the following theorems are provable.

If, in the following general form

$$\begin{bmatrix} \dot{x}_1 \\ \dot{x}_2 \end{bmatrix} = \begin{bmatrix} f(x_1) & g(x_2) \\ r(x_1) & h(x_2) \end{bmatrix} \begin{bmatrix} x_1 \\ x_2 \end{bmatrix} \tag{28}$$

the functions g and r are sign definite but opposite (skew), there is no change imposed on the basic nature of the system response if these functions are replaced by constants with the same sign properties.

If the irrotational component of the system field is unstable in a certain region, the system is totally unstable in that region.

If all the elements of the irrotational component are positive semi-definite the system is unstable.

If both the irrotational and solenoidal components are stable in a region enclosing a stationary point, the system is stable in that region.

If a solenoidal component has sign definite, skew-symmetric elements, and if the irrotational component is stable in a region, the system is totally stable in that region.

If the irrotational component has elements which are sign negative semi-definite, and if the solenoidal component is stable, then the system is stable.

The Harmonic Oscillator

Let us consider the simple second order linear problem given by the equations

$$\begin{bmatrix} \dot{x}_1 \\ \dot{x}_2 \end{bmatrix} = \begin{bmatrix} 0 & 1 \\ -2 & -3 \end{bmatrix} \begin{bmatrix} x_1 \\ x_2 \end{bmatrix}$$

The eigenvalues for this system are -2 and -1, and the eigenvector matrix U is

$$U = \begin{bmatrix} 1 & 1 \\ -2 & -1 \end{bmatrix}$$

and therefore we have

$$U^{-1} = \begin{bmatrix} -1 & -1 \\ 2 & 1 \end{bmatrix}$$

The x-variables can be transformed into y variables by

$$y = U^{-1}x$$

then we have

$$\dot{y} = \Lambda y$$

in which

$$\Lambda = \begin{bmatrix} -2 & 0 \\ 0 & -1 \end{bmatrix}$$

We now have two possible interpretations given by

a)

$$\begin{bmatrix} \dot{x}_1 \\ \dot{x}_2 \end{bmatrix} = \begin{bmatrix} 0 & -1/2 \\ -1/2 & -3 \end{bmatrix} \begin{bmatrix} x_1 \\ x_2 \end{bmatrix} + \begin{bmatrix} 0 & 3/2 \\ -3/2 & 0 \end{bmatrix} \begin{bmatrix} x_1 \\ x_2 \end{bmatrix}$$

$= -\text{grad}_x\, \emptyset_x$ + rotational vector V

in which $\emptyset_x = x_2(x_1+3x_2)/2$

b)

$$\begin{bmatrix} \dot{y}_1 \\ \dot{y}_2 \end{bmatrix} = \begin{bmatrix} -2 & 0 \\ 0 & -1 \end{bmatrix} \begin{bmatrix} y_1 \\ y_2 \end{bmatrix} = -\text{grad}_y\, \emptyset_y$$

in which $\emptyset_y = 2y_1^2 + y_2^2$

As there is only one set of trajectories in time, these can be generated in two ways as is shown in figure 5.

4. SOME EXAMPLES OF STRUCTURAL ANALYSIS

By relating the condition expressed in equation 3 to the network form of the state space equations 11 it can be clearly seen that the following generalization

$$x'.\dot{x} = x'.A^+ x + x'.A^- x$$

$$= x'.A^+ x = \xi \quad \text{(say)} \tag{29}$$

can be interpreted as a form of power statement, for the RHS of this equation can be taken to represent a generalized dissipation function. Therefore the surface given by

$$\xi = x'.A^+ x = 0$$

divides the x-space into regions of positive and negative dissipation.

Example

$$\dot{x}_1 = x_1 - \frac{x_1^3}{3} + x_2 \tag{30}$$

$$\dot{x}_2 = -x_1$$

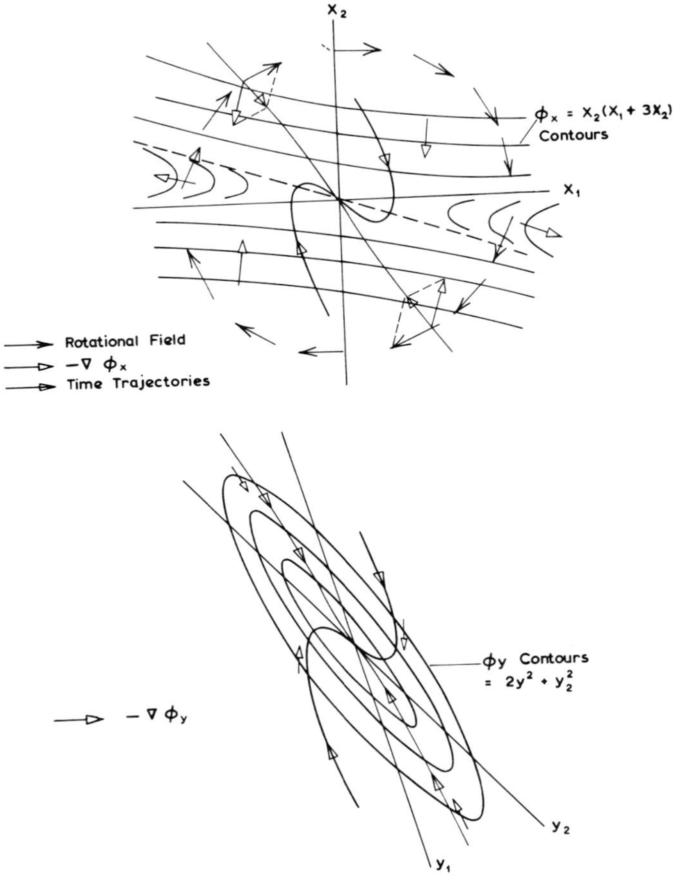

Fig. 5.

which in state space form gives

$$\begin{bmatrix} \dot{x}_1 \\ \dot{x}_2 \end{bmatrix} = \begin{bmatrix} 1 - \frac{x_1^2}{3} & 1 \\ -1 & 0 \end{bmatrix} \begin{bmatrix} x_1 \\ x_2 \end{bmatrix}$$

then we obtain

$$\xi = \langle x', \dot{x} \rangle$$

$$= x_1^2 - \frac{x_1^4}{3}$$

and, associated with this, the other functions

$$\emptyset = \frac{x_1^2}{2} - \frac{x_1^4}{12} \quad \text{and} \quad \nabla\phi = x_1 - \frac{x_1^3}{3}$$

These can all be represented on a common diagram as in figure 6. Within the inner shaded area, where $\emptyset$ is positive; the region $-\nabla\emptyset = 0$ will coincide with this and the horizontal arrows indicate the direction of $\nabla \emptyset$. Now if we imagine superimposed on this configuration the rotational field, then the resultant x vector must exhibit a 'divergence' in the inner region, and a 'convergence' in the outer region. There must therefore exist a limit cycle which represents the trajectory around which the passage is such that the negative dissipation which occurs as it passes through the inner region is exactly balanced by the positive dissipation which occurs in the outer region.

These concepts will be further extended in the next example.

Example

$$\begin{bmatrix} \dot{x}_1 \\ \dot{x}_2 \end{bmatrix} = \begin{bmatrix} -0.5 & -1.0 \\ 1.0 & -0.5x_2^2+2 \end{bmatrix} \begin{bmatrix} x_1 \\ x_2 \end{bmatrix} \qquad (31)$$

$$\emptyset = 0.25x_1^2 - x_2^2 + 0.125x_2^4$$

$$\sigma = 2.5x_1x_2 - x_1x_2^3/2 + 4x_2^2 + 0.25x_2^6 - 2x_2^4 + 0.25x_1^2$$

Examination of the $\emptyset$ contours in the region of the origin indicates a stationary point which is also a saddle point (fig. 7). The $\emptyset = 12.5$ contour is closed and exists in a positive σ region, and hence the motion around it is inwards and outside this contour all motion is globally inwards. However, as stated above, the origin is a saddle point and there must, therefore, exist a trajectory of equilibrium between the $\emptyset = 12.5$ contour and the

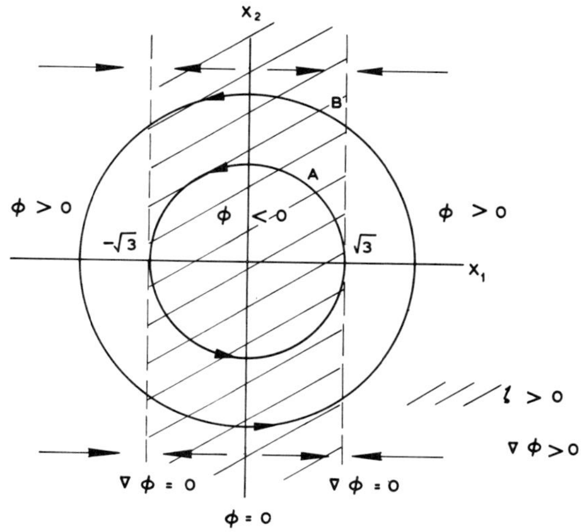

A. Trajectory in completely unstable region.

B. A trajectory spanning stable and unstable regions

Fig. 6.

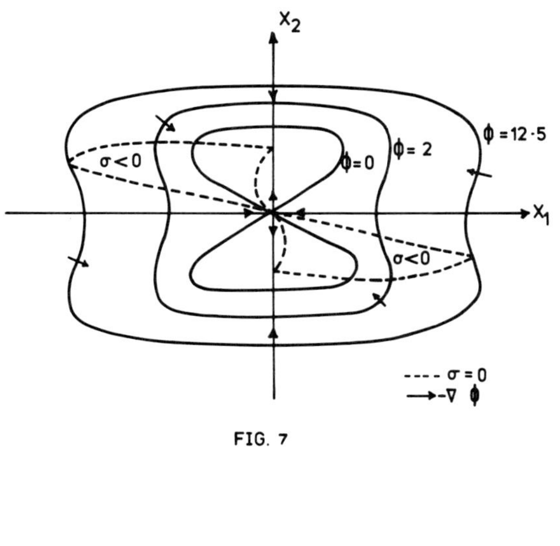

FIG. 7

origin i.e. a limit cycle which is stable (c.f. Poincaré-Bendixon Theorem [3]), and this is shown in figure 8.

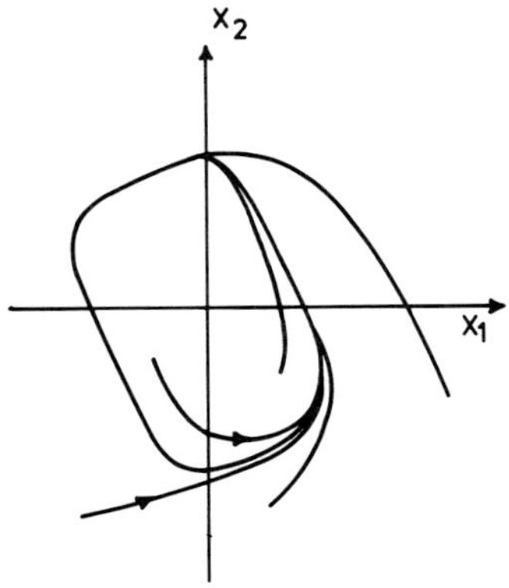

FIG. 8

A variation of the last example as follows

$$\begin{bmatrix} \dot{x}_1 \\ \dot{x}_2 \end{bmatrix} = \begin{bmatrix} -0.5 & -1.0 \\ 1.0 & 0.5x_2^2 - 2 \end{bmatrix} \begin{bmatrix} x_1 \\ x_2 \end{bmatrix} \tag{32}$$

suggests immediately that the limit cycle will now have inverse properties to that previously obtained.

The variations of $\emptyset$ and σ can be obtained, as in Fig. 9. Within the shaded circle, Theorem 2 applies and the system is stable and the limit cycle is unstable. An expansion of the cycle of Fig. 10 occurs due to the negative value of $f_1(x_1)$ and the existence of two equilibrium points which have an attractive effect on the system trajectories, as shown in Fig. 11.

Let us now consider the system

$$\dot{x}_1 = -x_2$$

$$\dot{x}_2 = x_1 - \frac{x_2^5}{10} + x_2^3 - x_2 \tag{33}$$

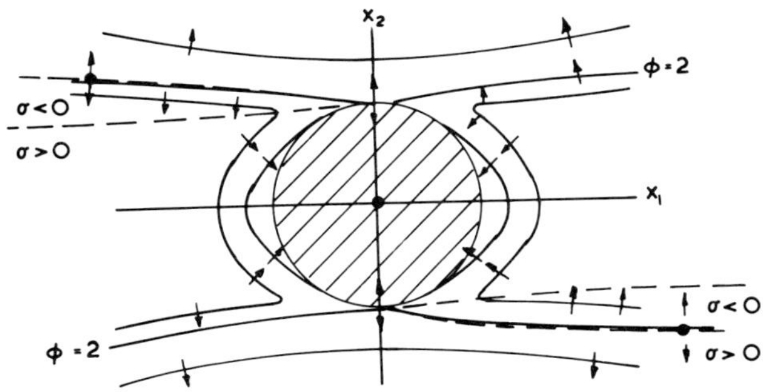

Fig. (9)

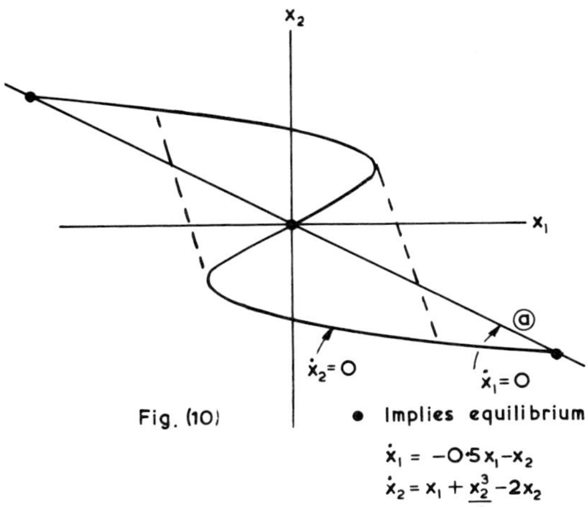

Fig. (10)

• Implies equilibrium points.

$$\dot{x}_1 = -0{\cdot}5x_1 - x_2$$
$$\dot{x}_2 = x_1 + \frac{x_2^3}{2} - 2x_2$$

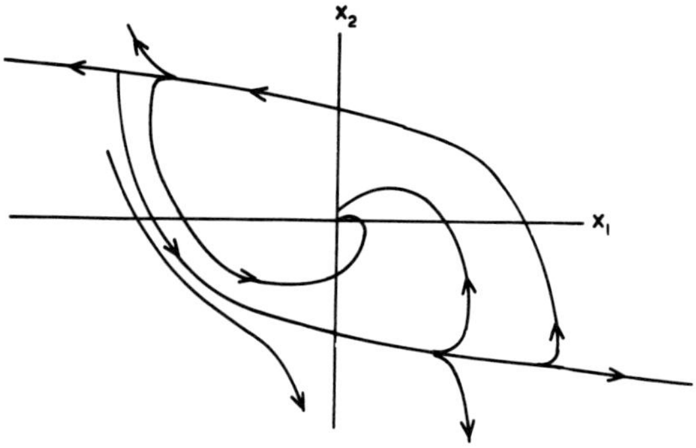

Fig. (11)

Trajectories for the system in Fig (10) @

which can be written in the form

$$\begin{bmatrix} \dot{x}_1 \\ \dot{x}_2 \end{bmatrix} = \begin{bmatrix} 0 & 0 \\ 0 & -0.1\, x_2^4 + x_2^2 + 1 \end{bmatrix} \begin{bmatrix} x_1 \\ x_2 \end{bmatrix} + \begin{bmatrix} 0 & -1 \\ 1 & 0 \end{bmatrix} \begin{bmatrix} x_1 \\ x_2 \end{bmatrix}$$

This exhibits two concentric limit cycles of which the inner is unstable and the outer is stable. The Ø contours are shown in figure 12 which indicates three separate stability regions, and a

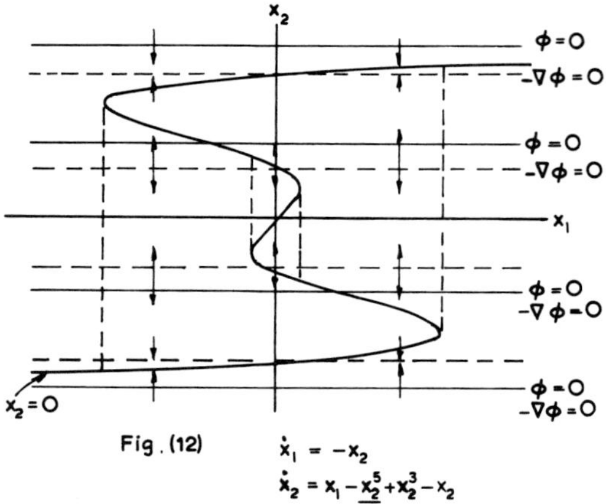

Fig. (12)

$$\dot{x}_1 = -x_2$$
$$\dot{x}_2 = x_1 - \frac{x_2^5}{10} + x_2^3 - x_2$$

stable equilibrium point. Changing the solenoidal field in this system does not affect the general stability, nor the existence of the cycles, providing its bounded rotational nature is preserved.

If we now consider the more general form of equation

$$\dot{x} = \begin{bmatrix} f_1(x_1) & g(x_2) \\ r(x_1) & f_2(x_2) \end{bmatrix} x \qquad (34)$$

but initially impose the constraint that $g(x_2) = -1$ and $r(x_1) = 1$, we can establish the equilibrium points by setting the vector $\dot{x} = 0$, and obtaining the points of intersection of

$$x_2 = -f_1(x_1)$$

$$-x_1 = -f_2(x_2)$$

For simplicity, if $f_1(x_1) = \pm 1$, and we have the form

$$\begin{bmatrix} \dot{x}_1 \\ \dot{x}_2 \end{bmatrix} = \begin{bmatrix} \pm 1 & -1 \\ 1 & \frac{-x_2^4}{10} + x_2^2 - 1 \end{bmatrix} \begin{bmatrix} x_1 \\ x_2 \end{bmatrix} \qquad (35)$$

we can see in figure 13 how this can be interpreted. This example also demonstrates how concentric limit cycles result from the presence of polynomials, as before, and these cycles are suggested by the vertical dotted lines. However, they only occur in reality if we also have compatible points of intersection (i.e. points of equilibrium). For a positive value of $f_1(x)$ two equilibrium points have been produced, as in Fig. 13 within the trajectory of a limit cycle. Examination of the Ø contours as in Fig. 14 shows that these points are unstable. Therefore they will repulse the trajectories of the stable external cycle and attract those of internal unstable one, causing the disappearance of both

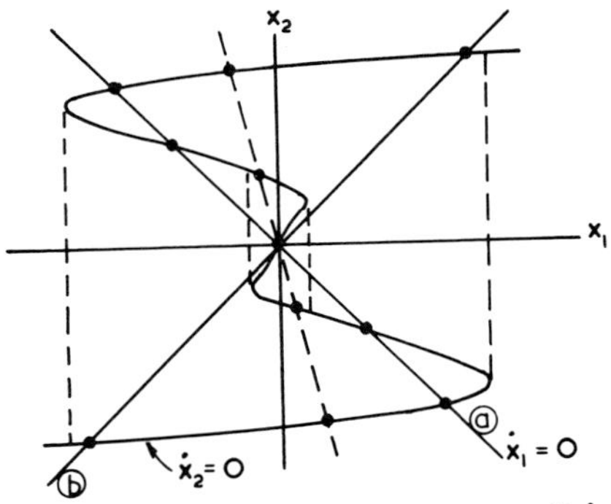

Fig.(13)

• Implies equilibrium points.

cycles, as shown in Fig. 15. The change in stability is well illustrated by the evaluation of the Ø contours as $f_1(x_1)$ increases positively in value. The unstable region will expand largely in the vicinity of the origin ($f_2(x_2)$ is small in magnitude), whereas the effect of $f_1(x_1)$ on the contours is small at large values of x_2, ($f_2(x_2)$ dominates) as in Fig. 14. However, the

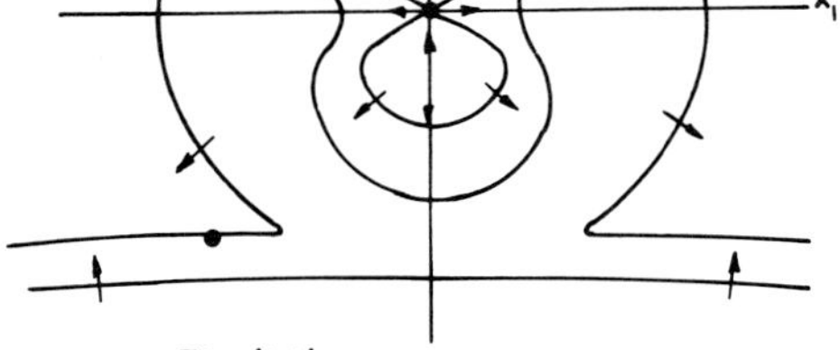

Fig.(14)

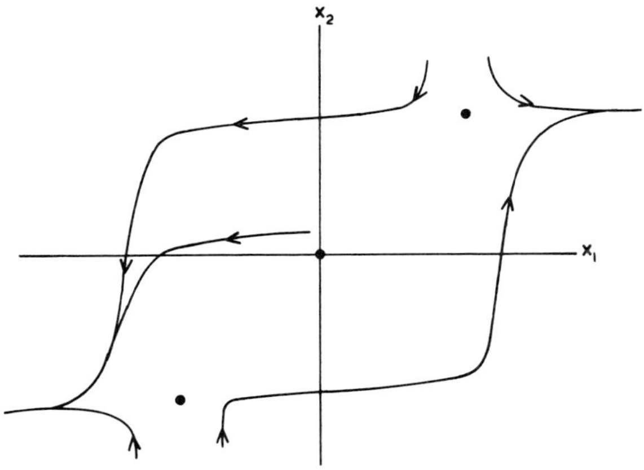

Fig.(15)

system trajectories will escape along the x_1 axis where the element of the Ø contours is more effective. If $f_1(x_1) = -1$, five equilibrium points exist, also shown in Fig. 13. The σ and the Ø contours in Fig. 16 show that three equilibrium points are stable and two are unstable. A distinct stability region surrounds the origin, and at some points around this region irrotational trajectories diverge almost into opposite directions. This is an outline of an unstable limit cycle, and it is part of the separatrices which identify the bounds of the regions of stability associated with each stable equilibrium point.

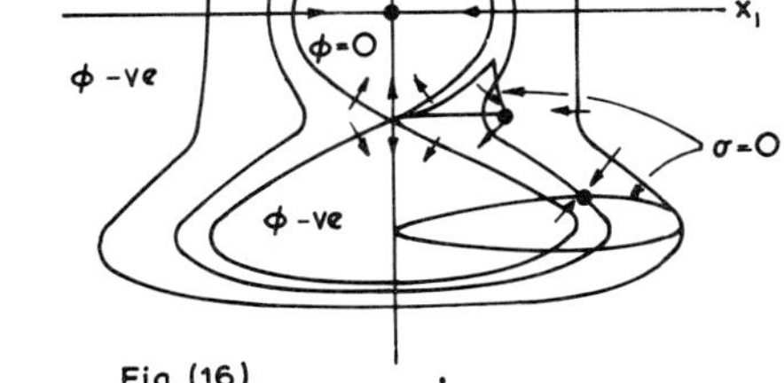

Fig.(16)

$$\dot{x}_1 = -x_1 - x_2$$

$$\dot{x}_2 = x_1 - \frac{x_2^5}{10} + x_2^3 - x_2$$

Coupled Non-linear Oscillators

Such systems can be represented, in their simplest form, by the equations

$$\begin{aligned} \ddot{x}_1 + \mu_1(x_1^2 - 1)\dot{x}_1 + ax_1 &= bx_2 \\ \ddot{x}_2 + \mu_2(x_2^2 - 1)\dot{x}_2 + cx_2 &= dx_1 \end{aligned} \tag{36}$$

in which b and d are the coupling coefficients, a and c are the frequency parameters, and μ_1 and μ_2 are the 'wave-shape' coefficients, which are not important if comparable in value in the qualitative analysis and can be assumed to be unity.

If the following variable transformations are adopted

$$\begin{aligned} \dot{x}_1 &= \int -(x_1^2 - 1)dx_1 + x_2 \\ \dot{x}_2 &= -ax_1 + bx_3 \\ \dot{x}_3 &= \int -(x_3^2 - 1)dx_3 + x_4 \\ \dot{x}_4 &= dx_1 - cx_3 \end{aligned} \tag{37}$$

we then have

$$\begin{bmatrix} \dot{x}_1 \\ \dot{x}_2 \\ \dot{x}_3 \\ \dot{x}_4 \end{bmatrix} = \begin{bmatrix} 1-x_1^2/3 & 1 & 0 & 0 \\ -a & 0 & b & 0 \\ 0 & 0 & 1-x_3^2/3 & 0 \\ d & 0 & c & 0 \end{bmatrix} \begin{bmatrix} x_1 \\ x_2 \\ x_3 \\ x_4 \end{bmatrix} \tag{38}$$

which is a more convenient form for analysis as the required conditions now exist for the integration of $\emptyset$, and both coupling and frequency coefficients appear in the off-diagonal elements, and therefore govern the rotational aspects of the behaviour. The analysis of this type of system is of considerable interest, particularly from the point of view of establishing the conditions for entrainment. It is also clear, as just suggested, that the rotational field is of central importance to the existence, stability and number of periodic solutions such systems possess. If the matrix of off-diagonal elements has a constant structure, then a fairly detailed analysis may be possible, but if it is of variable structure a number of outstanding problems appear to remain.

Volterra Type Equations

Equation of the form

$$\begin{aligned} \dot{x}_1 &= x_1 - x_1 x_2 \\ \dot{x}_2 &= x_1 x_2 - x_2 \end{aligned} \tag{39}$$

are simple examples of a broad class of equations that arises in chemical kinetics and in population dynamics. The difficulty associated with the analysis of this class of system resides in the fact that the divergence of the velocity vector is not zero. As in the previous type of problems a variable transformation can often be introduced to overcome this. If we define

$$\begin{aligned} y_1 &= \ln x_1 \\ y_2 &= \ln x_2 \end{aligned} \tag{40}$$

we then have

$$\begin{aligned} \dot{y}_1 &= 1 - e^{y_2} \\ \dot{y}_2 &= e^{y_1} - 1 \end{aligned} \tag{41}$$

and then div $\dot{y} = 0$.

Such a transformation may be far more difficult, if not impossible, in higher order and complex problems, although such transformations do have a physical interpretation in that the description has now been couched in terms of affinities rather than in concentrations, and the new form is more closely related to energy concepts than before. It is also interesting to note that the constant elements above demand a form of conservation within the system, as they possess values which are equal and opposite. This form of conservative system is well-known also in optimal control problems [11]. Furthermore as the non-linear elements occur in the off-diagonal elements, they are associated with the equivalence of non-linear storage devices, as opposed to the existence of non-linear dissipation properties which diagonal non-linearities would suggest. This class of problem still presents considerable difficulties in analysis, but the simple example given here suggests that energy based formulations may be more tractable.

Complex Biological Equations

In the mathematical modelling of the behaviour of biological systems dependent on complex chemical reactions, such as those associated with protein synthesis, equations of a highly 'irregular' form often result.* Examples of these could be

$$\dot{x}_1 = \frac{k_1 x_1^r}{1+k_a x_1^r + k_b x_2^s} - k_2 x_1 \tag{42}$$

$$\dot{x}_2 = k_2 x_1 - k_3 x_2$$

in which r and s have values ≤ 6, $k_1 \simeq k_2 \simeq k_3 \simeq 10^{-5}$, and $k_a \simeq k_b \simeq 10^{-3r}$. Unlike the two previous examples, this problem does not suggest

---

* We are indebted to Dr. B. Goodwin for initially suggesting this problem.

any useful transformation of variables, and it is highly unlikely that one could be found. So in the form presented the conditions for the extraction of the gradient scalar field and the solenoidal field certainly do not exist. However, it is still possible to exploit some aspects of the qualitative techniques developed, and a brief statement of some useful approaches will be given.

If $r \geq s$ and if r is also even, then for all possible combinations, the number of equilibrium points is constant, so it is possible to form some understanding of the nature of the solutions by assuming that $r = s = 2$. Consideration of the values of k_a and k_b, which are very small, suggest a first approximation in which these terms will be initially neglected. This overcomes the difficulty of determining the gradient field, then we can write

$$\begin{aligned} \dot{x}_1 &= k(x_1^2 - x_1) \\ \dot{x}_2 &= k(x_1 - x_2) \end{aligned} \tag{43}$$

and the ϕ and σ functions can now be determined and plotted. The justification for neglecting k_a and k_b in the first instance can be demonstrated.

The other possible conditions such as

$r \geq s$ and r being odd
$r < s$ with $r > 1$ and odd, with s even
with r even and s odd

can be assessed in similar ways, and a useful global view obtained of the possible variations in the system transient behaviour. Naturally such an approach is very specific to the problem in hand, and can in no sense be claimed as general, except that the underlying field concepts are common to other types of problem.

This problem and the previous Volterra equations suggest obliquely that difficulty of analysis may well be accentuated by unfortunate forms of equations. These may have resulted from certain approximations or assumptions made in the modelling process for ease of formulation, or from the fact that the sets of variables chosen to describe the system mask a more desirable underlying simplicity. Intuitively it is felt that in chemical problems, perhaps more than most others, the required potential functions that govern, or generate, the motion should be related to recognizable, naturally occurring, potentials. Whether this can, in fact, be reasonably expected has not yet been clearly determined. These points emphasize the need to consider physical structure in its own right as a tool of analysis.

Castastrophe Theory

In another chapter of this volume Sewell discusses aspects of this subject, first formulated by Rene Thom [12] as an extension of the earlier work on structural stability by Andronov [13]. This theory is proposed as an approach to an understanding of the evolutionary behaviour of natural processes, and the qualitative nature of the work suggests that there would be some value in a brief comparative study of the ideas of Thom and those expressed in this article.

Structural stability is concerned with the conditions, expressed in terms of the system parameters, for changed to occur in the general nature of the system behaviour. Some illustration of such changes, often manifest as sudden discontinuities, has been given earlier under the heading of bifurcation, which can occur when a periodicity is destroyed by parametric change. More generally Thom has proposed that a topological space can be defined from the system parameters, and that a potential surface be associated with this space across which the state of the system is considered to vary. Furthermore he has demonstrated that the variety of such surfaces is not unlimited.

The potential functions with which we have been concerned in this chapter have invariably been defined within the state-space of the system, and not, as in Catastrophe Theory, in some parametric space. We have also relied to a considerable extent on the structure presented by the describing equations, which itself is a reflection of the basic physics of the processes within the system. If such structure has not existed, a priori, due to approximation or other assumptions in the model, it has been possible, in some cases, to impose on the system some equivalent physics without destroying the validity of the analysis. Therefore the generation of the relevant potential functions can usually be seen as a natural outcome of the structure. Thom, on the other hand, offers a certain number of potential surfaces from which one must be selected as appropriate for the behaviour of a given problem, but appears to offer no systematic way by which such a choice can be made.

Time does not appear explicitly in the analysis of Thom, but it seems reasonable to suppose that some correlation must exist between the field analysis described here and that based on a parametric manifold on which a potential surface is defined, and this could provide an area for future study.

5. CONCLUSIONS

In a recent paper Aizerman [14] has discussed some of the problems still unsolved in control theory, and yet which are deceptively simple in their formulation. It is significant that in his paper he has discussed, as related topics, aspects of pathological behaviour (i.e. inconsistency with physical reality), the past insistence on the derivation of precise results, and possible linguistic methods (i.e. Fuzzy Set Theory) [15] of approaching those problems which has so far resisted other geometric and algebraic interpretations.

Contrary to Rutherford's dictum 'Qualitative is but poor quantitative' there is now much evidence that a qualitative approach may be the only one with any hope of success in many problems of non-linear behaviour that are presented in biology, economics and the social sciences.

The techniques briefly introduced in this article certainly fall into the category of qualitative dynamics, are firmly based in those properties concerned with the physical structure and mechanism. The recognition of these latter properties must not, however, encourage the analyst to dismiss models in which they are not present. Rather he should use his understanding of them as a basis on which to construct his analytical methods, and to adopt as his objective answering questions such as 'What is the *nature* of the dynamic behaviour?' or 'What is the structure of the phase space?'

REFERENCES

1. Branin, F.H., "The Algebraic Topological Basis for Network Analogies and the Vector Calculus", Symp. on Generalized Net works, Polytechnic Institute of Brooklyn, April 1966.
2. Kron, G., *Diakoptics*, Macdonald, London, 1963.
3. Jones, D.L., Holding, D.J. and Evans, F.J., *The Classification of Physical Variables in Network Theory and Mechanics with Applications to Variational Analysis, Physical Structure in Systems Theory*, van Dixhoorn and Evans (Eds), Academic Press, London, 1974, p. 143.
4. Korn, J., Evans, F.J. and Holding, D.J., *Thermodynamics and Field Theory as a Basis for Systems Dynamics*, ibid, p. 167.
5. Jones, D.L. and Evans, F.J., "A Classification of Physical Variables and Its Application to Variational Methods", *J. Frank. Inst.* 291, 449, 1971.
6. Callen, H.B., *Thermodynamics*, John Wiley, New York, 1960.
7. Prigogine, I., *Introduction to the Thermodynamics of Irreversible Processes*, Charles C. Thomas, 1955.

8. Brayton, R.K. and Moser, J.K., "A Theory of Non-Linear Networks", *Quart. Appl. Maths.*, 22, 1, 1964.
9. Flower, J.O. and Evans, F.J., "Irreversible Thermodynamics and Stability Considerations of Electrical Networks", *J. Frank. Inst*, 291, 123, 1971.
10. Abd-Ali, A., Fradellos, G. and Evans, F.J., "Structural Aspects of Stability in Non-linear Systems", Parts 1 and 2, *Int. J. Control*, 22, 4, 1975.
11. Seabrook, R. and Evans, F.J., "Some Questions Related to the General Variational Problem", *J. Frank. Inst.*, 296, 4, 1973.
12. Thom, R., *Structural Stability and Morphogenesis*, W.A. Benjamin, 1975.
13. Andronov, A.A. et al., *Theory of Oscillations*, Pergamon Press, 1966.
14. Aizerman, M.A., "Fuzzy Sets, Fuzzy Proofs and Some Unsolved Problems in Automatic Control", *Proceedings 6th Int. Congress IFAC*, Boston, 1975.
15. Zadeh, L.A., "Outline of a New Approach to the Analysis of Complex Systems and Decision Processes, *IEEE Trans., Man and Cybernetics*, Vol. SMC-3, January 1973.

ON MEASUREMENTS AND THEIR GROUP-THEORETICAL FOUNDATION

OLE IMMANUEL FRANKSEN
Electric Power Engineering Department
The Technical University of Denmark
Lyngby, Denmark

SUMMARY

Measurements and their scale-forms are described algebraically in terms of their group structures and geometrically in terms of directed graphs. On the basis of the resemblance between this description and Klein's Erlanger Program for the mathematical description of the different geometries, it is suggested that research into ways of crossing the disciplinary boundaries should derive from the concept of measurement as an extension of the Erlanger Program to the empirical sciences.

1. INTRODUCTION

Measurement is the fundamental element shared by the various disciplines of science. As a concept measurement transforms the empirical content of observations of reality into algebraic properties defining the abstract framework within which problem formulation and solution are confined. Hence, in crossing the disciplinary boundaries we may take advantage of the common structures imposed upon our models by the algebraic representations of measurements. A common structure of particular interest in this respect is the interpretation of scale-forms of measurements as mathematical groups interrelated in a similar manner as the different kinds of geometries have been interrelated in mathematics by Klein's Erlanger Program. It is suggested that systems science research should emulate Klein's approach in terms of group-like structures derived from the algebraic properties of measurements and, further, that these properties should be represented by a kind of directed graphs which complement the oriented graphs conventionally used to depict Kirchhoff's laws in electrical network theory or their analogous physical laws in other fields of engineering.

2. MEASUREMENT AND THE CONCEPT OF SCALES

The meaning of the concept of measurement is common sense. Yet, to come up with a definition of the term, acceptable in a wide range of human endeavour, is far more difficult. Measurements exist in a variety of forms. In physics, for example, we may agree upon forms of measurements not possible in psychology or sociology. Alternatively, forms acceptable in the social sciences may be rejected in physics as not coming up to acknowledged standards. The most widely, but not generally, accepted definition of measurement in its broadest sense, is perhaps the following (Campbell, 1921 [4]; Stevens, 1946 [16]:

Measurement is the assignment of numerals to observable qualities of objects or events according to rule.

In this definition the word *numeral*, rather than the common term *number*, is used to denote any representation of an algebraic system exhibiting some, but not necessarily all, of the properties pertaining to the common number system. The word *rule*, similarly, is intended to mean any consistent convention or direction.

The essence of the definition is that measurements establish an isomorphism between empirical aspects of objects or events and some one or more features of the algebraic number systems. In this connection it will be unfair not to disclose that different interpretations of the definition are possible. However, the formulation of the definition is such that it permits us to combine and discuss, in a meaningful and consistent manner, at least the more significant of the possible viewpoints (Franksen, 1969 [8]; 1974 [9]). Of particular interest among these interpretations is one known as the *viewpoint of symbolism*.

The main idea of this interpretation may be thus stated:

Measurements may be classified according to the set of algebraic properties which guarantees preservation of the empirical information under any mapping leaving these properties invariant.

Any representation of an algebraic system possessing a set of properties of this nature is called a *scale*. Any framework of algebraic properties, defining a one-one correspondence between a set of different scales, is called a *scale-form*.

The contested issue in the viewpoint of symbolism, is that the whole conception of measurement is based on something so ill defined as the preservation of empirical information. Essentially, the criticism on this point follows two lines of argument. First, any empirical operation is always attended by error. Secondly, how is the preserved empirical information specified in the first place. The answers, provided by this interpretation, to these questions may be given like this.

Pure reasoning tells us that, inherent in the notion of keeping the empirical information intact, is the fact that the scale-form will define a mathematical group of mappings. Indeed, one-one mappings among a set of scales, preserving the empirical information, can be defined only for mappings satisfying the algebraic postulates of a group: the associative, the identity, and the inverse laws. Further, any consistent set of properties defining a scale-form, cannot be proven empirically. To illustrate consider the questions. Is an observable quality deterministic or probabilistic. Can observable qualities of fundamentally different objects (e.g., astronomic, engineering, and atomic lengths) be classified together as the same quality in spite of the fact they are not determined empirically by the same rule or operation. Does the properties of continuity or differentiability automatically follow for any observable quality (e.g., weight or length) empirically satisfying a binary rule of addition. Clearly, theoretical concepts of this nature are given meaning only by being derived from certain postulates, the satisfaction of which cannot be proven, but which experience indicates are in agreement with observations (Franksen, 1969 [8]).

The second disconcerting element in the notion of empirical information lies in the difficulty of specifying the content matter to be preserved. Precisely what is preserved or not preserved in a particular circumstance depends on the empirical operations and the meaning we ascribe to their outcomes. For example, we may perform operations that allow us simply to identify or discriminate a particular property of an object or event. Sometimes we want to preserve nothing more than that type of simple outcome that identifies or classifies items of interest according to some observable quality. At other times we may want to go further, provided our empirical operations permit, and determine rank-orders, equal intervals, or equal ratios. The scale-form achieved depends upon the character of the basic empirical operations performed. These operations are limited ordinarily by the nature of the observable quality being scaled and by our choice of procedures. Once selected, however, the operations determine one or another scale the form of which defines the structural content to be preserved of the empirical information. Therefore, classification of measurements according to scale-form is but a systematizing device keeping intact such empirical information that we deem appropriate for description of the observable quality.

A classification of scale-forms that has been found useful in relation to engineering and economic measurements has been given in Table 1. The columns of the table list different properties of the four scale-forms given one in each row. The last column specifies the defining property, namely, the group of mappings under which the scale-form remains invariant. The designations of the groups originate in the mathematical literature (Birkhoff & MacLane, 1953 [3]).

SCALE	BASIC EMPIRICAL OPERATION	ALGEBRAIC RULE OF CLASSIFICATION	ALGEBRAIC STRUCTURE		GROUP STRUCTURE
			BINARY RELATIONS	BINARY OPERATIONS	
NOMINAL	Determination of equality	=	=		Symmetric group
ORDINAL	Determination of greater or less	<	= <		Monotonic group
INTERVAL	Determination of the equality of intervals or of differences	-	= <	+	Affine group: $x' = ax + b$, $a > 0$
RATIO	Determination of the equality of ratios	÷	= <	+ x	Similarity group: $x' = ax$, $a > 0$

Table 1. A Classification of Scales of Measurement.

The expression in economics that *measurements of some observable quality are defined up to a system of such-and-such transformations*, is another way of conveying the information of the last column (v. Neumann & Morgenstern, 1944 [14]; Alchian, 1953 [1]). The classification of scale-forms in relation to the group properties may briefly be explained as follows.

A *nominal scale* admits any permutation or one-one substitution of the assigned numerals since, basically, the scale-form partitions the measurements of the observable quality into *equivalence classes*. The scale-form for n classes is characterized by the socalled *symmetric group* of degree n that is the group of all permutations of n elements.

An *ordinal scale* introduces a rank-ordering that can be transformed by any *monotonic function* whether increasing or decreasing. In economic parlance we say that the observable quality may be measured up to any monotone transformation. The defining group has also been designated the *isotonic group* (Stevens, 1946 [16]).

An *interval scale* defines measurements of an observable quality up to any positive linear transformation or, with a loan from geometry, *affine mapping* involving addition of any constant and multiplication by any positive constant. The adding of a constant introduces a new reference or zero point whereas the multiplication by a factor changes the unit.

A *ratio scale* specifies measurements of an observable quality up to any positive linear transformation based on multiplication by any positive constant. Thus, the unit may change, but the reference or zero point is fixed or absolute. The name of the group originates in geometry where *similarity* is a one-one mapping that multiply all lengths or distances by a positive factor.

It is an interesting fact that each mathematical group defining a scale-form is a subgroup of the group listed above it in the table. For example, the affine group is also known as the "*extended" similarity group* because it consists of all affine mappings in which the change of unit is a similarity mapping. It follows as a consequence that *differences between measurements on an interval scale are measured on a ratio scale.* Another illustration is that the entire theory of finite groups can be developed indirectly as a branch of the theory of permutation groups. The subgroup property of the last column of the table originates in a dual property of the column recording the algebraic structure. The latter column is cumulative in the sense that any scale-form will exhibit the algebraic properties of the scale-form listed above it.

The physical reality of the world we live in, has made us all familiar with measurements on interval and ratio scales. Thus, we can all agree that dates on a calendar, temperatures in Celcius or Fahrenheit, positions in a coordinate system, and potentials of any kind are measured on interval scales, whereas periods of time, absolute temperature, distances, and flows or forces are measured on ratio scales. In contradistinction we find it far more difficult to identify measurements on nominal and ordinal scales. It is not because such measurements are rare, since truly they are abundant. Rather, it is because we do not recognize their use as measurements. To illustrate, telephone and street numbers are well-known concepts in our daily life, but in spite of their usefulness we hesitate to accept them as measurements though their importance stems from the fact that, usually, telephone numbers are defined on a nominal scale and street numbers on an ordinal scale.

It would be wrong to end up our discussion of measurements and scale-forms without considering those situations where we may find it difficult to ascertain the basic empirical operation defining the scale-form. In such cases we may find it useful to proceed in the following manner based on the dual cumulative properties of the last two columns of Table 1.

Assuming that the measurements are represented by numbers from the common number system we simply ask ourselves: In what ways can we transform the measured values and still have them serve all the functions originally fulfilled. We know that the values on any ratio scale can be multiplied by a constant, which changes the size of the unit. If, in addition, a constant can be added (or a new zero point chosen), it is proof positive that we are not concerned with a ratio scale. Then, if the purpose of the scale is still served when its values are squared or cubed, it is not even an interval scale. And finally, if any two values may be interchanged at will, the ordinal scale is ruled out and the nominal scale is the sole remaining possibility(Stevens, 1946 [16]).

3. ALGEBRAIC PROPERTIES OF SCALE-FORMS

Scales of measurement and their classification by scale-form are but means to an end. Obviously, we wish to go beyond a mere taxonomy to models that can explain and predict. Models are generalized conclusions derived from empirical information. They preserve this content of information in idealized forms we find useful. For the purpose of explanation we may transform the abstract representation of this information from complex or perhaps even perplexing forms to others more intelligible, but for the purpose of prediction such forms must keep intact the invariant structure of the empirical information. In truth the abstract properties of a model are confined within the algebraic framework assigned to the defining measurements by our choice of scale-forms. Consequently, we shall find it useful to investigate in more detail the algebraic framework on which models must be founded.

A natural starting point will be the cumulative property with respect to the algebraic structure common to the four scale-forms defined in Table 1. The implication of this property in relation to the partition into binary relations and binary operations, is the fact that all scale-forms will exhibit the structure of a binary relation whereas only two will admit the additional structure of a binary real number operation. Thus, in the process of establishing any model we are forced to consider the possible algebraic properties of binary relations. For ready reference let us briefly summarize these properties.

Formally, "R" denotes a *binary relation* on a given set S of elements if, given two elements x and y in the set S, either x stands in the relation R to y (in symbols, xRy), or x does not stand in the relation R to y (in symbols, $x \not R y$). Accordingly, R has the property that for any pair $x,y \varepsilon S$, either xRy or $x \not R y$. It should be emphasized that in this definition ordering is essential since, basically, the ordered pairs $\langle x,y \rangle$ and $\langle y,x \rangle$ of the two elements $x,y \varepsilon S$ are different. If, for all ordered pairs $\langle x,y \rangle$ in the set S, either xRy or yRx we say the relation R is *complete*. Alternatively, if the elements of one or more ordered pairs $\langle x,y \rangle$ in the set S are incomparable, $x \not R y$ and $y \not R x$, we say that the relation R is *partial*. Further, to any relation R we can define a *converse relation* R^* by writing yR^*x whenever xRy holds. Finally, concerning the possible algebraic properties, we say that a binary relation R in a set $S=\{x,y,z\}$ is:

- *Reflexive* if xRx for all $x \varepsilon S$;
- *Irreflexive* if xRx for no $x \varepsilon S$;
- *Symmetric* if xRy implies yRx for all $x,y \varepsilon S$;
- *Antisymmetric* if xRy and yRx imply $x=y$ for all $x,y \varepsilon S$;
- *Transitive* if xRy and yRz imply xRz for all $x,y,z \varepsilon S$.

Manifestly, a binary relation cannot possess all of the listed properties simultaneously since pairwise the reflexive and the irreflexive respectively the symmetric and the antisymmetric are mutually exclusive. Arbitrarily, but supported by the collective experience of science in general, only two of the remaining combinations are introduced as algebraic foundations of the acknowledged scale-forms: the equivalence and the partial order relations.

An *equivalence relation* is *reflexive, symmetric,* and *transitive.* By virtue of in particular the symmetric property it will partition the total set into non-overlapping subsets called *equivalence classes*. The fact that equivalence classes are either disjoint (no element in common) or identical, places the equivalence relation at the foundation of any classification scheme. The nominal scale-form submits empirical information to the algebraic structure of an equivalence relation. Thus, measurements of an observable quality are assigned the same numeral only if they belong to the same equivalence class, and, no measurements belonging to any other class is assigned that numeral. Clearly, this property remains invariant under any permutation among the classes of the assigned numerals.

A *partial order relation* is *reflexive, antisymmetric* and *transitive*. In general it admits incomparable elements in the sense of the set theoretical relation inclusion. Of course, if the relation is less-than-or-equal among real numbers all elements are comparable. The ordinal scale-form imposes upon empirical information the algebraic structure of a partial order relation. Relative to the cumulative properties of the scale-forms we may say that the ordinal scale submits measurements on a nominal scale to a relation $<$ (or $>$) such that, for two measurements on a nominal scale $a \varepsilon A$ and $b \varepsilon B$ with the two classes $A \neq B$, the statement $a<b$ is true only because for all members of the two classes $a \leq b$ and $a \neq b$. If the relation $a<b$ is satisfied we say that a *precedes* b or, referring to the converse relation $b>a$, that b *succeeds* a.

In algebra the binary relation $a<b$ is called a *strict partial order*. It differs from the partial order relation because it is *irreflexive* and *transitive*. Instead of deriving the strict partial order relation from the condition that $x \neq y$ for all $x,y \varepsilon S$, we may introduce, corresponding to the antisymmetric property, a third property called the *asymmetric* property: if xRy implies $y \not R x$ for all $x,y \varepsilon S$. Well-known illustrations of this relation is strict-inclusion among sets and less-than among real numbers. In practice, however, it is usually not necessary to make any particular distinction between a partial order and a strict partial

order. This explains why in establishing the ordinal scale-form we have not considered the strict partial order separately.

The fact that the algebraic structure of the ordinal scale-form is a partial order, reveals that because of incomparable measurements the assigned numerals may not necessarily submit to transformations by any monotonic function. We may assume, therefore, that the term *isotonic* transformation, introduced by Stevens at the suggestion of the mathematical physicist G.D. Birkhoff, is intended to designate mappings that take this possibility into account (Stevens, 1946 [16]).

As a tool algebra enables us to discern the patterns and symmetries of the empirical information represented by measurements. One of the most useful algebraic observations in this respect is the metamathematical *principle of duality*. It originates in the fact that any relation has a converse relation. It applies to any model based on measurements satisfying a partial order (Birkhoff & MacLane, 1953 [3]):

> *Any theorem which is true in every partially ordered system remains true if the partial order relation R and its converse R^* are interchanged throughout the statement of the theorem.*

A widely accepted misconception in relation to the four scale-forms, seems to be that measurements on nominal and ordinal scales cannot be operated on like measurements on interval and ratio scales. Definitely, the binary number operations of addition and multiplication are not defined for measurements on nominal and ordinal scales, but nothing prevents us from introducing Boolean or other equally meaningful operations that we may find profitable.

4. DIGRAPHS AS GEOMETRICAL REPRESENTATIONS OF ALGEBRAIC RELATIONS

Descartes's work in coordinate geometry introduced the idea that geometry is a special branch of algebra. Graph theory is a particular subject within combinatorial or algebraic topology which is concerned with continuity properties of geometric structures disregarding distances and angles. Algebraic determination of the latter properties presupposes the existence of binary operations on numbers. It is not surprising, therefore, that the theory of binary relations is the algebraic counterpart of graph theory. Graphs are simply geometrical or, to be specific, topological representations of binary relations with the important exception that they cannot represent relations on infinite sets or collections of elements. Yet, where graphs are applicable a **reasoning** based on their geometrical properties is an invaluable aid in the derivation of structural consequences of binary relations. Measurements on any scale-form exhibit the properties of an algebraic relation. Accordingly, we must place graph theory at the foundation of the formulation and solution of real world problems.

Earlier, we found it necessary to introduce a mathematical set of elements as the basis for the definition of a binary relation. The reason a set was needed rather than any arbitrary collection of elements, was that the establishment of ordered pairs $\langle x,y \rangle$ must be founded on the ability to distinguish between the elements. But this property is equally well preserved if in this definition indexed collections of elements are introduced instead of the unordered sets. Onwards in the discussion of binary relations and their graphs we shall assume this to be the case.

A geometrical image of the algebraic symbolism xRy must satisfy essentially three conditions:

(1) The binary relation R must be depicted by a geometrical object which differs from the objects taken to represent the two elements x and y.

(2) The fact that in the ordered pair $\langle x,y \rangle$ x is the first and y the second component, must be clearly indicated.

(3) The final geometrical figure must represent xRy as a structural continuity devoid of absolute measures which

like distances or angles presuppose the coexistence of compatible binary operations.

In graph theory these three conditions are complied with as follows:

(i) Two types of geometrical objects are introduced; viz., points called *vertices* or *nodes* and curved or straight-line segments designated *edges*, *arcs*, or *pointers*.

(ii) Each segment is *directed* or *oriented* by an arrow which introduces an order by its direction from tail to head.

(iii) The points have no dimension and the segments are continuous of arbitrary shape and length. Interconnection of two segments (points) can take place only via a point (segment).

A *directed graph*, also called a *digraph*, is a geometrical representation in terms of these building blocks of a binary relation R on an indexed, finite collection of elements a,b,c,...,n. Conventionally, a one-one correspondence is introduced between the elements and the points or vertices of the digraph. If we have xRy in the indexed collection, indicating that the relation R is satisfied for the ordered pair <x,y>, a directed edge is established from the vertex representing the first component x to the vertex representing the second component y of the ordered pair. Thus, relative to the directed edge the tail of the arrow indicates the vertex of the first component x, called the *initial vertex*, and the head of the arrow the vertex of the second component y, named the *final vertex*. If, on the other hand, we have $x \not R y$, indicating that the relation R is not satisfied for the ordered pair <x,y>, no directed edge is introduced from vertex x to vertex y. Obviously, since the digraph represents by its directed edges a binary relation, each edge must be bounded always by two vertices depicting the ordered pair in xRy. Conversely, some vertices may remain unconnected by edges, i.e. *isolated*, by virtue of the fact that the corresponding elements are not in the relation R with any other elements of the indexed collection.

5. CONNECTEDNESS AND REACHABILITY

A consequence of the one-one correspondence between a digraph and a binary relation, is that the algebraic properties of a relation must have unique geometrical interpretations. In truth, the merit of a digraph is that it offers a simple geometrical picture of all the logical consequences that follow from the intrinsic algebraic properties of a given relation. The structural properties of a digraph in terms of which the geometrical illustrations may be given, form a subject matter of their own. As a separate branch of mathematics graph theory is based on a rather elaborate and special set of concepts. Fortunately, most of these concepts are intuitively obvious and easy to memorize because of their lucid, almost common sense geometrical explanation. A generally accepted or standardized terminology for these concepts does not exist. In the following introduction of some of the more important of these concepts we shall adhere to such terms which seem to prevail in engineering applications of graph theory. To elucidate the meaning of these concepts we shall use the digraph of Fig. 1 as an illustrative example. For this purpose the nine vertices of this digraph are labelled alphabetically from A to I while the twelve directed edges are numbered consecutively from 1 to 12. Let it be emphasized that the two sets of labels are quite arbitrary and that any set of numerals may be used to define the two index-sets.

Incidence is the continuity property that a given directed edge, disregarding its orientation, is bounded by a given node. Thus, in Fig. 1 edges 7, 8, and 9 are all *incident* on vertex G. But continuity is also a matter of traversing a digraph following an unbroken route along the edges from vertex to vertex. It is useful in this connection to distinguish between continuity properties of a digraph which are independent of the orientation of the directed edges and those which are dependent. We may describe

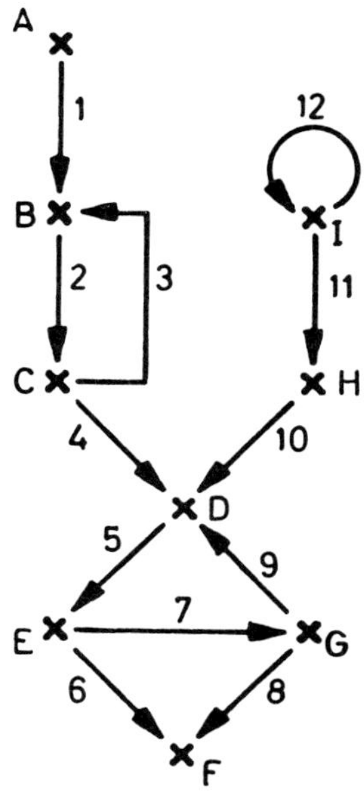

Fig. 1 - A Cyclic Digraph

these continuity properties of a digraph by relating the former to the connectedness by the edges and the latter to the reachability among the vertices. The two sets of continuity properties will be discussed in that order.

Connectedness is a term used to generally describe continuity properties of a digraph which are independent of the individual orientations of the directed edges. To fix ideas we shall describe any unbroken *route* along the edges from vertex to vertex of a digraph by an alternating sequence of vertex- and edge-indices beginning and ending with a vertex-index. Apart from the continuity property that any edge-index must be preceded by its one vertex-index and succeeded by its other vertex-index, no further restriction is placed on the concept of a route. Thus, the alternating sequence of indices may repeat vertex- as well as edge-indices. Examples of routes in Fig. 1 are sequences like B3C2B3C4D,

I12I11H, D9G8F, and D5E7G9D. Several important concepts may be derived from the notion of a route. A *path* is a route in which no edge is repeated. Hence, among the four routes illustrated previously, only the last three are paths. An *open path* is a route in which no vertex is repeated. Spontaneously, this definition may be interpreted as a dual of a path rather than as a special path. However, a little reflection will soon convince anyone that in this definition we have included implicitly the path concept. Indeed, in an index-sequence defining a route it is impossible to repeat an edge without also simultaneously repeating a vertex. The route D9G8F, quoted earlier, is an open path. A *closed path* is a path beginning and ending with the same vertex-index. Consequently, a closed path may intersect itself at some of the vertices. This makes this concept somewhat ambiguous. Therefore, we shall find it useful to introduce a further refinement. A *mesh* is a closed path in which only the beginning vertex-index is repeated. Hence, the sequences D5E6F8G9D and D5E7G9D are meshes. A rather special mesh is a *loop* which is a single edge with an identical initial and final vertex. A typical illustration of a loop is the sequence I12I.

A description of the connectedness of a digraph may be based on these definitions. Thus, a *connected digraph* is characterized by the existence of an open path between any pair of vertices. Thence, the digraph of Fig. 1 is connected. If a graph is not connected it will fall into two or more *separate parts* or *components* each of which is connected. For this reason a separate part may be called also a *connected component*. An isolated vertex is a special example of a connected component. A connected digraph in which any pair of vertices are connected by a directed edge is called a *complete* digraph. If the number of vertices is N it is readily seen that the total number of edges in a complete digraph is:

$$\frac{N \times (N-1)}{2}$$

A *tree* is a connected digraph containing no closed paths. Of course, a connected component of a digraph may also form a tree. A digraph separated into connected components all of which are trees, is called a *forest*. If a tree is selected in a connected digraph containing closed paths, it partitions the edges into two equivalence classes: the tree and the set of edges not belonging to the tree. The latter class is called the *tree-complement* or, more briefly, the *co-tree*. A partition of the connected digraph of Fig. 1 into a tree and a co-tree is illustrated in Fig. 2.

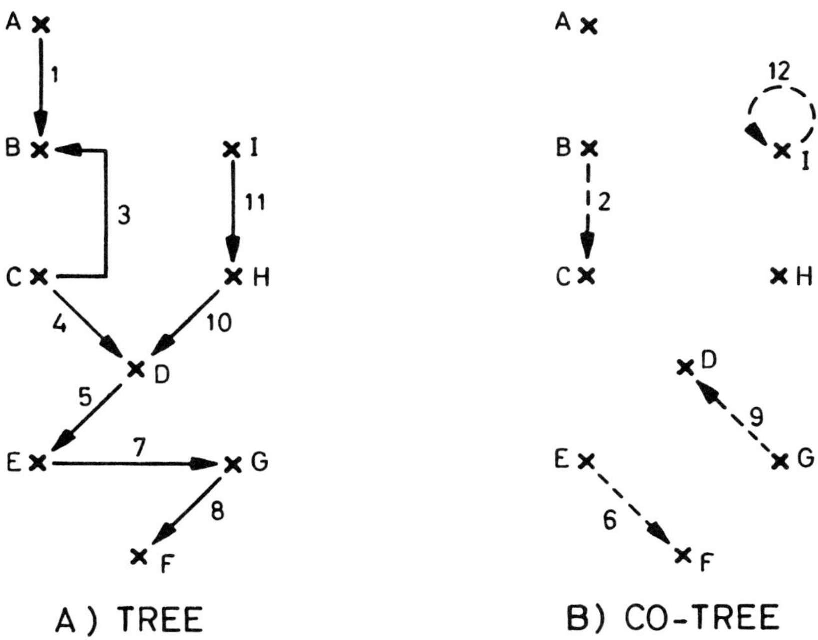

Fig. 2 - Partition of a Digraph into a Tree and a Co-Tree

The edges of a tree are called *tree-branches* or just *branches*. The edges of a co-tree are named *links* because when added one at a time to a tree they will each together with the tree define a mesh called a *basic mesh*. A loop is a special type of basic mesh involving no branches.

For a tree with N vertices it is obvious that the number of branches is (N-1). In a forest of K trees we find correspondingly that the total number of branches is (N-K). Now, consider a connected digraph with N vertices and B directed edges. If we partition this digraph into a tree and the corresponding co-tree, assuming that the number of links is M, we clearly have that:

$$B = (N-1) + M$$

This relation is a special case of the so-called *Euler-Poincaré condition* which for a digraph separated into K connected components may be written:

$$B = (N-K) + M$$

Thus, relative to this condition we may say that a connected digraph is made up of K=1 connected components. A *cut-set* of a digraph is a set of directed edges which, if removed from the digraph, will increase the value of K by 1:

$$K \leftarrow K + 1$$

For a connected digraph it is possible to reformulate this definition such that each branch of an arbitrarily selected tree will define a so-called *basic cut-set* together with an appropriate subset of links of the corresponding co-tree. For a given choice of tree a dual relationship will exist between the set of basic meshes and the set of basic cut-sets. Further extensions of this kind of duality can be introduced if the digraph is *planar*, that is, it can be drawn on a sphere (or a plane) such that edges intersect only in vertices. However, we shall refrain from a discussion of such properties since, in the present context, they are of less interest (Franksen, 1965 [7]).

Reachability is a term used to generally describe continuity properties of a digraph which depend upon the individual orientations of the directed edges. The concepts describing these properties are derived from those of connectedness in a straightforward manner. A *directed path* is a path for which the alternating sequence of vertex- and edge-indices has the property that the vertex-index immediately adjoining to the left (right) of any edge-index is the initial (final) vertex of that edge. We say that vertex y *is reachable from* vertex x if there is a directed path from x to y. Evidently, even if two vertices are connected by a path neither or only one of the vertices may be reachable from the other. To illustrate, in Fig. 1 vertex D is reachable from vertex I along the directed path I11H10D but vertex I is not reachable from vertex D. A *simple path* is a directed open path. Thus, the directed path just quoted is a simple path. A *cycle* is a directed mesh. Hence, the sequences B2C3B and D5E7G9D are cycles. The designation a *loop* is maintained as a description of the special kind of cycles exemplified by I12I.

Digraphs containing no cycles have reachability properties fundamentally different from digraphs containing cycles. To distinguish between these two types of digraphs the former are called *acyclic digraphs* and the latter *cyclic digraphs*. The fact that a digraph is acyclic does not preclude that it may contain meshes. To illustrate, the digraph of Fig. 3 is acyclic. By definition all trees must be acyclic. A certain kind of trees, called *directed trees* have distinctive reachability properties of particular interest. An example of such a tree is given in Fig. 4A. The directed tree in Fig. 4B is a special case known as a *binary tree*.

The characteristic reachability properties of cyclic digraphs derive from those of a cycle. Basically, *a cycle is a mesh in which every pair of vertices are mutually reachable*. If any pair of vertices in a connected digraph are mutually reachable the digraph is said to be *strongly-connected*, or more briefly *strong*, because of the similarity between its properties of connectedness

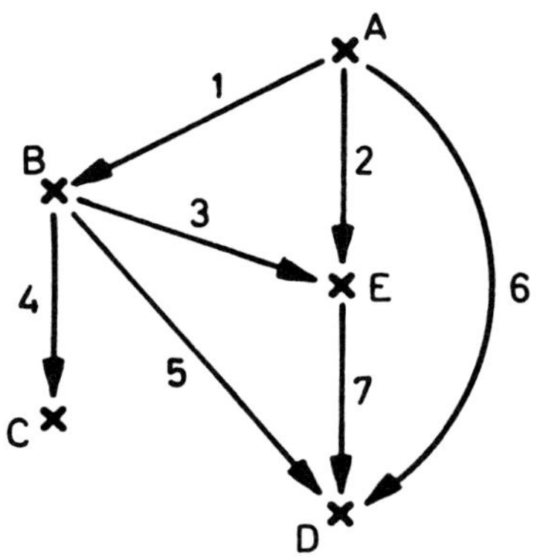

Fig. 3 - An Acyclic Digraph

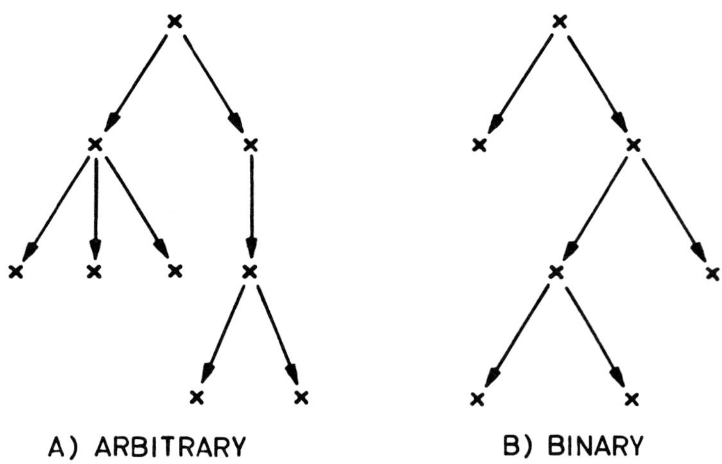

Fig. 4 - Directed Trees

and of reachability. If a digraph is disconnected into separate parts one or more of its components may be strongly-connected. Thus, the digraph of Fig. 5 consists of three strongly-connected components.

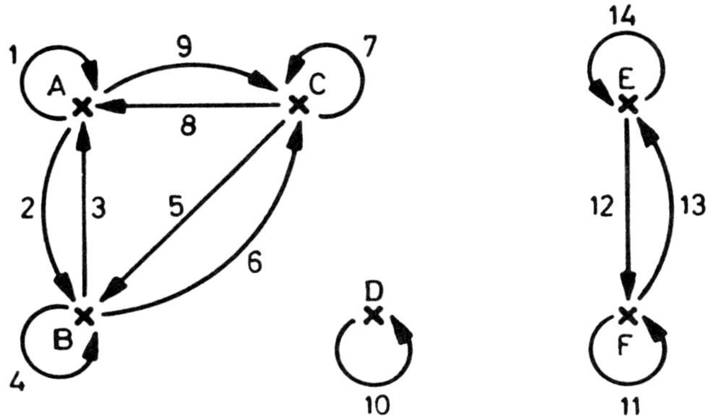

Fig. 5 - A Digraph with Strongly-Connected Components

Of course, such cases are exceptional. To discuss the more general situation let us introduce the concept of a *subgraph of a digraph* D to mean a digraph whose vertices and edges form a subset of the vertices and edges of the original digraph D. Using this terminology we will find in general that only subgraphs of connected cyclic digraphs are strongly-connected. An unambiguous description may have its problems. For example, imagine that we have isolated the strongly-connected subgraph of Fig. 6. It obviously contains two distinct cycles: A1B5D4A and A1B2C3D4A.

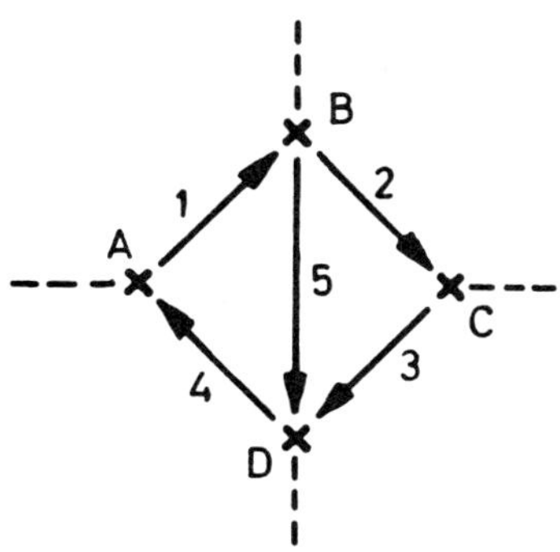

Fig. 6 - A Strongly-Connected Subgraph

Yet, the two cycles have vertices and edges in common. It may be quite complicated, therefore, to recognize and identify the strongly-connected subgraphs of a cyclic digraph from a mere listing of its cycles. A more reasonable description of each strongly-connected subgraph may be a specification of its vertices adjoined, perhaps, by a listing of its edges. To admit a unique description along these lines we must sharpen the notion of a strongly-connected subgraph. In general, we shall say that a subgraph with a special property is *maximal with respect to that property* if no larger subgraph (i.e., with more vertices or edges) contains it as a subgraph and has the property. It follows that the distinguishing feature of cyclic digraphs relative to acyclic, is that cyclic digraphs contain maximal, strongly-connected subgraphs called *strongly-connected components* or, more briefly, *strong components*. In Fig. 1, to illustrate, we have three strong components: B,C with 2,3; D,E,G with 5,7,9; and I with 12.

In principle, if the strong components are removed from a cyclic digraph the remainder subgraphs will be all acyclic. We may say, therefore, that the strong components partition a cyclic digraph into two equivalence classes of subgraphs, say, a cyclic and an acyclic. Of course, in practice we just cannot remove the strong components since, possibly, we may end up with acyclic digraphs containing edges unbounded by vertices. For the purpose of representing binary relations unbounded edges are devoid of meaning. Hence, we are forced to adopt another approach which in a meaningful manner will achieve the required partition.

The characteristic property of any strong component is that all vertices are mutually reachable. A loop is the simplest possible strong component. In relation to strong components in general we may say that, as a concept, a loop has condensed the set of mutually reachable vertices into a single vertex that is reachable from itself. Evidently, if in a cyclic digraph we shrink each strong component into a loop the resulting condensed digraph will uniquely identify the class of acyclic subgraphs. All we then have to do to produce the acyclic subgraph is simply to remove from the condensed digraph, called the *cyclic condensed digraph*, the set of edges forming loops. This process which achieves a partition of a digraph into two equivalence classes of respectively cyclic and acyclic digraphs, is called the *condensation of a digraph with respect to its strong components*. Clearly, the condensed digraph with all the loops removed will be an *acyclic condensed digraph*. In the literature, usually, the latter is designated simply *the condensed digraph* with the tacit understanding that it is acyclic (Harary, Norman, & Cartwright, 1965 [10]). As a special case we note that the condensation of an acyclic digraph leaves the digraph invariant.

The process of condensation of a cyclic digraph with respect to its strong components is exemplified in Fig. 7. Comparison of the cyclic condensed digraph with the given digraph will illustrate a couple of facts which, though evident, may be of some interest. First, if the given digraph is connected then also the condensed digraph will be connected. Secondly, two or more edges outside any strong component may be condensed in particular situations into a single edge in the same manner as all edges of a strong component are condensed into the single edge of a loop. Thus, edges 6 and 8 of Fig. 7A are condensed into the single edge 6,8 of Fig. 7B. It is not difficult to see that such situations will occur whenever two or more edges are bounded such that their initial (final) vertices belong to a strong component A and their final (initial) vertices either are a common vertex outside A or the vertices of a strong component different from A.

Comparison of the continuity properties of connectedness and reachability reveals a structural similarity of particular importance:

Connectedness and reachability display a common pattern of bilateral symmetry which partitions the associated continuity properties of a digraph into two dual sets pertaining respectively to the concepts of open and closed paths.

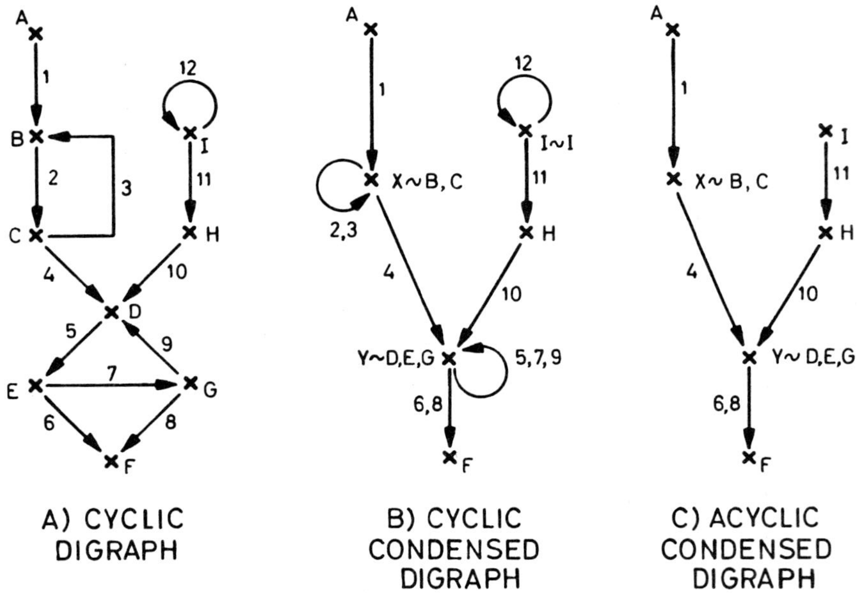

Fig. 7 - Condensation of a Cyclic Digraph

To illuminate the meaning of this statement let us focus for a connected digraph on two aspects in relation to a scheme of property classification common to connectedness and reachability.

First, any digraph may be partitioned into two equivalence classes of subgraphs exhibiting disjoint continuity properties. Thus, we have, in relation to connectedness, a tree and a co-tree defining the basic meshes and, in relation to reachability, an acyclic condensed digraph and a set of condensed loops defining the strong components.

Secondly, a unique correspondence in patterns of property exists between the disjoint subgraphs of connectedness and those of reachability. In particular, three things may be noted. Trees are special kinds of acyclic digraphs. Strong components are particular combinations of basic meshes. And, each link of a co-tree defines a basic mesh as, similarly, each edge of the set of condensed loops defines a strong component. It is because of this correspondence that the terminology of open and closed paths has been introduced to describe the structural duality of the continuity properties.

6. REACHABILITY PROPERTIES OF BINARY RELATIONS

A binary relation xRy on an ordered pair <x,y>, we recall, was represented in a digraph by a directed edge oriented from a vertex x to a vertex y. The very fact that the orientation of the directed edge identifies the first component x and the second component y of the ordered pair <x,y>, enables us to establish unique digraph representations of the different algebraic properties of a binary relation on a finite collection of indexed elements. In virtue of the latter fact a digraph will contain a finite number of vertices and, hence, directed edges (Ore, 1963 [15]).

The *converse* relation R^* to any relation R was defined by writing yR^*x whenever xRy holds. Thus, if there is a directed edge from vertex x to vertex y in the digraph corresponding to R, then there will be a directed edge with the opposite orientation from vertex y to vertex x in the digraph representing R^*. Clearly, the digraph defined by R and the digraph defined by R^* will be identical apart from the fact that their edges are of opposite orientation. In other words, the *converse digraph to a given digraph* is the latter but with oppositely directed edges.

The *reflexive* and the *irreflexive* algebraic properties are based on the ordered pair $\langle x,x \rangle$ of the same element x. In this case, therefore, the ordered pair is represented by a single vertex x. Hence, a loop depicts the reflexive property whereas no loops correspond to the irreflexive property.

The *symmetric* and the *antisymmetric* algebraic properties are based on two ordered pairs $\langle x,y \rangle$ and $\langle y,x \rangle$ of two elements x and y. In a digraph these two elements are represented by two vertices x and y. The symmetric properties implies that x and y are interconnected by two edges of opposite directions forming together a cycle. The antisymmetric property is represented in identically the same manner if $x=y$. On the other hand, if $x \neq y$, then the antisymmetric property is tantamount with the fact that only directed edge is bounded by the two vertices x and y. The purpose of the *asymmetric* property is to single out the lastmentioned digraph representation as a distinct counterpart to the symmetric interpretation.

The *transitive* algebraic property is based on three ordered pairs $\langle x,y \rangle$, $\langle y,z \rangle$, and $\langle x,z \rangle$ of three elements x, y, and z. In a digraph these three elements are represented by three vertices x, y, and z. A representation of this property will be that the existence of one directed edge from vertex x to vertex y together with another directed edge from vertex y to vertex z, implies the existence of a *resultant* edge directly from vertex x to vertex z. By repeated use of the property of transitivity it may be concluded that when there is a directed path from one vertex x to another z then there is also a directed edge from x to z. Alternatively, if the relation is *intransitive* no such resultant edges exist between the vertices on the directed path.

Transitivity is a cornerstone on which the logical processes of deduction are founded because without this property conclusions cannot be inferred. Therefore, the scale-forms of measurements are defined only in terms of transitive binary relations. But, if we confine ourselves to digraphs corresponding exclusively to transitive relations, it is readily visualized that our manner of representing transitivity will result in digraphs cluttered with edges. Not only is the handling of such digraphs quite impractical but, even worse, we destroy their utility as recognizable distinct structures. For this reason we find it useful to represent the algebraic properties of transitive binary relations by the continuity properties of digraphs rather than by the geometrical objects per se.

Reachability of a digraph and transitivity of a binary relation establish together the desired correspondence in a unique and concise way. Indeed, transitivity as well as all the remaining other algebraic properties may be rephrased distinctly in terms of reachability properties. To demonstrate this it will be advantageous to introduce a further refinement of the reachability concept. Reachability in a digraph of vertex y from vertex x, it will be remembered, was defined by the existence of a directed path from x to y. We shall say that vertex y *is immediately reachable from* vertex x, if x is the initial and y the final vertex of a directed edge. In terms of this and the other concepts of reachability the algebraic properties may be given a graph-theoretical interpretation as follows.

A relation is reflexive or irreflexive on an element x depending upon whether or not at vertex x of the corresponding digraph there is a loop along which x is immediately reachable by itself. A relation is symmetric or asymmetric on two different, but "comparable" elements x and y depending upon whether the two vertices x and y of the corresponding digraph are mutually or only unilaterally immediately reachable from each other. A relation is transitive on three elements x, y, and z when in the corresponding digraph vertex y is immediately reachable from vertex x and vertex z immediately reachable from vertex y such that there is a directed path from x to y and from y to z making z reachable from x. By repeated use of this rule it is seen that transitivity of a binary relation and reachability along directed paths of a digraph are concepts in a one-one correspondence.

The observation that the reachability of a digraph provides a simple geometrical illustration of the logical processes of deducing conclusions founded on the transitivity of a binary relation, is thought-provoking. It implies that such processes may be turned into well-defined computational schemes based on reachability properties. From a computational point of view, essentially, transitivity between vertices of a directed path is based on combinations of ordered pairs such that in each pair one element is immediately reachable from the other. This indicates that the desired computational schemes must derive from operations on ordered pairs of immediately reachable vertices. A set of concepts that turns out to be particularly helpful in this respect may be defined as follows.

The *topological length* or, more briefly, the *length* of a directed path is its number of directed edges. If a simple path is defined from vertex x to vertex y, then the *topological distance* or, simply, the *distance* from vertex x to vertex y is the length of the simple path from x to y. Since a cycle may be described as a simple path returning to its starting vertex, it is seen that the "size" of a cycle may be described as the distance along the cycle from one of its vertices to itself. Quite naturally, this brings us to the question of the distance in a simple path from a vertex to itself. Special statements or statements with a conditional clause are avoided in mathematics. Therefore, we generalize the concept of a distance by saying that *a single vertex is a simple path of zero length*. A path of length 0 is sometimes designated a *trivial* path. This generalization is consistent with the definition of distance in the sense that any vertex is reachable from itself along a directed path containing a zero number of directed edges.

Often, we find in a digraph that *more than one* simple path exist from vertex x to vertex y. In such cases, obviously, we must uniquely specify the simple path defining the distance from x to y. In practice, particularly important examples in this regard are distances defined along simple paths of minimum or maximum lengths. To identify the distance along a simple path of minimum length a terminology like the *minimum distance*, the *shortest distance*, or the *geodesic*, may be applied. Similarly, the distance along a simple path of maximum length may be designated the *maximum distance*, the *longest distance*, or the *detour*. Introduction of a terminology along these lines normally implies that we are not interested per se in any specific simple path or in the fact that two or more simple paths may exhibit the same minimum or maximum length. Further, it is important to note that whereas minimum as well as maximum distances are defined in general for acyclic digraphs, *only minimum distances are defined for cyclic digraphs*. In fact, the distance along any cycle from one of its vertices to itself must be a minimum distance since otherwise we had to traverse the directed path defining the cycle indefinitely.

In a digraph, usually, the distance from vertex x to vertex y will not equal the distance from y to x if at all x is reachable from y along any simple path. The very fact that in any digraph we may have a pair of vertices x and y such that *no directed paths exist*, say from y to x, merits a further generalization of the definition of a distance to also include a distance of a non-existent, socalled *void* or *empty simple path*. The first idea that comes naturally to mind is that a void simple path is a path of zero distance. But this concept has been used already to signify the common sense notion that the distance from a vertex to itself is zero.

An alternative idea, therefore, may be to say that since vertex x is inaccessible from vertex y the *distance* from y to x *is infinite* (Harary, Norman, & Cartwright, 1965 [10]). Of course, this choice raises several problems relative to the definition of suitable reachability operations representing consequences of transitivity. To begin with, joining simple paths end to end into a single directed path, it is quite obvious that the length of the resulting path is obtained by a mere addition of the distances of the individual simple paths. Hence, if one of the contributions is infinite the desired result will be obtained that the sum total will also be infinite. However, if we want to compare several simple paths of different distances from one vertex x to another y to find the minimum or maximum distance, then we have a problem at hand of the sign of the infinity. The fact is that, to be consistent, a comparison for minimum distance must be based on a plus infinity, whereas a comparison for maximum distance requires the introduction of a minus infinity if we wish to define the distance of a non-existent simple path by infinity. Of course, the basic assumption underlying these applications of a signed infinity distance, is the fact that *the distance of any finite simple path is non-negative.*

The dependence upon type of comparison, whether for maximum or minimum distance, of the sign of the distance of a void simple path, serves to emphasize that with respect to these comparisons the void distance should behave like a zero under addition. Similarly, the desired property that the resulting directed path should be void when joining together a void simple path with other simple paths of finite distances, indicates that in this type of operation a void distance should behave like a zero under multiplication. Thus, even if the void distance is not zero it certainly shares some common properties with that number.

In mathematics the symbol zero usually plays a dual role of a numerical quantity and an empty or null element. Indeed, as a symbol zero was originally introduced into our positional number system to indicate empty spaces on an imaginary abacus. In the algebra of unordered sets this role of the symbol zero is played by the *empty* or *void set* $\emptyset$. To illustrate this idea of $\emptyset$ as a "spaceholder" like zero, just consider the fact that the set $\emptyset$ defines any set of no elements, but that $\{\emptyset\}$ is a set of only one element, namely $\emptyset$. In relation to indexed collections of elements it is certainly feasible, corresponding to any index or set of indices, to introduce the empty set ∅ to play the role of a "space-holder" whenever no element exists relative to the given index. The role of the *space* in a set of alphabetic characters exemplifies the usefulness of this concept. Basically, since topological distances of finite simple paths are scalar numbers all operations defined on topologically distances must be scalar operations. Hence, we are forced to introduce the definition that *the distance of a void simple path is the empty scalar* indicated by the symbol ∘. Evidently, to make this definition useful we must also define the necessary binary operations between any pair of scalar numbers including zero and the empty scalar. Denoting the pair of scalar numbers A,B and minimum and maximum respectively of A,B by $A \lfloor B$ and $A \lceil B$ respectively. Table 2 extends these two operations to include the empty scalar. It is readily seen that this table is isomorphic to addition and that the operations are associative and commutative.

$A \lfloor B$ or $A \lceil B$	*B EMPTY* $B = \circ$	*B NON-EMPTY* $B \neq \circ$
A EMPTY $A = \circ$	∘	*B*
A NON-EMPTY $A \neq \circ$	*A*	$A \lfloor B$ *or* $A \lceil B$

Table 2. Minimum and Maximum Extended to the Empty Scalar

Based on a similar table, but isomorphic to multiplication we may extend the conventional binary addition of two scalars *A* and *B* to also include the empty scalar ∘. However, we shall find it useful to simultaneously include also multiplication of the two scalars one of which may be the empty scalar. Consider, for example, that we wish to determine the total length of a directed path made up of given number A of simple paths all of the distance B. Clearly, in this case the total length will be AxB. The extension of both of these associative and cummutative operations are defined in Table 3.

$A + B$ or $A \times B$	B EMPTY $B = \circ$	B NON-EMPTY $B \neq \circ$
A EMPTY $A = \circ$	$\circ$	$\circ$
A NON-EMPTY $A \neq \circ$	$\circ$	$A + B$ or $A \times B$

Table 3. Addition and Multiplication Extended to the Empty Scalar

The advantage of introducing the empty scalar, defined by the operations in Tables 2 and 3, as a measure of the distance of a void simple path, is that it conforms operationally with the fact that any other finite simple path, including a single vertex, has a numerically non-negative distance. Hence, in this generalized form the concepts of topological length and distance provide a computational tool for description of the consequences of transitivity of a binary relation. The question, of course, is how these operations are influenced by the remaining other algebraic properties. To answer this problem let us consider the three kinds of transitive binary relations: the equivalance, the strict partial order, and the partial order relation.

The fact that an equivalence relation partitions an indexed collection of elements into disjoint equivalence classes implies that the corresponding digraph will be partitioned similarly into an equal number of connected components. That is, each equivalence class is represented geometrically by a connected component. The reflexive property means that in the corresponding digraph there is a loop at every vertex. The symmetric property tells us that within each connected component of the digraph any pair of vertices will be mutually immediately reachable. Hence, each connected component may be conceived to have been produced by a superposition of a complete digraph and its converse. Consequently, a connected component representing an equivalence class will be always strongly-connected. In conclusion, therefore, we have that *a cyclic digraph made up exclusively of strongly-connected components represents an equivalence relation.* Fig. 5 illustrates a digraph depicting an equivalence relation that partitions the indexed collection of elements into three equivalence classes.

A distinguishing feature of the order relations we are about to consider, is that they are partial rather than complete. The difference between a complete and a partial relation, we remember, is that in the complete all elements are comparable with one another whereas in the partial all elements need not be comparable, though each must be with itself. Accordingly, a complete ordering is not excluded by a strict partial order or a partial order relation. A complete ordering is simply considered a particular case that merits no special measures.

The algebraic properties defining a strict partial order relation may be established from either of two viewpoints. Both approaches share the common property that they originate in the notion of a *quasi-order* relation defined as a binary relation satisfying only the irreflexive and transitive algebraic properties (Donellan, 1968 [6]; Harary, Norman, & Cartwright, 1965 [10]). The difference between the two viewpoints derives from a difference in interpretation on how to supplement the quasi-order with a non-symmetric algebraic property. On the one hand we may assume the preexistence of an equivalence relation on the collection of indexed elements such that the collection is partitioned into equivalence classes with exactly one element in each class. On the other hand, no previous relation of equality is supposed with the result that antisymmetry cannot be defined. Hence, in this case we must introduce explicitly a non-symmetric condition in the form of the asymmetric algebraic property defined previously. In either case we end up with the same digraph representation of non-symmetry, namely, that each pair of vertices may bound only a single directed edge. Combination of this quality with the characteristic of the irreflexive property that no vertex will have a loop, reveals that cycles or strong components cannot be produced without violating the transitive property. Consequently, we see that *an acyclic digraph represents a strict partial order relation.* The acyclic digraph of Fig. 3 and the directed trees of Fig. 4 exemplify a strict partial order relation on indexed collections of elements.

Measurements on a scale-form based on a strict partial order relation usually originate in some causality condition. For example, an event at some point in time cannot be the cause of an event at an earlier point in time. Or, a manufacturing product is normally not used as a component or assembly in its own production. On the other hand, measurements on scale-forms based *solely* on a strict partial order relation are inconceivable. It will be possible always to classify an indexed collection of measurements by some equivalence relation that partitions the measurements into equivalence classes each containing one or more measurements. Consequently, as stated for the scale-forms in Table 1, a strict partial order relation on measurements must be accompanied by an equivalence relation. For this reason it was stated earlier, in connection with the discussion of binary relations defining scale-forms of measurement, that in practice the strict partial order relation is not considered separately but is joined together with the equivalence relation into a combined relation of partial order.

The only dissimilarity between a digraph representation of an equivalence relation and of a partial order relation is that the partial order relation substitutes the symmetric property by one of antisymmetry. Hence, in contradistinction to the irreflexive property of the strict partial order relation, the reflexive property of the partial order relation will introduce a loop at each vertex. Further, by virtue of the antisymmetric property of the partial order relation the corresponding digraph will contain some pairs of vertices bounding a single edge only and other pairs of vertices bounding two edges of opposite direction forming together a cycle. Combination by repeated use of the transitivity property of these characteristics result in the conclusion: *A cyclic digraph that may be partitioned into subgraphs, defining strongly-connected components and being interconnected by single edges, represents a partial order relation.*

Comparison of the digraph representation of a partial order with those of equivalence and strict partial order, furnishes the self-evident observation that any cyclic digraph, representing a partial order, may be produced by superposition of two digraphs corresponding respectively to an equivalence and a strict partial order. An extremely important consequence of this fact is that, computationally, we may determine the reachability properties of any digraph representing a partial order relation by some kind of "addition" (e.g., the set operation union) of the reachability properties of the two digraphs representing respectively the corresponding relations of equivalence and strict partial order. For example, if the equivalence relation partitions the indexed collection of measurements into equivalence classes containing each but a single measurement, the reachability properties of the partial order relation are determined by combining those of an acyclic digraph with those of a cyclic digraph defined by a loop at each vertex of the acyclic digraph. Obviously, in a case like this we may find it advantageous to organize the indexed collection of measurements as if they were submitted to only a strict partial order relation. Indeed, it is always beneficial to attempt some organization according to a strict partial order of any indexed collection of measurements.

Basically, the reachability properties of acyclicness and cyclicness of a digraph derive from the combination of transitivity with other algebraic properties of binary relations. It is noteworthy in this respect that the algebraic properties, apart from transitivity, define the digraph in terms of its vertices and directed edges. Transitivity, in the context of a digraph, are conclusions concerning the continuity properties of reachability. A computational determination of the consequences of transitivity is carried out quantitatively in terms of topological lengths and distances. However, these concepts do not permit us, explicitly and quantitatively, to take into account the continuity properties relating vertices with directed edges. But it is exactly by these continuity properties of a digraph that the binary relation algebraic properties other than transitivity are represented. Consequently, it will be useful to extend the quantitative measure of reachability in terms of number of directed edges to also include the continuity properties of vertices.

The *degree of a vertex* x is the number of directed edges incident on x. Clearly, this is a measure of the connectedness of a digraph. To change this concept into a measure of reachability we must be able to distinguish between the outgoing and incoming edges incident on a vertex. The *outdegree*, denoted *OD*, of a vertex x is the number of directed edges having x as the initial vertex. Similarly, the *indegree*, called *ID*, of a vertex x is the number of directed edges having x as the final vertex. Obviously, for any digraph the sum total of all outdegrees must equal the sum total of all indegrees since the said sum, by the definition of a directed edge, is the number of edges. The combination of outdegree and indegree provides a useful classification of the vertices of a digraph. Thus, a *source* is a vertex with indegree zero and a *sink* a vertex with outdegree zero. Also, the notion of a degree may be used to uniquely define certain kinds of digraphs such as the directed trees illustrated in Fig. 4. To illustrate, a *directed tree* is an acyclic digraph in which exactly one vertex has indegree 0 and every other vertex has indegree 1. Correspondingly, a *binary tree* is a directed tree in which every vertex has outdegree 0 or 2. Often, in relation to directed trees the source is called a *root* and the sinks *terminals* or *leaves*. A valuable property of directed trees is the fact that the topological distance between any pair of vertices is uniquely defined. Another is that any vertex is reachable from the source. Incidentally, the distance in a directed tree from the source to any vertex is called the *level* of the vertex.

The definition of a directed tree was based on the concept of an acyclic digraph. In this connection the question naturally arises whether it is possible to specify, using the degree of a vertex as a measure, some general property of acyclic digraphs which enables us to distinguish between acyclic and cyclic digraphs. The answer to this question is affirmative. An obvious characteristic of a directed tree is that it contains one source and at least one sink. Generalizing this observation to any acyclic digraph the decisive *property of acyclicness* may be stated:

> *An acyclic digraph has at least one vertex of outdegree zero and at least one vertex of indegree zero.*

Assuming that the digraph is acyclic the proof proceeds by contradiction. Suppose that the digraph has no sink. Then every vertex has a positive outdegree ≥ 1. Consequently, starting with an arbitrary vertex x, we can establish a directed path through the vertices x,y,z,... ad infinitum. But the number of vertices in the digraph is finite. It follows that in the sequence of vertex- and edge-indices defining the directed path, sooner or later, some vertex-index must appear twice. But this implies that the directed path contains a cycle. Hence, by contradiction the digraph must contain at least one sink. Paraphrasing the principle of duality stated previously the same argument may be applied to the converse digraph. Consequently, by what may be called *directional duality* an acyclic digraph must contain at least one source.

Another conclusion that may be drawn from this proof is the fact:

> *If every vertex of a digraph has a positive outdegree or, conversely, a positive indegree, the digraph will contain a cycle.*

Now, clearly, it is easy to dream up a cyclic digraph which contains at least one source and one sink. Thus, the cyclic digraph of Fig. 1 has vertex A as source and vertex F as sink. The importance of this property as a computational tool for determination of acyclicness of a digraph must be seen in relation to the unique fact that *a vertex in any cycle is neither a source nor a sink.* Accordingly, an exhaustive search of a digraph, verifying the existence or non-existence of a source and a sink for the conceivable subgraphs, will reveal whether or not it is acyclic.

The concept of *directional duality,* introduced in the course of the proof of the property of acyclicness, is based on a comparison of a digraph with its converse, that is, the given digraph with the direction of every edge reversed. From this definition it is evident that the converse of the converse of a digraph is the digraph itself. A generalization of this idea of taking the converse is that of a *dual operation.* The meaning of this concept is any operation on a digraph which, when applied twice, results in the original digraph. For each dual operation there is an associated collection of *dual concepts.* In particular, a concept is *self-dual* with respect to a dual operation if it remains invariant after performance of the operation. If the dual operation is that of taking the converse the associated concepts are called *converse concepts.* Associated self-dual concepts may in this connection be called self-dual directional. It should be emphasized that in the context of digraphs *the converse of a concept* is one which results in place of the concept, when the operation of converse is applied to a digraph. That is, a converse concept is a directionally dual concept. Statements about digraphs are given in terms of concepts and their duals. *The dual of a statement* about a digraph refers to that statement obtained upon replacing each concept in the statement by the dual concept. Consequently, the dual of a dual of a statement is the original statement. Rephrasing the principle of duality stated previously for partially ordered systems we obtain the *directional duality principle:*

> *Any theorem which is true for a digraph remains true if each concept and its converse concept are interchanged throughout the statement of the theorem.*

An immediate consequence of this principle, used in the proof of the property of acyclicness, is that any digraph and its converse will exhibit converse or directional dual reachability properties. On the other hand, properties of connectedness for a digraph and its converse are self-dual directional. In practice we often find it useful simultaneously to consider the converse set of reachability properties for a given digraph and its converse. Evidently, it will suffice in such cases simply to derive the reachability properties for the given digraph and then apply the directional duality principle without bothering about introducing the converse digraph at all. In the approach of this kind, based on a single digraph and the principle of directional duality, we need some terminology which, relative to the given digraph, classifies the reachability properties such that they are uniquely identified as belonging to the given digraph or its converse. A commonly accepted way of doing this is based on a terminological loan from the concept of strict partial ordering. Thus, the relation represented by the given digraph is designated a *predecessor relation* whereas that depicted by the converse digraph is called a *successor relation*. Accordingly, referring solely to a directed edge from vertex x to vertex y of the given digraph, the given relation and its converse are described respectively in terms of immediate reachability as follows:

(1) x *is the immediate predecessor of* y, or
x *precedes* y *immediately*.

(2) y *is the immediate successor of* x, or
y *succeeds* x *immediately*.

We may say, therefore, that the first statement is a *direct* interpretation of the given digraph while the second statement represents a *converse* interpretation of the digraph.

It is important to realize that the sole purpose of this terminology is identification of directional dual reachability properties of a given digraph and its converse. The corresponding connectivity properties are directional self-dual and, hence, uninteresting. However, if the converse digraph is substituted by a *geometrical dual* digraph, this digraph and the given will exhibit geometrical self-dual reachability properties. Consequently, the roles of reachability and connectedness mirror each other relative to the two kinds of duality. Terminologically, to identify geometrical dual connectivity properties of a given digraph and its geometrical dual the former may be called the *primal* and the latter the *dual* digraph. It should be noted that the concept of geometrical duality is restricted to *planar* digraphs. That is, digraphs which can be drawn in a plane so that their edges intersect only in their vertices. The edges of a planar digraph form a set of adjoining polygons in the plane, dividing it up into polygonal pieces designated *regions* or *faces*. The reason that geometrical duality is defined only for planar digraphs, is that this kind of duality is based upon a one-one correspondence: vertices $\leftrightarrow$ regions and branches $\leftrightarrow$ links, between the primal and the dual digraphs.

7. THE ERLANGER PROGRAM OF SYSTEMS SCIENCE

The introduction in the early 17th century of *coordinate* or, as it has also been called, *analytical geometry* made mathematics a double-edged tool. By expressing geometrical curves and shapes in algebraic forms, from which quantitative knowledge can be derived, algebra brings to geometry a classification by logical structure and a system by methods of solution that turn geometry into a mechanized tool of reasoning. Conversely, by interpreting algebraic statements geometrically, geometry endows algebra with an intuitive content of meaning suggesting novel interpretations and deduction of new conclusions.

Previously in this chapter, to obtain insight into the algebraic properties of the scale-forms associated with measurements, a geometrical interpretation was given representing the measurements and their properties in terms of digraphs. Now, to derive conclusions in a systematic and exhaustive manner from measurements with given properties, the reverse process of expressing digraphs on algebraic form must be undertaken.

It is a well-known fact that digraphs and their properties are represented algebraically by Boolean matrices and operations upon such matrices. Due to lack of space we shall have to refrain from a discussion of this topic referring the reader instead to some pertinent literature (Harary, Norman, & Cartwright, 1965 [10]; Bertztiss, 1971 [2]). Thus, in the following we shall confine ourselves to the more principal aspects involved in the algebraic representation of digraphs.

The transformation of the properties of binary relations from an algebraic form into geometrical digraphs and back again to another algebraic form, does not seem to exactly minimize the work in problem solving. Intuitively, a more rational approach will be to go directly from one algebraic form to another. True enough and, as a matter of historical fact, this was precisely what C.S. Peirce did about 1870 when in his pioneering work he applied matrix notation to the logic of relations (Copilowish, 1948 [5]). The situation in engineering and economics, however, is radically different. The empirical world does not present us with a well-defined set of algebraic properties. All we have are some observations of intrinsic properties, represented by our choice of measurement, of an object or event. To identify the abstract properties of such empirical relationships, calls primarily for geometry. The Greeks established geometry as the first empirical science achieving rigour and consistency. All other empirical sciences are modelled on the axiomatic method of geometry. Truly, human beings perceive empirical reality as a geometrical reality. Objects are basically geometrical figures and events are points on the paths of moving bodies. Therefore, geometry is a natural first step in the representation of measurements of empirical qualities. Once a geometrical picture, say, a digraph is established, we may use it to verify the consistency of the empirical information by identification of the isomorphic algebraic properties. Furthermore, after having expressed the geometrical properties on algebraic form and determined their algebraic consequences, we may turn back to the geometrical picture to facilitate the interpretation of the empirical meaning of these consequences.

Coordinate geometry is based on the fundamental assumption that the properties of a collection of geometrical elements, defining some space, may be represented by algebraic expressions. A *space*, taking the notion of a geometry in its most general sense, is simply a mathematical set of elements together with some geometrical concepts which, like distance or continuity, bind the elements together. Conventionally, the elements are geometrical entities or objects such as points, lines or vectors, but any conceivable interpretation is in fact possible. To achieve an algebraic representation of the properties of the geometrical elements a structure, called a *frame of reference* or *coordinate system,* is superimposed upon the space. The purpose of the superimposed structure is two-fold. On the one hand, it provides a systematic identification of the geometrical elements. On the other, it associates each distinct geometrical element with one or more numerals, the socalled *coordinates*, possessing algebraic properties derived from or functionally related to the fundamental concept (distance or continuity) defining the space geometrically.

Sometimes the coordinates will possess properties that permit their use for identification of the geometrical elements. In general, however, it is necessary to introduce one or more index-sets for this purpose. The number R of index-sets necessary for the algebraic specification of some property of the distinct geometrical elements, is known as the *rank*, *order* or *valence* of the corresponding collection of coordinates. By virtue of the fact that the rank R is a cardinal number, usually finite, it may be used to define a *ranking index-set* the R indices of which assign some arbitrary ordering to the originally given number of index-sets. The finite or transfinite cardinal number N of indices of each of the latter index-sets is known as its *dimension*. We may designate, therefore, the latter index-sets as the *dimensional index-sets*. Together, the ranking index-set and the dimensional index-sets organize the corresponding collection of coordinates in an *R-array*. Thus, a *scalar* has no dimensional index-set, a *vector* a single dimensional index-set, and a *matrix* two dimensional index-sets. Alternatively, we may say that a scalar is a single coordinate, a vector an N-tuple of coordinates, and a matrix a (MxN)-table of coordinates.

Basically, the organization by index-sets of collections of coordinates into R-arrays must be separated conceptually from the notion of the *dimension* of the space. In a conventional Cartesian coordinate system we tacitly assume that the dimension of a space is the minimum number of coordinates needed to fix a point in space. However, this assumption breaks down if the binding geometrical concept, defining a space on a set of points, is not the Euclidean distance function. Thus, in topology the notion of dimension is introduced by a recursive definition which, based on the concept of continuity, applies to abstract topological spaces yet grant to line and plane the dimensions usually assumed for them. Accordingly, if the dimension of an abstract space is N, then any geometrical object in this space must be represented algebraically relative to the selected reference frame of index-sets by an R-array the dimensional index-sets of which all have the dimension N. It follows that R-arrays with dimensional index-sets of different dimensions represent mappings, not on the space itself, but on some subspaces. For example, a square matrix may be either the component of a geometrical object in a space of the corresponding dimension or a mapping of the space on itself, but a rectangular matrix will be a mapping of a space on another space or subspace.

The central idea in coordinate geometry, we have stated previously, is the use of algebraic equations to represent and study geometrical figures of arbitrary shape and dimensions like curves and surfaces. It is a common experience that, say, a physical object may be represented in a variety of different coordinate systems or reference frames. To the human eye the geometrical images may be only vaguely similar although they are equally valid descriptions of the "true" shape of the object. The fact that coordinate geometry associates algebraic equations with geometrical shapes, implies that distinct geometrical properties, in order to be "true" in any coordinate system, must be represented under change of coordinates by an *invariant form* of the corresponding algebraic expression. That is, going from one coordinate system to another, invariancy of algebraic properties depicts preservation of geometrical properties. To illustrate, the equality in one coordinate system of two algebraic expressions each representing an invariant geometrical property, implies equality in all coordinate systems. In particular, if an algebraic expression of invariant form vanishes in one coordinate system it will vanish in all.

Historically, the impact of this observation upon mathematical thinking was decisive. It gave coordinate geometry an entirely new perspective which fundamentally changed the theoretical basis of those empirical sciences, in particular physics, that relied on mathematics for abstract structures of modelling. The mathematician Felix Klein who, like Fourier, discarded mathematics beyond such structures that may serve in investigation of nature, formulated the new viewpoint in 1872 in an inaugural address upon his appointment to the University of Erlangen. For some odd reason, perhaps related to the intricacy of German grammar, his formulation is known today in English literature as the *Erlanger program* rather than as the Erlangen Program. Klein's basic definition of a geometry may be rendered as follows (Klein, 1893 [11]):

> *A geometry is the study of those properties of a space which remain invariant when the elements of the space are subjected to the transformations of some group of transformations.*

Before Klein each geometry was derived axiomatically like Euclid's. Indeed, the Euclidean geometry was the sole geometry until the early 19th century when Gauss, Lobatchevsky, and Bolyai introduced a Non-Euclidean geometry based on Euclid's but with a different formulation of the parallel axiom. The work on affine and, in particular, projective geometry had progressed significantly due to the efforts of Möbius and respectively von Staudt and Cayley when Klein entered the scene bringing with him a profound knowledge of group theory from his student days in Paris together with his friend Sophus Lie.

Klein's basic idea was that each geometry can be characterized by a mathematical group of transformations and that a geometry is really concerned with invariants under this group. Moreover, because some groups contain others as subgroups, some geometries will embrace others. That is, a subgroup of the transformation group of some geometry will define a subgeometry such that all theorems of the original geometry continue to be theorems in the subgeometry. Clearly, since the transformation group of a geometry may contain more than one subgroup, it may be possible to define alternative subgeometries of a given geometry.

Traditionally, each geometry had been treated in its own right as a separate theory derived from a fundamental set of independent axioms. In Klein's Erlanger program this view was subsumed as a more general viewpoint based on the invariancy of geometrical objects or properties under groups of transformations. The classification of the different geometries according to defining groups of transformations imposes upon them a partial order. We may conceive Klein's scheme of classification, therefore, as a directed tree with a geometry at each vertex. Placing the most general geometry at the root and the most specific at the leaves, the tracing of a path from one of the leaves to the root will reveal a vanishing of the differences of the configurations simultaneously as the deeper lying, more general, characteristic features are coming to the fore instead. The more comprehensive the group is which we take up, the more we penetrate into the depths: the details flow together and we see only the larger and more general concepts. A directed tree which illustrates Kline's conception of the principal geometries at his time, is given in Fig. 8 (Kline, 1972 [12]).

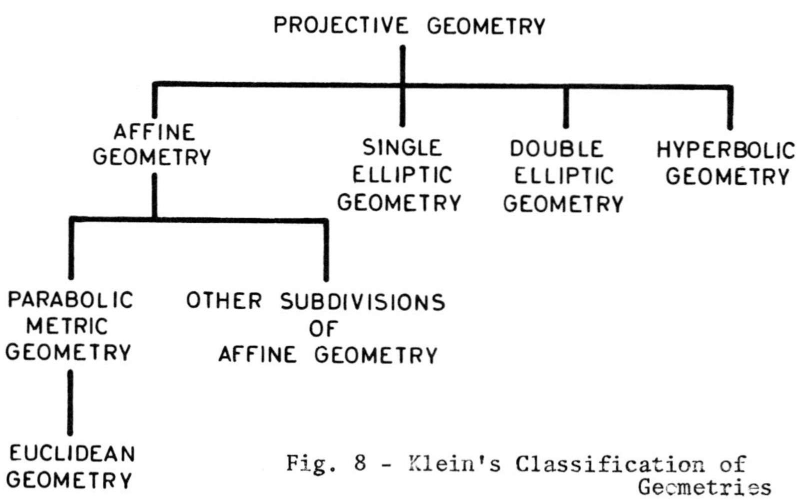

Fig. 8 - Klein's Classification of Geometries

Evidently, some of the most important geometries are missing in this figure. However, it is not because they cannot be included if we are willing to interpret Klein's definition of a geometry in its broadest sense. Take, for example, differential geometry or tensor analysis. Tensors are geometrical objects generalizing the vector concept. In any coordinate system the components of a tensor are N-arrays of coordinates or functions of the coordinates. The fundamental invariant property is not distance per se, but the differential distance between two neighbouring points. Now, consider a transformation from one coordinate system to another as a functional relationship satisfying the ordinary regularity conditions (i.e., single-valued, continuous, differentiable, and with a non-zero Jacobian). Under this assumption, no matter what functional relations exist between the two sets of coordinates, their differentials are always linearly dependent. Consequently, expressing in general the components of a tensor in terms of the differentials of the coordinates rather than in terms of the coordinates themselves, tensor analysis may be described as a geometry the invariant properties of which are defined by the affine group of transformations on the differentials of the coordinates. The affine group contains as sub-groups the full linear group and the group of translations. Hence, we have here a scheme in terms of coordinate differentials which parallels Klein's in terms of coordinates. Alternatively, since the admissable transformations in tensor analysis include the transformations of Klein's scheme as a special case, we realize that a classification of geometries by groups of transformations must take the form of an acyclic digraph interconnecting the two schemes and rising from a source-vertex representing a geometry more general than projective geometry.

The advances in physics and engineering achieved in our century by using tensor analysis to express the physical laws in a form invariant under change of coordinates, illustrates the impact upon the empirical sciences of Klein's view. To extend his idea to a broader field of applications we shall find it useful to include topology into his scheme as the most general geometry. To this end we shall begin by establishing the link between tensor analysis and topology.

Essentially, the structure of a generalized space in tensor analysis is determined by two non-metric properties, representing algebraic properties of binary relations, and two metric properties, depicting algebraic properties of binary operations on numbers. Considering geometry as an empirical science we may say that the non-metric properties of the generalized space are "measured" on nominal or ordinal scale-forms while the metric properties are "measured" on interval or ratio scale-forms. The non-metric properties of the space are its topology and the reference frames or coordinate systems in terms of which it is described. The metric properties are the metric tensor, associating coordinates and their direction vectors with measures of lengths and angles, and the differential equation between the absolute and the covariant derivative measuring the curvature or non-linearity of the space.

The distinctive feature characterizing this partition of properties into, what here has been called, metric and non-metric properties, is the fact that in the former we have included the reference frame or coordinate system alongside the topological or continuity properties. Consequently, using continuity as the geometrical concept binding together the elements of the space, we may elide the metric properties in order to obtain a more general geometry. That is, solely on the basis of the topology of the space and the reference frames we may establish a coordinate geometry in complete analogy with conventional tensor analysis, but based on a more comprehensive and general group structure (Mautner, 1946 [13]).

Having extended Klein's Erlanger Program in this way to include tensor analysis and topology, we may note that even then his definition is not all embracing. Indeed, geometries have been introduced based on the idea of an abstract space with a superimposed structure that may not necessarily be definable in terms of some transformation group. In many branches of engineering, for example, we find it useful to introduce more general abstract objects which, like semigroups and monoids, possess only some of the properties of groups. A *semigroup*, it may be recalled, is a set together with an associative binary operation. If this operation is also commutative the semigroup is called *commutative*. A *monoid* is a semigroup which has an identity element. Thus, since a group is a monoid in which each element has an inverse, we see that all groups are, in particular, semigroups. Therefore, to include these more general structures of coordinate transformations, we may *substitute the word "group" by the word "semigroup" in Klein's definition of a geometry.*

8. CONCLUSION

With this proviso we may say that aim of systems science as a theory, is to establish an "Erlanger Program" for the empirical sciences conceiving each distinct model of reality as a geometry in the abstract sense of Klein's definition. Earlier in this chapter, in agreement with the viewpoint of symbolism, we classified measurements according to the set of algebraic properties, the so-called scale-form, which guarantees preservation of the empirical information under any mapping leaving these properties invariant. The consequence that follows from implementing the general idea of the Erlanger Program is that:

> *Modelling of reality from measurements on some scale-form, preserves objectivity with respect to the empirical information by being established in terms of invariant properties of that geometry which is defined by the transformation group of the scale-form.*

Thus, analogies between different fields of endeavour originate in isomorphisms between the defining group structures just as dualities derive from automorphisms within the same group structure. It is interesting to note that implicitly the idea of invariant form has governed the whole evolution of physics since Newton's days. The transition of ideas and models from one physical field to another was based on invariancies forming analogous natural laws. Also, the development of classical economic theory stems from analogies explaining economic behaviour in mechanistic terms (Franksen, 1969 [8]; 1974 [9]).

Evidently, the four scale forms introduced in this chapter comprise only a subset of the possible scale-forms that may and have been used. Thus, in psychology a logarithmic interval scale has been introduced as a measure of utility (Stevens, 1959 [17]). The selection of the four scale-forms discussed here, is simply a convenient choice for our purpose. Further, if we deem it appropriate, the definition of a scale-form may be based on the concept of a semigroup, say, rather than on that of a group. So far, however, experience does not seem to indicate the necessity of such a step apart from purposes of aggregation.

The very fact that the algebraic structure of the different scale-forms of Table 1 is cumulative, implies that a digraph may be used to represent the properties of equivalence and order of any scale-form. Essentially, this may be done in either of two ways. If we depict these properties in terms of the *connectivity* of the digraph the orientations of the edges are introduced only as a matter of reference. Clearly, in this case we end up with the *oriented* or *linear graph* approach typically used for the description of natural laws (such as Kirchhoff's) in fields like electrical, mechanical, hydraulic, and civil engineering. On the other hand, if we depict these properties in terms of the *reachability* of the digraph the orientations of the edges are introduced to represent some intrinsic, often causal, relationship between the measurements. In this case we end up with the *directed graph* typically used in computer science, production engineering, and the social sciences. Thus, dependent upon the manner in which we formulate and categorize measurements in general there seems to be, at least from an overall point of view, two preferred, essentially divergent ways in which graph theory may be used to represent measurements in the mathematical modelling of natural and man-made systems.

Concluding these more general remarks concerning systems science as an empirical counterpart of the extended Erlanger program, a computational advantage of this viewpoint should be pointed out. Group theory and the related algebraic structures form the core of the computational approaches especially suitable for digital computer solutions. Hence, the development of efficient computational solutions for use in practice, considered as an important aspect of systems science, may greatly benefit from an explicit recognition of measurements as the coordinates of some geometrical space conceived within the framework of the extended Erlanger Program.

REFERENCES

1. Alchain, A., "Utility Theory and Consumer's Surplus", *American Economic Review*, March 1953, 26-50.
2. Berztiss, A.T., *Data Structures - Theory and Practice*, Academic Press, Inc., N.Y., 1971.
3. Birkhoff, G. & MacLane, S., *A Survey of Modern Algebra*, Rev. Ed., The MacMillan Co., N.Y., 1953.
4. Campbell, N., *What is Science?* 1921. (Reprinted by Dover Publications, N.Y., 1952).
5. Copilowish, I.M., "Matrix Development of the Calculus of Relations", *The Journal of Symbolic Logic*, Vol. 13, No. 4, Dec. 1948, 193-203.
6. Donnellan, T., *Lattice Theory*, Pergamon Press, Ltd., London, 1968.

7. Franksen, O.I., "Kron's Method of Tearing", *Fourth Power Industry Computer Application Conference*, Clearwater, Florida, May 1965.

8. Franksen, O.I., "Mathematical Programming in Economics by Physical Analogies. Part I: The Analogy between Engineering and Economics", *Simulation*, no. 6, June 1969, 297-314. "Part II: The Economic Network Concept", *Simulation*, no. 1, July 1969, 25-42. "Part III: System Equilibrium and Mathematical Programming", *Simulation*, no. 2, August 1969, 63-87.

9. Franksen, O.I., "Basic Concepts in Engineering and Economics", In *Physical Structure in Systems Theory - Network Approaches to Engineering and Economics*, by J.J. van Dixhoorn & F.J. Evans (eds.), Academic Press, London, 1974, 247-278.

10. Harary, F., Norman, R.Z., & Cartwright, D., *Structural Models - An Introduction to the Theory of Directed Graphs*. John Wiley & Sons, Inc., N.Y., 1965.

11. Klein, F., "Vergleichende Betrachtungen über neuere geometrische Forschungen", *Matematische Annalen*, Bd. 43, 1893, 63-100.

12. Kline, M., *Mathematical Thought from Ancient to Modern Times*, Oxford University Press, N.Y., 1972.

13. Mautner, F.I., "An Extension of Klein's Erlanger Program: Logic as an Invariant-Theory", *American Journal of Mathematics*, Vol. 68, 1946, 345-384.

14. von Neumann, J. & Morgenstern, O., *Theory of Games and Economic Behaviour*, Princeton University Press, 1944, 3.Ed. (Reprinted by John Wiley & Sons, Inc., N.Y., 1964).

15. Ore, O., *Graphs and Their Uses*, Random House, Inc., N.Y., 1963.

16. Stevens, S.S., "On the Theory of Scales and Measurements", *Science*, Vol. 103, No. 2684, June 1946, 677-680.

17. Stevens, S.S., "Measurement, Psychophysics, and Utility", In *Measurement: Definitions and Theories*, by C.W. Churchman & P. Ratoosh (eds.), John Wiley & Sons, Inc., N.Y., 1959, 18-63.

A QUALITATIVE MODEL OF THE QUALITY OF RANGELAND OF THE SAHELIAN ECOSYSTEM

DANIEL J. HORNBACH
Department of Biology
University of Dayton
Dayton, Ohio

RAYMOND FITZ
School of Engineering
University of Dayton
Dayton, Ohio

SUMMARY

The Interpretive Structural Modeling (ISM) technique is described and this methodology is applied to modeling the Quality of Rangeland subsystem of the complex Sahelian ecosystem. The impetus for examing the Sahelian ecosystem is a famine which began in 1973, and resulted in the death of 100 thousand people and 5 million cattle.

The Interpretive Structural Modeling methodology is an emerging methodology which appears to be very useful as an aid to individuals and small groups in developing an understanding of complex situations. It is especially useful for decision makers for it allows them to gain an understanding of the complexity of issues with which they are forced to deal.

1. INTRODUCTION

In 1973, a famine struck six West African countries, just south of the Sahara, that resulted in the death of some 100 thousand people and 5 million cattle [10]. The immediate cause of the famine at first seems to be a drought which began in 1968, but in reality the drought was only a trigger that lead to the upset of the delicate ecological balance. This balance will be explored further in section 3 - Description of the Sahel.

This paper illustrates the use of Interpretive Structural Modeling (ISM) as a qualitative model methodology which has been developed to help people think and communicate more clearly about complex issues, the complex issue here being the famine in the Sahel. The methodology uses a computer to collect a series of subjective judgements made by the participants and displays the structure of the system formed by these judgements in the form of a directed graph. The theory behind this qualitative modeling technique will be examined further in section 2 - Interpretive Structural Modeling.

Experience to date [2] has indicated that interpretive structural modeling enhances collaborative decision making, which allows policy planning to be implemented as social learning, and allows ethical and value judgements to be part of a policy planning process. ISM allows the integration of knowledge about a system's structure and resultant behavior so that system interventions can be made which are efficient in changing undesirable system behavior (Figure 1).

The policy planning process can be viewed as involving four major tasks (Figure 2), all of which can be approached through the ISM technique [2]. The first is critical issue analysis, where the decision making unit must develop an appreciation for how the focal system has evolved up until the present, and how it might be expected to evolve in the future if no action is taken. The second task is normative planning, where the decision making unit must develop desirable alternative futures for the focal system under study. A third task is strategic planning, where the decision making unit develops a set of objectives which are needed to realize the particular chosen future or ideal for the system. The fourth task is operational planning, where the decision making unit delineates goals, and means to these goals, which can be obtained over a particular period of time later in the future. It is the first task, that of critical issue analysis, for which a qualitative model of the quality of rangeland has been developed (see section 4 - Model of the Quality of Rangeland).

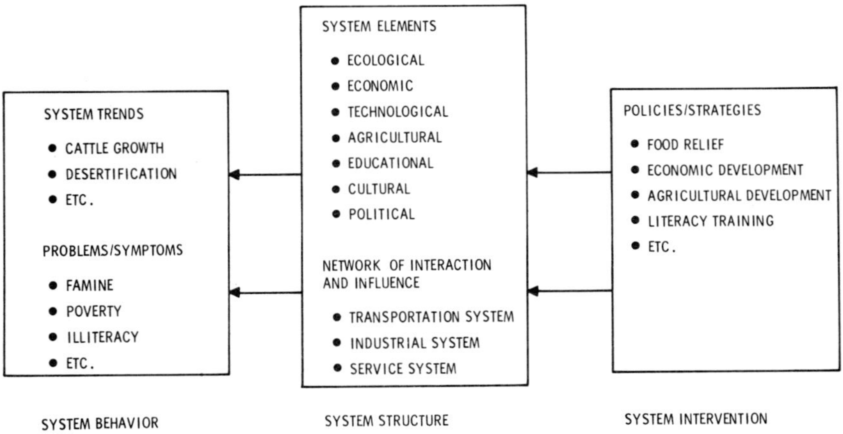

Fig. 1 - Interventions on System Structure by which to Correct Undesirable System Behavior

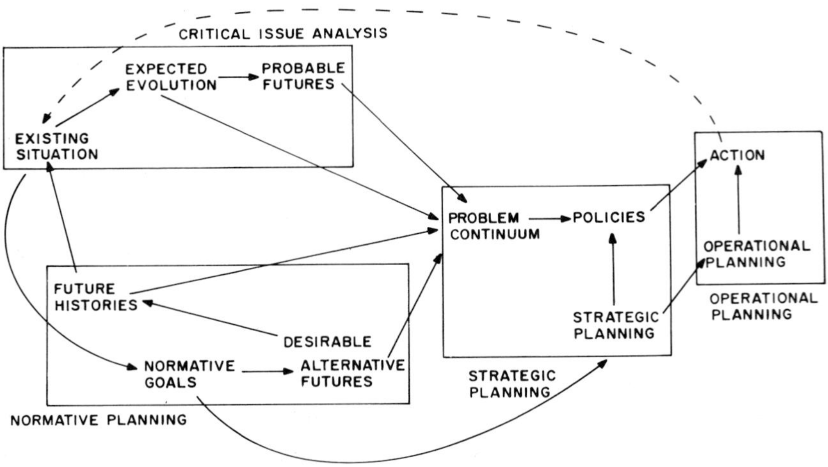

Fig. 2 - The Four Tasks of the Policy Planning Process and Their Interactions

2. INTERPRETIVE STRUCTURAL MODELING

A system ζ is an entity which consists of two or more elements (ξ) and a non-empty set of relationships ($\mathcal{R}$) among the elements, i.e., $\zeta = (\xi, \mathcal{R})$ [9]. A model S_M is a descriptive approximation of the system which consists of a set of contextual elements ($\hat{R}$) i.e., $S_M = (S, \hat{R})$. The elements of S, e_i, are descriptions of the elements of the system in colloquial language. The contextual relation, $\hat{R}$, is a phrase in colloquial language describing a relation that could exist among the elements.

A model may take one of many forms (Figure 3). A mental model is merely a person's conceptual map of certain reality. A mental model contains beliefs about what factors or elements are important and also beliefs about important relations among the factors or elements (Figure 4). A prose model is a mental model that has been put into words. If the prose model is rigorously constructed it should be a logical extension of the mental model. These first two types of models are available and readily used by many persons. The remaining three types of models fall into the domain of the analyst.

A matrix model allows for the incorporation of both structural and contextual properties of the system into a model. The structural properties are determined by examining the relationship between the member of a set of contextual elements. A binary matrix model $\mathcal{M}$ is a binary matrix and two associations being indicated by colons, i.e.,

$$\mathcal{M} = \{N, S:I, \hat{R}:R\}$$

where N is a binary matrix, S is a set of contextual elements associated in a one to one manner with the index set I, and $\hat{R}$ is a contextual relationship associated with the binary relationship R induced on IxI by N that is if $n_{ij} = 1$, then e_i R e_j and if $n_{ij} = 0$ then e_i $\bar{R}$ e_j, ($\bar{R}$ = not related). Figure 5 shows a matrix model that can be constructed from a set of system elements $S = \{e_1, e_2, e_3, e_4, e_5, e_6, e_7, e_8, e_9\}$ (Note this is a subset of the quality of rangeland model and will be used in exemplifying the ISM technique). Once the adjacency matrix has been systematically filled with zeros or ones according to the definitions given for a binary matrix model, the relationships between the elements of a system have been explicitly stated. If then a graphical illustration of the binary matrix can be derived, one has a structural model where the elements correspond to the nodes of a network model and the presence of a relationship between any two elements is denoted by a directed line connecting those two elements (nodes).

If a digraph is to be constructed from a matrix, the relationship used to determine the structural properties must be transitive, i.e., e_i R e_j and e_j R e_k *implies* e_i R e_k. There is a correspondence between the notion of transitivity and reachabil-

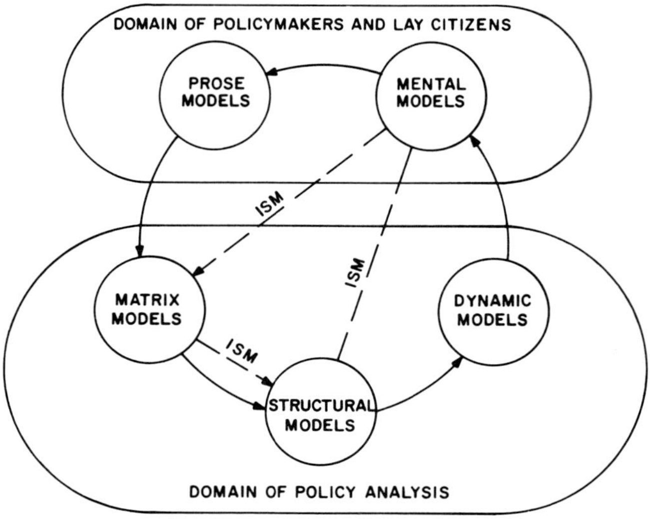

Fig. 3 - Relationships Between Different Types of Models and the Two Domains of Policy Planning. The Dotted Lines Show the Role Interpretive Structural Modeling can Play in Bridging the Communication Gap

MENTAL MODELS CONTAIN

- BELIEFS about what FACTORS OR ELEMENTS are important
  - Quality of Soil
  - Cattle Population
  - Western Development Aid
- BELIEFS about important RELATIONS among the factors of elements

  An increase in Western

  Development Aid causes

  an increase in the

  Cattle Population

Fig. 4

e_1 - CATTLE POPULATION

e_2 - CATTLE BIRTH RATE

e_3 - CATTLE FERTILITY

e_4 - CATTLE DEATH RATE

e_5 - CATTLE LIFE EXPECTANCY

e_6 - QUALITY OF LIFE FOR CATTLE

e_7 - QUALITY OF LIFE FOR CATTLE FROM FOOD

e_8 - QUALITY OF LIFE FOR CATTLE FROM WATER

e_9 - QUALITY OF LIFE FOR CATTLE FROM LACK OF DISEASE

RELATIONSHIP R DOES AN INCREASE IN (FACTOR e_i) CAUSE AN INCREASE IN (FACTOR e_j)?

A

INFLUENCING ↓ / INFLUENCED →	e_1	e_2	e_3	e_4	e_5	e_6	e_7	e_8	e_9
e_1	1	1	0	1	0	0	0	0	0
e_2	1	1	0	0	0	0	0	0	0
e_3	0	1	1	0	0	0	0	0	0
e_4	0	0	0	1	0	0	0	0	0
e_5	1	0	1	0	1	0	0	0	0
e_6	0	0	1	0	1	1	0	0	0
e_7	0	0	0	0	0	1	1	0	0
e_8	0	0	0	0	0	1	0	1	0
e_9	0	0	0	0	0	1	0	0	1

Fig. 5 - Original Adjacency Matrix

ity in graph theory. Reachability is defined in the following way; element e_i is said to be *reachable* from element e_j if a path can be traced on a digraph D from e_j to e_i. By convention an element is said to be reachable from itself with a path length of 0. The reachability matrix M, has element $m_{ij} = 1$ if element e_j is reachable from element e_i, otherwise $m_{ij} = 0$. The reachability matrix M of the adjacency matrix (Figure 5) is shown in Figure 6. The reachability matrix is also called the transitive

$M =$

	e_1	e_2	e_3	e_4	e_5	e_6	e_7	e_8	e_9
e_1	1	1	0	1	0	0	0	0	0
e_2	1	1	0	1	0	0	0	0	0
e_3	1	1	1	1	0	0	0	0	0
e_4	0	0	0	1	0	0	0	0	0
e_5	1	1	1	1	1	0	0	0	0
e_6	1	1	1	1	1	1	0	0	0
e_7	1	1	1	1	1	1	1	0	0
e_8	1	1	1	1	1	1	1	1	0
e_9	1	1	1	1	1	1	1	0	1

Fig. 6 - Reachability Matrix

closure of D due to the transitive nature of the relationship between the elements.

The reachability matrix can be utilized to develop a hierarchial restructuring of the digraph. A reachability set R_i is defined as all of those elements which are reachable from e_i, i.e.,

$$R_i = \{e_j | m_{ij}=1\} .$$

Further an antecedent set A_i is defined as all those elements which reach e_i, i.e.,

$$A_i = \{e_i | m_{ij}=1\} .$$

The intersection of A_i and R_i $(A_i \cap R_i)$ consists of those elements common to both sets. Those elements e_i for which $R_i \cap A_i = R_i$ are not reachable from any of the remaining elements and thus constitutes top level elements. An equivalent condition, which is easier to use in programming is $R_i \subset A_i$. By striking the column and row corresponding to the top level elements of the reachability matrix M, it is possible to repeat the process and determine the top level elements of the reduced system. These are in essence second level elements, and as this process continues, other level elements are determined. In the example being developed, the element set S can be partitioned into six hierarchial sets:

$$S = \{L_1; L_2; L_3; L_4; L_5; L_6\}$$
$$L_1 = \{e_4\}$$
$$L_2 = \{e_1, e_2\}$$
$$L_3 = \{e_3\}$$
$$L_4 = \{e_5\}$$
$$L_5 = \{e_6\}$$
$$L_6 = \{e_7, e_8, e_9\}$$

This process has reordered the original element set into a set of hierarchial components L_H. Note that the hierarchial reordering process could proceed from the bottom up by identifying the elements for which $A_i \cap R_i = A_i$. The reachability matrix M (Figure 6) can be rearranged by interchanging the rows and columns according to the hierarchial components L_H and the hierarchial reachability matrix M_H can be formed (Figure 7). This rearrangement and partitioning of the original adjacency matrix has provided more information about the system in that it shows the presence and content of levels. In large or complex systems, hierarchial rearrangement greatly enhances the communicability of the digraph.

Each of the levels, L_H can further be partitioned into *disjoint* and *strong* subsets;

$$L_H = \{T_H; U_H\}$$

M_H =

	e_4	e_1	e_2	e_3	e_5	e_6	e_7	e_8	e_9
e_4	1	0	0	0	0	0	0	0	0
e_1	1	1	1	0	0	0	0	0	0
e_2	1	1	1	0	0	0	0	0	0
e_3	1	1	1	1	0	0	0	0	0
e_5	1	1	1	1	1	0	0	0	0
e_6	1	1	1	1	1	1	0	0	0
e_7	1	1	1	1	1	1	1	0	0
e_8	1	1	1	1	1	1	1	1	0
e_8	1	1	1	1	1	1	1	0	1

Fig. 7 - Hierarchial Reachability Matrix

where T_H is the disjoint subset of all elements in L_H which are not related to any other elements in L_H, and U_H is the strong subset of all other elements in L_H not in T_H. In our example then, $E = \{L_1; L_2; L_3; L_4; L_5; L_6\}$ thus $E = T_1; U_2; T_3, T_4; T_5; T_6$, where

$$T_1 = \{e_4\}$$

$$U_1 = \emptyset$$

$$T_2 = \emptyset$$

$$U_2 = \{e_1, e_2\}$$

$$T_3 = \{e_3\}$$

$$U_3 = \emptyset$$

$$T_4 = \{e_5\}$$

$$U_4 = \emptyset$$

$$T_5 = \{e_6\}$$

$$U_5 = \emptyset$$

$$T_6 = \{e_7, e_8, e_9\}$$

$$U_6 = \emptyset$$

Each strong subset can be partitioned into blocks;

$$U_H = \{W_H^1;\ W_H^2;\ \ldots W_H^p\}.$$

A block W_H^r has the property that for a given e in W_H^r any other element in that block is both reachable form and antecedent to it. In our example;

$$U_2 = W_2 = \{e_1, e_2\}.$$

The standard form of the reachability matrix, the standard form matrix S_H (Figure 8), is obtained by rearranging the rows and columns within each level by allowing the elements of the disjoint set to precede each of the blocks of the strong subset. The individual blocks are called cycles.

The condensation matrix C (Figure 9) is obtained by replacing a cycle with the first element in the cycle. Notice that the condensation matrix is always a lower triangular form.

There are many digraphs that have the same reachability matrix. In most modeling activities it is helpful to start with the minimum edge digraph, i.e., the digraph with the least number of edges. The binary matrix corresponding to the minimum edge digraph if called the skeleton matrix Q (Figure 10). The skeleton matrix is calculated by determining the precedent set for each element of the condensation matrix. The precedent set is defined by;

$$P(e_i) = \{e \mid \text{there exists no } b \in E \text{ such that } e\ R\ b \text{ and } b\ R\ e_i\}.$$

In the skeleton matrix Q;

S =

	e_4	e_1	e_2	e_3	e_5	e_6	e_7	e_8	e_9
e_4	1	0	0	0	0	0	0	0	0
e_1	1	1	1	0	0	0	0	0	0
e_2	1	1	1	0	0	0	0	0	0
e_3	1	1	1	1	0	0	0	0	0
e_5	1	1	1	1	1	0	0	0	0
e_6	1	1	1	1	1	1	0	0	0
e_7	1	1	1	1	1	1	1	0	0
e_8	1	1	1	1	1	1	1	1	0
e_9	1	1	1	1	1	1	1	0	1

Fig. 8 - Standard Form Matrix

C =

	e_4	e_1	e_3	e_5	e_6	e_7	e_8	e_9
e_4	1	0	0	0	0	0	0	0
e_1	1	1	0	0	0	0	0	0
e_3	1	1	1	0	0	0	0	0
e_5	1	1	1	1	0	0	0	0
e_6	1	1	1	1	1	0	0	0
e_7	1	1	1	1	1	1	0	0
e_8	1	1	1	1	1	0	1	0
e_9	1	1	1	1	1	0	0	1

Fig. 9 - Condensation Matrix

Q =

	e_4	e_1	e_3	e_5	e_6	e_7	e_8	e_9
e_4	1	0	0	0	0	0	0	0
e_1	1	1	0	0	0	0	0	0
e_3	1	0	1	0	0	0	0	0
e_5	0	0	1	1	0	0	0	0
e_6	0	0	0	1	1	0	0	0
e_7	0	0	0	0	1	1	0	0
e_8	0	0	0	0	1	0	1	0
e_9	0	0	0	0	1	0	0	1

Fig. 10 - Skeleton Matrix

$$q_{ij} = 1 \text{ if } e_j \in P(e_i)$$

$$q_{ij} = 0 \text{ if } e_j \in P(e_i).$$

The completed digraph of the example developed is shown in Figure 9.

Interpretive Structural Modeling (ISM) uses the technique described above, except that it does not require an a priori knowledge of a binary matrix A. Instead ISM uses an "embedding" process to take advantage of the transitivity of the relationship $\hat{R}$. Transitive inference can be made within the reachability matrix to guide the systematic interrogation of the user with regard to the presence or absence of the relationship between the elements. A digital computer is used to keep track of the responses made by the user, to calculate inference, and to generate an efficient ordering of subsequent queries. The formation of the standard form, condensation, and skeleton matrices is termed the "extracting process" for a minimum edge digraph has been "extracted" from the hierarchial reachability matrix. Figure 12 shows a symbolic digraph of the principle operations in ISM.

The extracting process has removed all redundant links from the digraph, but these redundant links may be essential to convey the full meaning and pattern of the relations, therefore, ISM also employs a "comparison" operation. This operation allows the user to compare the results of the mathematical operations with his mental model and then to introduce corrections or modifications of the digraph. Finally a substitution operation consists of elaboration of text and interpretive symbols or additional graphical embellishments which will make the final "interpretive structural model" a more complete picture of the system that it represents.

ISM is intended for use when assaulting a complex system, so that a qualitative model of the structure of the system can be composed. ISM expedites the creation of a digraph which gives rise to a structural model. The entire process has been designed for a man/machine interaction, so that the user can concentrate on substantive concerns in order to make subjective judgements regarding the relationship between the elements. Figure 13 gives a graphical representation of this man/machine interaction.

A variation of the embedding technique is used in this exercise. Instead of constructing the adjacency matrix A, a computer program has been developed to construct the reachability matrix M directly. The complete algorithm for this process can be found in Warfield [11].

Suppose that a reachability matrix for the set of elements $E_n = \{e_1, e_2, \ldots e_n\}$ has been constructed (Figure 14). Now the modeler wishes to determine the relationship between element e_{n+1} and the other element of set E_n. The computer constructs the Bordering Reachability matrix B by "bordering" the original reachability matrix with the row and column e_{n+1} (Figure 15). Next the element set E_n is divided into five subsets:

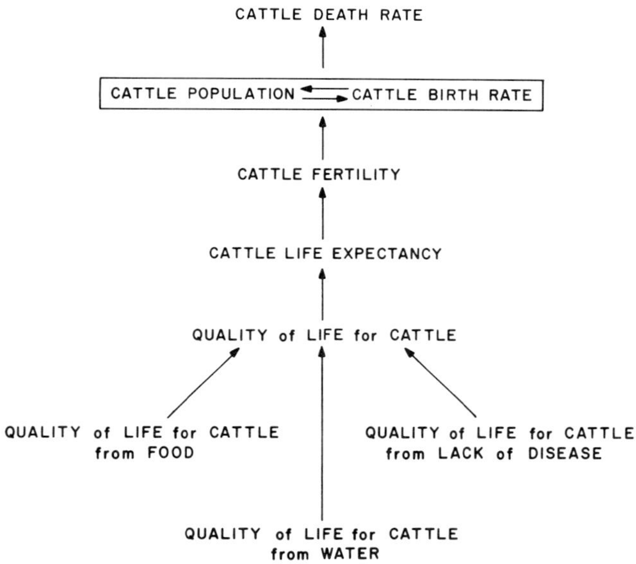

Fig. 11 - Digraph from Skeleton Matrix Q

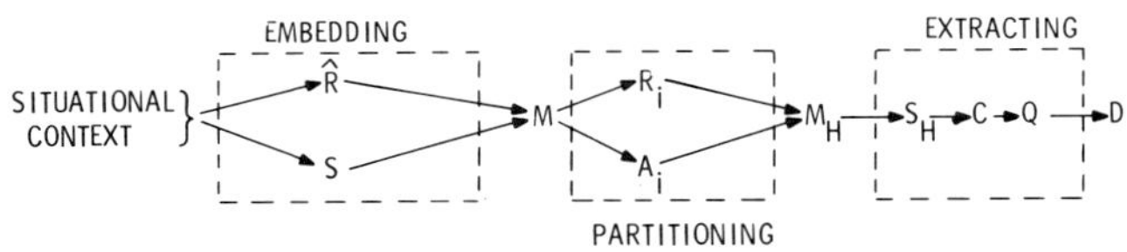

$\hat{R}$ - CONTEXTUAL RELATIONSHIP

S - SET OF SYSTEM ELEMENTS

M - REACHABILITY MATRIX

R_i - REACHABILITY SET

A_i - ANTECEDENT SET

M_H - HIERARCHIAL REACHABILITY SET

S_H - STANDARD FORM MATRIX

C - CONDENSATION MATRIX

G - SKELETON MATRIX

D - MINIMUM EDGE DIAGRAPH

Fig. 12 - Symbolic Digraph of the Operations in ISM

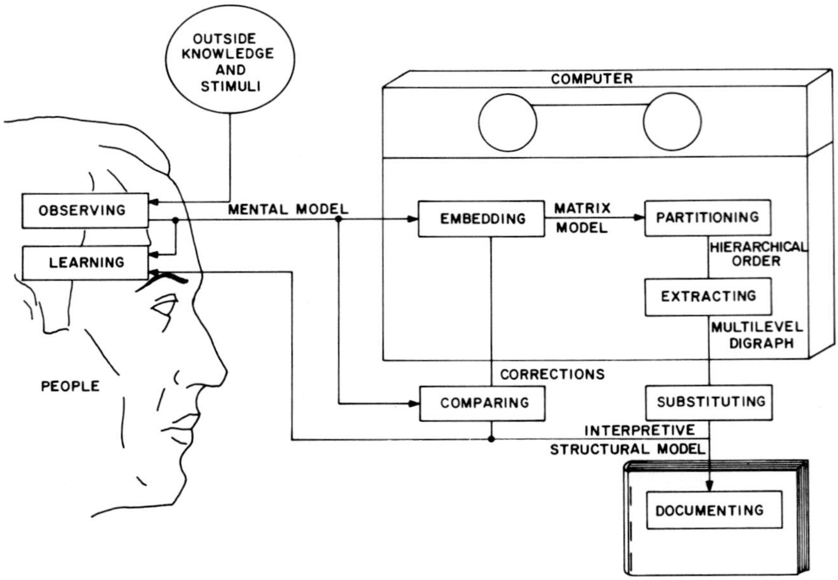

Fig. 13 - Functional Representation of Interpretive Structural Modeling, Showing Potential for Man/Machine Symbiosis

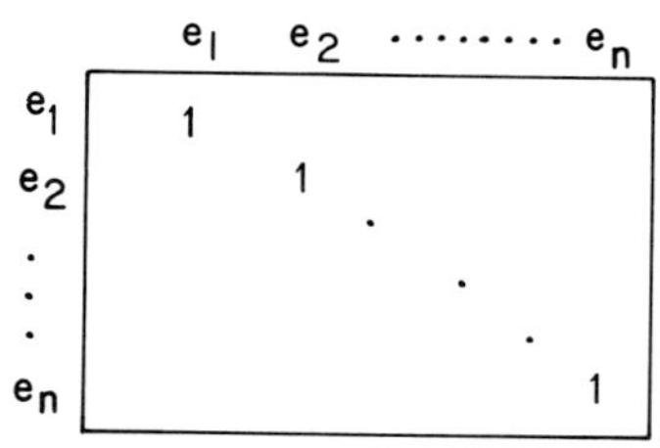

Fig. 14 - Reachability Matrix

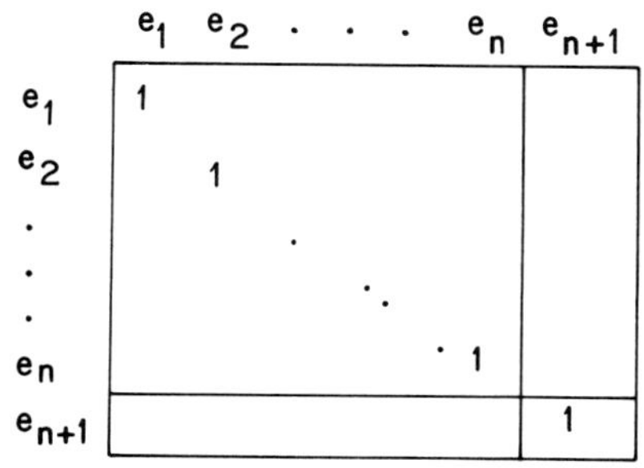

Fig. 15 - Bordering Reachability Matrix

1. e_j - The Pivot Element

2. $L(e_j)$ - The Lift Set of e_j

3. $D(e_j)$ - The Drop Set of e_j

4. $F(e_j)$ - The Feedback Set of e_j

5. $V(e_j)$ - The Vacancy Set of e_j

Thus $E_n = e_j; L(e_j); D(e_j); F(e_j); V(e_j)$. The definitions of these sets are given below and their interrelations can be seen diagramatically in Figure 16.

1. $L(e_j) = \{e | e \varepsilon E_n;\ e \neq e_j;\ e_j\ R\ e;\ e\ \overline{R}\ e_j\}$

2. $D(e_j) = \{e | e \varepsilon E_n;\ e \neq e_j;\ e_j\ \overline{R}\ e;\ e\ R\ e_j\}$

3. $F(e_j) = \{e | e \varepsilon E_n;\ e \neq e_j;\ e_j\ R\ e;\ e\ R\ e_j\}$

4. $V(e_j) = \{e | e \varepsilon E_n;\ e \neq e_j;\ e_j\ \overline{R}\ e;\ e\ \overline{R}\ e_j\}$

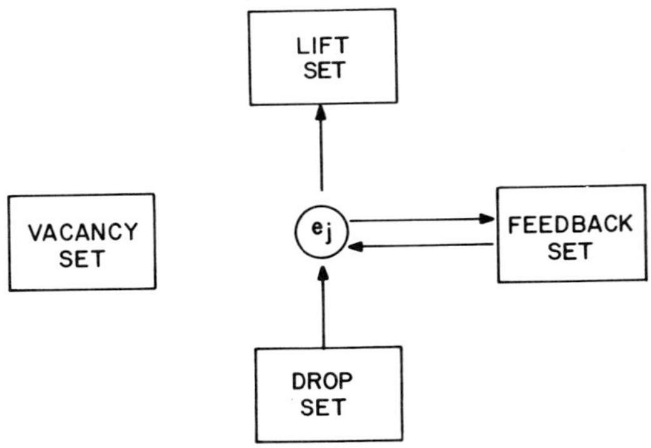

Fig. 16 - Relationships Between the Five Subsets of E_n

There are four types of questions that can be asked concerning the relationship between the pivot element e_j and the new element e_{n+1}. These four questions are: $e_j \; R \; e_{n+1}$, $e_j \; \overline{R} \; e_{n+1}$, $e_{n+1} \; R \; e_j$, and $e_{n+1} \; \overline{R} \; e_j$. There are inferences that can be made due to the transitivity of the relationship R.

1. Assume that the pivot element e_j is related to the new element e_{n+1}. There are four types of inference that can be made.

(i) If element e_k is a member of the drop set of e_j, $D(e_j)$ then the element e_k is related to the new element e_{n+1}.

$$e_k \in D(e_j) \text{ implies } e_k \; R \; e_{n+1}$$

(ii) If element e_k is a member of the drop set of e_j, $D(e_j)$, then the new element e_{n+1} is not related to element e_k.

$$e_k \in D(e_j) \text{ implies } e_{n+1} \; \overline{R} \; e_j$$

(iii) If element e_p is a member of the feedback set of

e_j, $F(e_j)$ then the element e_p is related to the new element e_{n+1}.

$$e_p \in F(e_j) \text{ implies } e_p \; R \; e_{n+1}$$

(iv) If element e_m is a member of the vacancy set of e_j, $V(e_j)$ then the new element e_{n+1} is not related to element e_m.

$$e_m \in V(e_j) \text{ implies } e_{n+1} \; \overline{R} \; e_m$$

2. Assume that the pivot element e_j is not related to the new element e_{n+1}, then there are two types of inferences that can be made.

(i) If element e_n is a member of the lift set of e_j, $L(e_j)$, then element e_n is not related to element e_{n+1}.

$$e_n \in L(e_j) \text{ implies } e_n \; \overline{R} \; e_{n+1}$$

(ii) If element e_p is a member of the feedback set $F(e_j)$, then element e_p is not related to element e_{n+1}.

$$e_p \in F(e_j) \text{ implies } e_p \; \overline{R} \; e_{n+1}$$

3. Assume that the new element e_{n+1} is related to the pivot element e_j, then there are four types of inference that can be made.

(i) If element e_n is an element of the lift set $L(e_j)$, then the new element e_{n+1} is related to e_n.

$$e_n \in L(e_j) \text{ implies } e_{n+1} \; R \; e_n$$

(ii) If the element e_n is a member of the lift set $L(e_j)$, then the element e_n is not related to the element e_{n+1}.

$$e_n \in L(e_j) \text{ implies } e_n \; \overline{R} \; e_{n+1}$$

(iii) If element e_p is a member of the feedback set $F(e_j)$, then the new element e_{n+1} is related to element e_p of the feedback set.

$$e_p \in F(e_j) \text{ implies } e_{n+1} \, R \, e_p$$

(iv) If element e_m is a member of the vacancy set of e_j, $V(e_j)$, then the element e_m is not related to element e_{n+1}.

$$e_m \in V(e_j) \text{ implies } e_m \, \overline{R} \, e_{n+1}$$

4. Assume the new element e_{n+1} is not related to the pivot element e_j. Two types of inference can be made with this assumption:

(i) If element e_k is a member of the drop set $D(e_j)$, then the element e_{n+1} is not related to e_k.

$$e_k \in D(e_j) \text{ implies } e_{n+1} \, \overline{R} \, e_k$$

(ii) If element e_p is a member of the feedback set $F(e_j)$, then element e_{n+1} is not related to element e_p.

$$e_p \in F(e_j) \text{ implies } e_{n+1} \, \overline{R} \, e_p$$

The computer picks the pivot element by examining the four assumptions shown above for each element of the set E_n and uses the element which could give the maximum inference as the pivot element.

After the entire reachability matrix is completed, the hierarchial matrix and digraph are formed as described above.

Figure 3 indicate that from a structural model one can construct a dynamic model. If one examines the work of Forrester [3] or Meadows et al. [8] for example, one will note that their causal loop diagrams are a type of structural model. The mental models of these research groups were translated into this graphical form without the intermediate matrix model being constructed. ISM gives a systematic way of constructing these causal diagrams and thus insures that the relationship between all structural elements have been made explicit. Thus the qualitative ISM modeling exercise is a starting point for the development of a dynamic model. Once the causal diagram has been constructed one can transform it to a flow digraph (e.g. a DYNAMO flow graph [3, 4, 8] and then translate the flow graph to equations. The equation will be time-dependent, with the structure of the system derived from the ISM exercise being constant, but with a flux of energy, matter or information through the structural components.

3. DESCRIPTION OF THE SAHEL

The Sahelian region consists of six nations; Chad, Mali, Mauritania, Niger, Senegal and Upper Volta, all of which lie approximately between 10° and 20° north latitude (Figure 17). The total area is about 1.7 times that of western Europe and about equal to 67 percent of the area of the United States [7].

To characterize the Sahel in one word would be to say that the Sahel is hot. Maximum temperatures in the northern areas exceed 50°C and seldom do readings drop below 0°C. Solar radiation is intense and there is cloudy weather only in the rainy season. Much of the year the dry north-easterly trade winds influence the region, and because of the high temperatures, evaporation rates are high. Rainfall is highly variable in this region, with some regions experiencing no rain for several years. There is only one rainy period in the Sahel, the summer, and often early in this period there are thunderstorms with associated violent winds. The heavy rains from these thunderstorms allow for sheet erosion to occur and the violent winds often carry sands from the adjacent desert allowing arable or grazable land to be buried. Figure 18 shows several climatic zones, the associated amount of rainfall, and the duration and season of rainfall. As a rule of thumb, there is one day of growing period (wet season) for each 5mm. annual precipitation [7].

Within the Sahel, droughts are frequent and severe, with estimates of drought lengths since the 16th century giving the following results [7]: of the 22 droughts listed, 6 lasted for 1

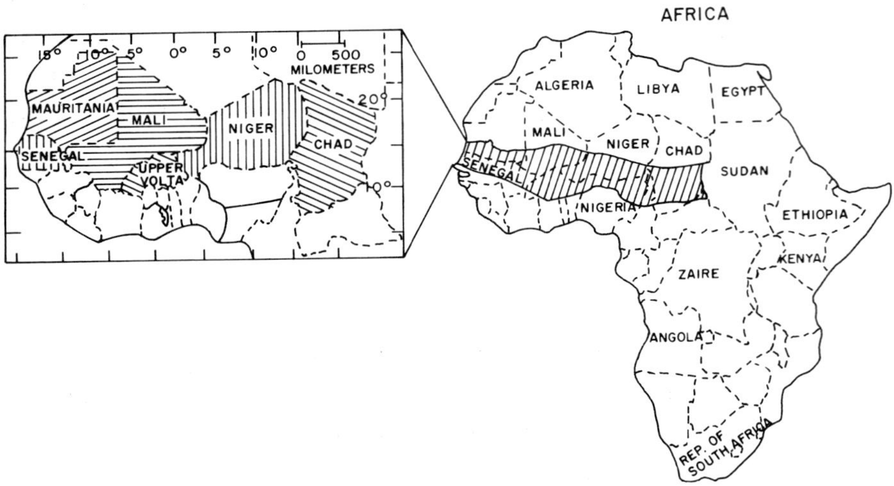

Fig. 17 - The Sahelian Region of West Africa

year, 8 for 2 years, 2 for 3 years, 1 for 4 years, 4 for 5 years and 1 for 18 years.

In general the Sahel is flat although there are a few mountainous regions. Soils are readily erodable by wind and water and in numerous areas the amount of soil covering is minimal with bare rock outcrops throughout the region. In the Sahel only 7.9 percent of the total land area is classified as arable in contrast to the 24.3 percent for all of Africa and 24.1 percent for all continents.

Various evidence suggests that much of north Africa formerly supported grazing cattle [1]. These indicators suggest that herds existed as early as 6000 to 7000 B.P. (before present), not only in areas that are grazable now, but also in areas now essentially bare rock and drifting sand [5].

It is generally agreed that the encroachment of the desert into usable lands is primarily the result of misuse of the ecosystem by man [6]. This encroachment, or process of desertifi-

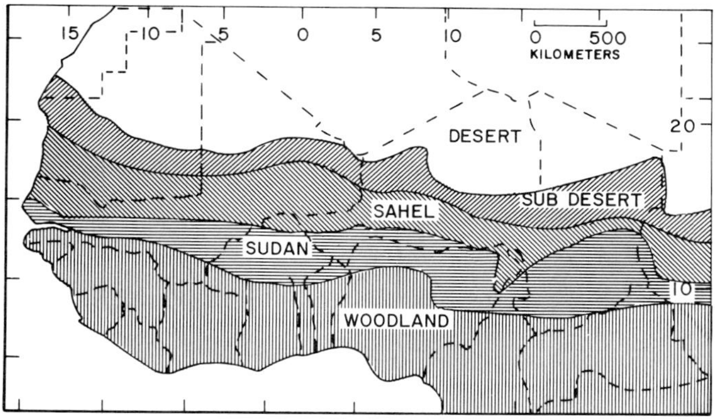

ZONE	MILLIMETERS ANNUAL PRECIPATION	No. RAINY DAYS PER YEAR	RAINY SEASON	PERCENT OF REGION
DESERT	100	10		36
SUBDESERT	100-300	10-25	JULY-AUG.	28
SAHEL	300-650	26-46	JUNE-SEPT.	17
SUDAN	650-900	50-65	MAY-OCT.	10
WOODLAND	900	65-80	MAY-OCT.	9

Fig. 18 - Major Vegetation Zones of West Africa

cation, in subdesert and savanna regions takes on the following characteristics: (a) a reduction of perennial plant cover, (b) impoverishment (reduction of species diversity) of flora, (c) soil erosion by wind and water action, and (d) the formation of mobile sand dunes and areas of bare rock [6].

This process is initiated by one or more human activities such as: (a) removing vegetation for firewood at a rate faster than normal regrowth, (b) over-grazing by livestock, or (c) use of lands for cash crops thus forcing food crops to be grown in regions of marginal rainfall. Add to these degrading activities the fragility of the ecosystem with its recurrent droughts, and the result is lowered agricultural production and thus an even higher pressure of population on the ecosystem. This vicious cycle can be seen in Figure 19.

Not only are there ecological constraints on development of the Sahel (variability and fragility of the ecosystem along with present population pressures) but there are also time, economic and social-political constraints. Time as a constraint for implementation of agricultural strategies is a fairly slow process. The land in the Sahel is degraded to the point that the first step must be regeneration of the soil to a better state. Due to the relatively few natural resources found in the Sahel, funding for new technologies may have to be external. Planning must fit human material needs but it must also be cognizant of human social systems. There are often competing social needs and thus certain strategies may be ill-suited to the sedentary fraction of the population and at the same time doom the nomads to extinction. Thus social constraints may be the most difficult to overcome.

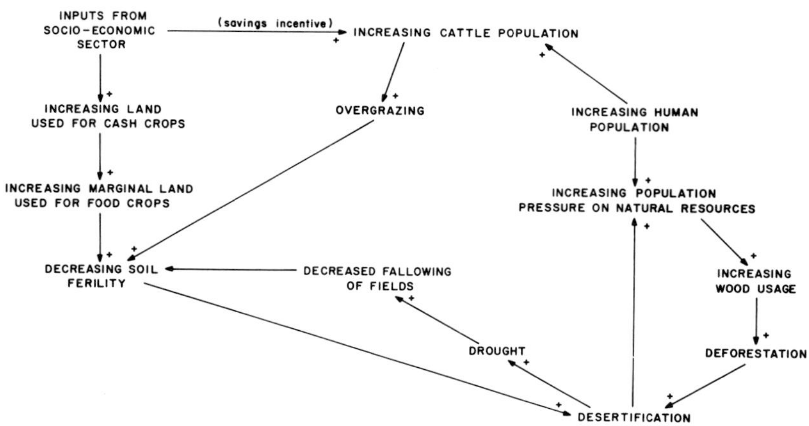

Fig. 19 - Major Feedback Loops of the Sahelian Ecosystem

4. MODEL OF THE QUALITY OF RANGELAND

A completed structural model of the dynamics of the ecosystem of the Sahelian region is very complex and contains many interacting feedback cycles. The approach used to develop a model of the Sahelian ecosystem was based on use of the geodesic cycles as prescribed by Warfield [11]. Although the geodesic notion was not used exactly it helped to identify certain subsystems within the model which could be developed one at a time. It was reasoned that once these subsystems where analyzed they could be integrated into a total system, which would give an appreciation of the Sahel ecosystem and its complexity.

The initial construction of subsystem models for the Sahel ecosystem is around eight sectors, namely, cattle population, quality of rangeland, soil erosion, sheep and goat population, human population, relief and development aid, sedentary villages, and water resources. There is a great overlapping of the sectors, but this overlapping is an asset when subsystem integration takes place. For each of the subsystems, historical data which indicates trends, both positive and negative, and a series of problems involved in each of these subsystems is being developed. From this trend and problem data, along with the physical and biological laws known to function in the ecosystem, a list of elements for each subsystem model is developed and a structural model is built.

An example of one subsystem model is that of cattle population. Figure 20 illustrates the growth and decline of cattle population in the country of Mali over a twenty year period. It is easy to identify three distinct periods of growth in this particular graph. During the first period cattle population is growing at a very small rate and is approximately a steady state condition. Around 1960, with the introduction of veterinary medicine and animal husbandry after the period of independence, the cattle

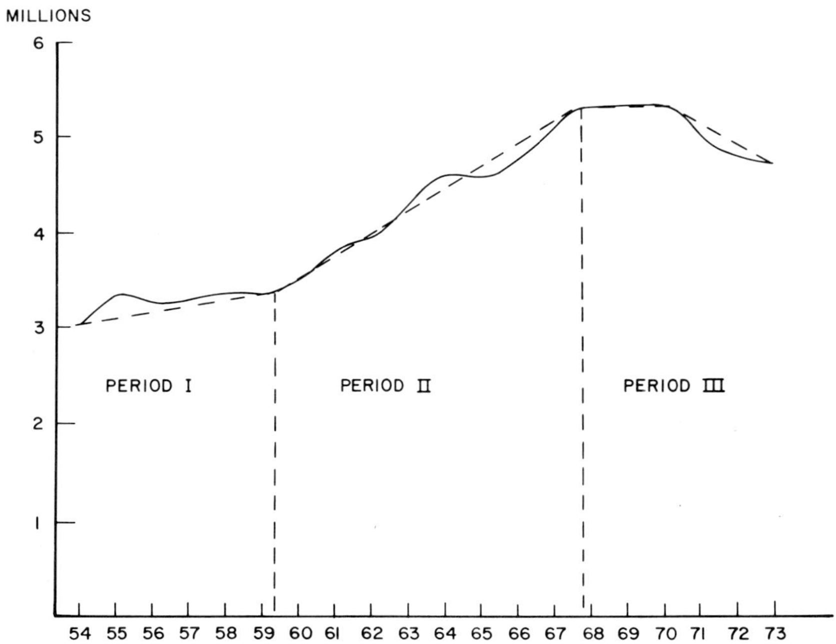

Fig. 20 - Cattle Growth in Mali (1954-1973)

population increases rapidly. Around the beginning of the drought in 1968, the dynamics of the cattle population begins to stabilize once again, and towards the end begins to decline. Viewing this graph, and having a knowledge of population dynamics, makes it fairly easy to identify what was occurring in each of the three periods outlined. Figure 21 shows the structural model of the cattle population subsystem using the relationship, R;

Does a change in
FACTOR A
cause a change in
FACTOR B.

A somewhat larger subsystem which overlaps greatly with the other subsystems is that of Quality of Rangeland. For modeling

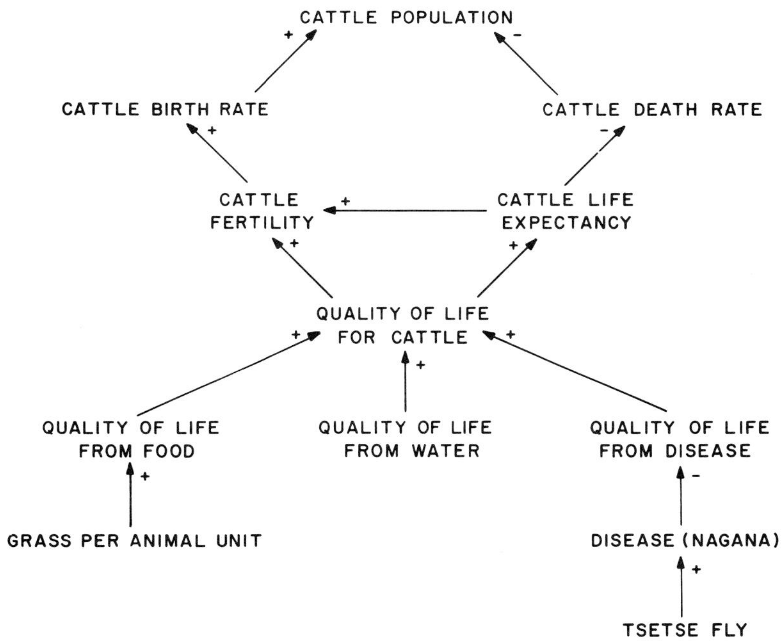

Fig. 21 - Digraph of Cattle Population Subsector

this subsystem two relationships were used so that a quasi-quantitative model would be produced. The relationships are;

1. Does an increase in
FACTOR A
cause an increase in
FACTOR B

and

2. Does an increase in
FACTOR A
cause a decrease in
FACTOR B.

A directed graph for the quality of rangeland subsystem has been constructed and is shown in Figure 22.

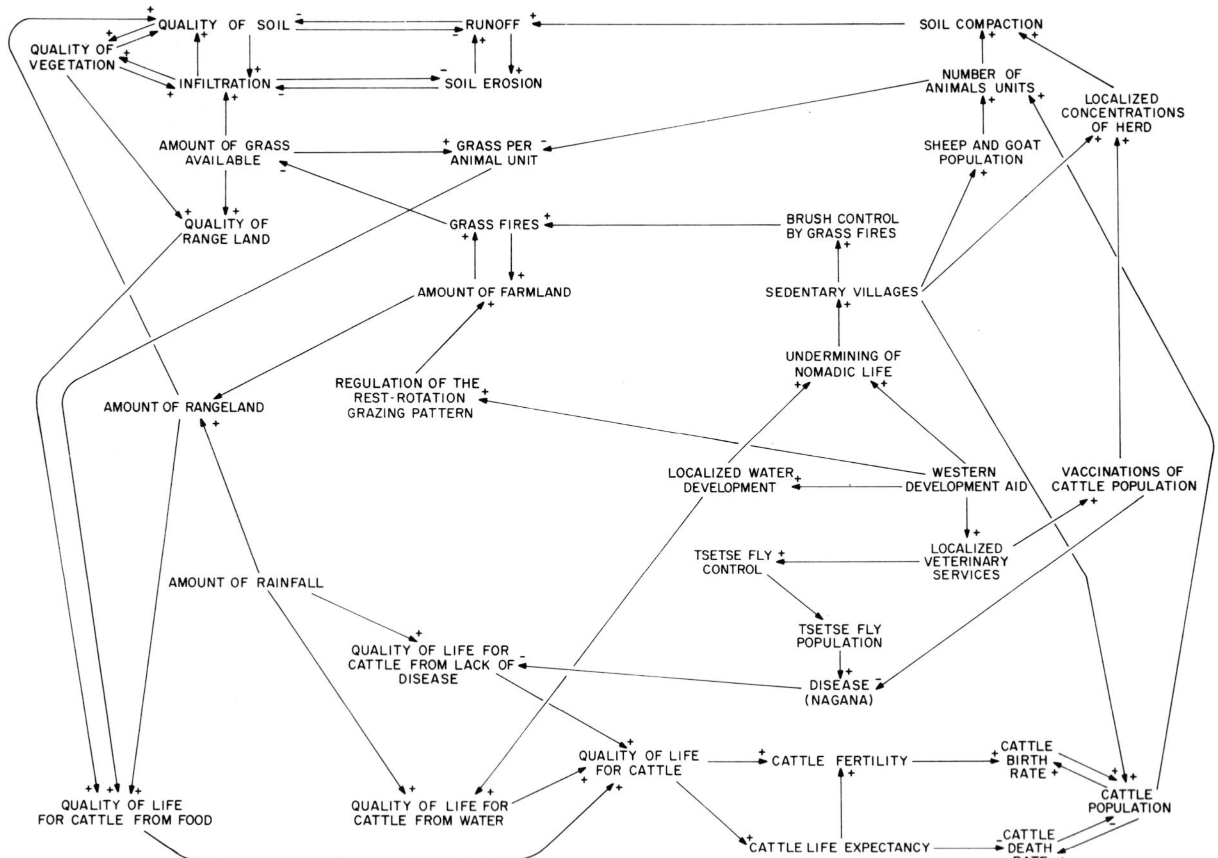

Fig. 22 - ISM Model of the Quality of Rangeland

5. SUMMARY AND ACKNOWLEDGEMENTS

In this paper, an application of interpretive structural modeling for the complex Sahelian ecosystem has been developed. Interpretive structural modeling seems to be a useful tool in helping to delineate the structure of complex systems. Preliminary indications and interactions with decision makers indicate that the ISM approach has sufficient potential that they are willing to further explore its use in policy decision making [2].

Don Geiger and Ed Zamierowski of the University of Dayton Biology Department have helped develop the content of the models presented, and Marie Zamiska has aided in the construction and review of the paper.

REFERENCES

1. Davidson, B., "Africa in Historical Perspective" in *Africa South of the Sahara,* (1973), pp. 3-16, Europa Publications, London.
2. Fitz, R., "Interpretive Structural Modeling and the Policy Planning Process" in *Modeling and Simulation, Vol. 6, Proceedings of the Sixth Annual Pittsburg Conference*, (W.G. Vogt and M.H. Mickle, eds.), (1975), Instrument Society of America, Pittsburg, Pa.
3. Forrester, J.W., *World Dynamics*, Wright-Allen Press, (1974), Cambridge, Mass.
4. Goodman, M.R., *Study Notes in System Dynamics*, Wright-Allen Press, (1974), Cambridge, Mass.
5. Hufnagl, E., *Libyan Mammals*, The Oleander Press, (1972), Stoughton, Wisconsin.
6. Kassas, M., "Desertification versus Potential for Recovery in Cirsum-Saharan Territories" in *Arid Lands in Transition*, (H.E. Dregne, ed.), (1970), AAAS, Washington.

7. Matlock, W.G. and E.L. Cockrum, *A Framework for Evaluating Long-Term Strategies for the Development of the Sahel-Sudan Region, Volume 2, A Framework for Agricultural Development Planning*, Center for Policy Alternatives Report No. CPA 74-9, (1974), Massachusetts Institute of Technology.

8. Meadows, D.L., W.W. Behrens, D.H. Meadows, R.F. Naill, J. Randers, and E.K.O. Zahn, *Dynamics of Growth in a Finite World*, Wright-Allen Press, (1974), Cambridge, Mass.

9. Sagasti, F.R., "A Conceptual 'Systems' Framework for the Study of Planning Theory, *Technological Forecasting and Social Change 5*, (1973), pp. 379-393.

10. Wade, N., "Sahelian Drought: No Victory for Western Aid", *Science 185*, (1974), pp. 234-237.

11. Warfield, J.N., *Structuring Complex Systems, Battelle Monograph No. 4*, Battelle Memorial Institute, Columbus, Ohio.

THE MULTIPLE-PARAMETER STABILITY THEORY AND ITS RELATION TO CATASTROPHE THEORY

K. HUSEYIN
Department of Systems Design
University of Waterloo
Waterloo, Ontario, Canada

SUMMARY

The connection between the multiple-parameter stability theory and catastrophe theory is discussed with reference to distinct critical points (elementary catastrophes of dimension one). It is demonstrated analytically that the *general* and *singular general* points, which are associated with smooth equilibrium surfaces of order two and three, correspond to fold and Riemann-Hugoniot catastrophes respectively. Similarly, it is shown through algebraic transformations only (without resorting to topology) that *general* point of order four (swallowtail) can be described by three significant parameters, and an analytical pattern is thus established for obtaining higher order surfaces (butterfly, etc.). Several examples illustrating the applications of the theory in various disciplines are presented.

1. INTRODUCTION

Consider a gradient system characterized by the potential function

$$V = V(Q_i, \Lambda) \qquad i=1,2,\ldots,N$$

where the Q_i describe the state of the system and Λ is an independently assignable parameter. At every given value of the *control* parameter Λ the system has a potential in terms of the *behaviour variables* Q_i, and in a slowly evolving process the equilibrium states of the system satisfy the relation

$$\text{grad}_Q V = 0$$

which yields N scalar equations, defining certain *equilibrium paths* in the $(Q_i - \Lambda)$ *configuration space* (which is also referred to as the load-deflection space). In Mechanics, for example, the Q_i are the generalized coordinates, describing the deflections and Λ is the loading parameter. Poincaré's bifurcation theory identifies mainly two types of critical points on the equilibrium paths at which the Hessian,

$$\left| \partial^2 V / \partial Q_i \, \partial Q_j \right| \quad ,$$

vanishes. Namely, the *limit* point which represents a local extremum on an equilibrium path and the point of *bifurcation* at which two distinct paths intersect each other resulting in an exchange of stabilities. If the critical point involves coincident eigenvalues, then, multi-furcations rather than bifurcation may occur*. In the development of the general theory of elastic stability in Mechanics, bifurcation and limit points played a central role, and even the behaviour of imperfect structural systems, which essentially involve an additional control parameter(s)

---

* It has been shown [1,2] that, under some circumstances, trifurcation can also occur at simple (distinct) critical points.

(imperfection parameter), was referred to these points [3,4,5].

Systems are often under the influence of several parameters, and it was first pointed out in Ref. [1] that bifurcation and limit points cannot describe the behaviour of multiple-parameter systems fully and that a reclassification of critical conditions based on equilibrium surfaces rather than paths is needed. Evidently, if the system is characterized by a potential function

$$V = V(Q_i, \Lambda^j) , \qquad i=1,2,\ldots,N \quad j=1,2,\ldots,M ;$$

the equilibrium equations $\partial V/\partial Q_i = 0$ define an M dimensional surface(s) in the configuration space spanned by the Q_i and Λ^j, and it was proposed [1] that a classification which distinguishes between a proper and improper (degenerate) surface would be most appropriate. Thus, *general* and *special* critical points, describing the former and latter situations respectively, were introduced. A general nonlinear multiple-parameter theory of stability was then developed [6,7,8] on the basis of these new concepts. *General* points associated with simple eigenvalues can be *singular* or *non-singular* and the latter can further be classified as *elliptic*, *hyperbolic* or *parabolic* [7]. The *stability boundary* is defined as the projection of the *critical zone* of the equilibrium surface (where the Hessian vanishes initially) into the load-space (control space).

It is remarkable that catastrophe theory developed by René Thom [9] essentially parallels the above stability theory and gives a full topological classification of instability phenomena (catastrophes) which can arise when $M \leq 4$. The Riemann-Hugoniot catastrophe and fold catastrophe, for example, correspond to singular and nonsingular general points respectively. Similarly, the bifurcation set (stability boundary) is the projection of the fold line (critical zone) into the control space.

Using topological theorems, all catastrophes are given in the simplest canonical forms and the "universal Unfoldings" (potential functions associated with each form) contain only the critical behaviour variable(s). To achieve this in the stability theory, certain appropriate transformations were introduced, and in particular the transformation of control parameters, $\Lambda^i = \beta^{ij} \phi^j$, contributed to identifying significant parameters [7]. Here, one notes the quantitative nature of the stability theory. The systems of practical interest can rarely be described by one behaviour variable adequately, and when more than one variable is involved it is normally impossible to identify a priori the critical variable(s). The systematic methods of analysis underlying the stability theory, however, enables one to maintain a link between the basic and transformed variables throughout the analysis and automatically produces the significant ones. The catastrophe theory, on the other hand, while providing a full qualitative description of phenomena, may not be directly applicable as will be seen in an example presented in Section 4(i). Simplest forms of catastrophes may not be adequate from other points of view as well. Topological arguments, for example, lead to the conclusion that the fold catastrophe is a one-parameter phenomenon. While this is satisfactory in a qualitative discussion, the convexity of the stability boundary, for instance, can only be established by incorporating an additional control parameter (as in the case of hyperbolic general point), and in a quantitative analysis such properties of the boundary have a vital practical as well as theoretical value. Furthermore, although a certain form of the equilibrium surface (a particular catastrophe) can be expressed in terms of critical variables only, as remarked earlier, the latter can be determined through appropriate transformations and the effect of the non-critical variables is reflected in the final equation of the surface in terms of curvatures and other quantities.

Another interesting point is concerned with the concept of "structural stability" used in catastrophe theory. An equilibrium point is said to be structurally stable if and only if the

Hessian does not vanish at that point. This implies that an equilibrium point which is actually unstable in terms of the stability theory (in the sense of Lyapunov) may be called structurally stable. A structurally unstable point of "finite codimension", on the other hand, serves as the "organizing centre" of the so-called elementary catastrophes, which are formed in a structurally stable way and are associated with smooth surfaces. It follows that the elementary catastrophes are related to *general* points. *Special* points may arise when an appropriate control parameter is set zero. While it is true that the nature, with all its imperfections, will normally produce *general* points, the practical value of *special* points, nevertheless, is enormous. An example from Mechanics may be cited to illustrate this point. Consider a flat thin plate subjected to two independent uniformly distributed axial compression forces along its perpendicular edges. It is expected, of course, that this system will have small imperfections. However, an idealized model which initially ignores such imperfections is invaluable in a quantitative analysis. In fact, the most convenient method of analysis concerning the effect of small imperfections is based on the perfect model, and a thorough knowledge pertaining to the latter is almost a prerequisite for exploring the behaviour of the imperfect system and not vice-versa.

In the following sections attention will be restricted to simple critical points or to elementary catastrophes of dimension one. Coincident critical points, including the umbilics, are discussed elsewhere [10]. Several examples from various fields are presented and some experimental results are given.

2. CLASSIFICATION OF CRITICAL POINTS

Consider a system whose behaviour can be described by a finite set of behaviour variables Q_i $(i=1,2,\dots,N)$ and suppose that the behaviour of the system is influenced by M independent control parameters Λ^j. Then, several fundamentally important questions present themselves: How does the system behave when the parameters are varied gradually? Starting from a known stable state, for example, will the system lose its stability at one stage of the process? What implications do such instabilities have? Can they be predicted? How? These problems have been studied in the general theory of stability and are also the subject of the catastrophe theory.

Suppose now that the system under consideration is a *gradient system* (conservative), possessing a real, single-valued and continuously differentiable (at least in the region of interest) potential which is a function of the Q_i and Λ^j,

$$V = V(Q_i, \Lambda^j). \tag{1}$$

The total potential energy of an elastic conservative system in structural mechanics is a typical example for the all-embracing potential (1) where the Q_i are the generalized coordinates describing the deformations of the system, and Λ^j represent loads, imperfections, etc. [1,2]. In other disciplines, one can construct certain other appropriate potentials capable of describing a given system, providing the latter is a gradient system.

The condition

$$\text{grad}_Q V = 0 \tag{2}$$

yields N equilibrium equations of the form

$$\frac{\partial V}{\partial Q_i} = 0 \tag{3}$$

which evidently define an M dimensional *equilibrium surface* in the $N + M$ dimensional configuration space spanned by the Q_i and Λ^j. An equilibrium state on this surface is called critical if the Hessian

$$\det \left[\frac{\partial^2 V}{\partial Q_i \ \partial Q_j}\right] \tag{4}$$

vanishes at that point.

Interest now focusses on the characteristic properties of the equilibrium surface. In order to explore these properties, consider an arbitrary equilibrium point F on the surface and introduce the local coordinate system (u_i, λ^j), through the transformation

$$\begin{aligned} Q_i &= Q_i^F + \alpha_{ij}\, u_j \\ \Lambda^i &= \Lambda_F^i + \lambda^i \end{aligned} \qquad \text{where } \alpha_{ij}\, \alpha_{jk} = \delta_{ik} , \tag{5}$$

to diagonalize the quadratic form (in the $q_i = Q_i - Q_i^F$) of the potential function around F. Here the summation convention is adopted, and the u_i are the principal coordinates. The potential function takes the form

$$H(u_i, \lambda^j) \equiv V(Q_i^F + \alpha_{ij}\, u_j \ , \ \Lambda_F^i + \lambda^j) \tag{6}$$

where the Hessian matrix $[H_{ij}]_F$ is diagonal.

<u>Noncritical points</u>.

If F is noncritical, all the eigenvalues of the Hessian matrix are different from zero. The solution of the equilibrium equations $\partial H/\partial u_i = 0$ can be expressed in the parametric form

$$u_i = u_i(\varepsilon^k) \ , \quad \lambda^j = \lambda^j(\varepsilon^k) \quad \text{where } k=1,2,\ldots,M \tag{7}$$

for a multiple-parameter perturbation technique aimed at determining sufficient surface derivatives which can be used to construct the asymptotic equation of the equilibrium surface around F. In this case ε^k can be chosen as λ^k in order to simplify the procedure. The details of the technique as well as the rules concerning the appropriate selection of the ε^k have been discussed in Refs. [2,11] and will not be repeated here.

Substituting the assumed solution $u_i = u_i(\lambda^k)$ back into the equilibrium equations results in the fundamental identities

$$H_i\ [u_j(\lambda^k),\lambda^k] \equiv 0 \tag{8}$$

where $H_i \equiv \partial H/\partial u_i$.

Differentiating with respect to λ^k yields

$$H_{ij}\ u_j^k + H_i^k = 0 \tag{9}$$

where superscripts are used to denote differentiation with respect to the control parameters.

Evaluating (9) at F gives

$$u_j^k = -\ H_i^k/H_{ii}\ , \tag{10}$$

and the first order equation of the equilibrium surface takes the form

$$H_{ii}\ u_i + H_i^j\ \lambda^j = 0 \tag{11}$$

describing an M dimensional plane and indicating a one-to-one correspondence between a given set of λ^i and u_i . This is the most general equilibrium behaviour of the system, as in the linear theory of elasticity.

3. SIMPLE CRITICAL POINTS

Suppose F is a critical point at which $H_{11} = 0$ and $H_{ss} \neq 0$ (for all $s\neq1$). The critical point F is *special* or *general* according to whether or not

$$\mathrm{grad}_\lambda \ H_1(0,0) \tag{12}$$

vanishes respectively. As remarked earlier, the former situation may be a product of our models which, nevertheless, represents an invaluable concept from an analytical as well as a practical point of view, and has been presented in Ref. [2]. Attention in this paper is focussed on the general points which are closely related to elementary catastrophes.

i) General critical points (Fold catastrophe).

Considering now the general critical point F at which

$$H_{11} = 0, \quad H_{ss} \neq 0 \text{ (for all } s\neq 1) \text{ and at least } H_1^1 \neq 0, \tag{13}$$

and following the multiple-parameter perturbation procedure [2], one selects the perturbation parameters ε^k as u_1 and λ^m (m=2, ...,M) and expresses the solution of the equilibrium equations as

$$u_s = u_s(u_1,\lambda^m) \ , \quad \lambda^1 = \lambda^1(u_1,\lambda^m) \tag{14}$$

The first perturbation of the identities

$$H_i[u_s(u_1,\lambda^m) \ , \ \lambda^1(u_1,\lambda^m) \ , \ u_1,\lambda^m] \equiv 0 \tag{15}$$

with respect to u_1 and λ^m (m=2,...,M) yields

$$H_{is} \, u_{s,1} + H_i^1 \ \lambda_1^1 + H_{i1} = 0 \tag{16}$$

and

$$H_{is} \, u_s^m + H_i^1 \, \lambda^{1,m} + H_i^m = 0 \tag{17}$$

respectively, which upon evaluation at F result in the surface derivatives

$$\left.\begin{aligned}
&\lambda^1_1 = 0\,, \quad u_{s,1} = 0 \\
&\text{and} \\
&\lambda^{1,m} = -\frac{H^m_1}{H^1_1}\,, \quad u^m_s = -\frac{1}{H_{ss}}\left[H^m_s - H^1_s\frac{H^m_1}{H^1_1}\right]
\end{aligned}\right\} \tag{18}$$

A second perturbation with respect to u_1 gives

$$\lambda^1_{11} = -\frac{H_{111}}{H^1_1}\,, \quad u_{s,11} = -\frac{1}{H_{ss}}\left[H_{s11} - H^1_s\frac{H_{111}}{H^1_1}\right]. \tag{19}$$

Using (18) and (19) one can construct the first order equations of the equilibrium surface in the vicinity of F as

$$H^i_1\,\lambda^i + \frac{1}{2}H_{111}(u_1)^2 = 0 \tag{20}$$

$$H_{ss}\,u_s + \left[H_{s11} - H_{111}\frac{H^1_s}{H^1_1}\right](u_1)^2 + \left[H^m_s - H^1_s\frac{H^m_1}{H^1_1}\right]\lambda^m = 0 \tag{21}$$

which describe the *general* critical point of order two.

Equation (20) represents the projection of the equilibrium surface into the $u_1 - \lambda^i$ subspace, involving the critical behaviour variable only. It describes a curved surface, and was first obtained in Ref. [1,6]. If arbitrary control paths of the form

$$\lambda^i = \lambda^i(\xi) \tag{22}$$

are considered, Equation (20) reduces to

$$H'_1\,\xi + \frac{1}{2}H_{111}(u_1)^2 = 0\,, \tag{23}$$

describing the corresponding equilibrium path on the surface, and is immediately recognized as the "*fold catastrophe*". Clearly, the behaviour of the system remains the same qualitatively for all control paths (22) and it may, therefore, be suggested that the general critical point of order two (fold catastrophe) is essentially a one-parameter phenomenon (i.e., of co-dimension one [9]). This, however, is misleading in the sense that one may be led to believe that the equilibrium surface (20) is always *parabolic* and there is no need for a second parameter at all. Actually, the properties of the critical line (fold line) as well as the equilibrium surface can only be explored fully by retaining the parameters and introducing an appropriate transformation. Thus, introduce [1,7] the transformation

$$\lambda^i = \beta^{ij} \phi^j \qquad \beta^{ij} \beta^{jk} = \delta^{ik} \tag{24}$$

such that when (24) is applied, the resulting transformed function

$$\Pi(u_i, \phi^j) \equiv H(u_i, \beta^{ij} \phi^j) \tag{25}$$

has the properties

$$\Pi_1^1 \neq 0 \quad \text{and} \quad \Pi_1^m = 0 \qquad \text{for all } m . \tag{26}$$

Using the perturbation technique as before one can construct the first order equation of the equilibrium surface as (Figure 1)

$$\Pi_1^1 \phi^1 + \frac{1}{2} \Pi_{111} u_1^2 + c^m u_1 \phi^m + \frac{1}{2} d^{mn} \phi^m \phi^n = 0 \tag{27}$$

where c^m and d^{mn} are constants and involve derivatives of Π with respect to u_s as well as u_1. The surface is synclastic, anticlastic, or parabolic according to whether the matrix $[c^m c^n - \Pi_{111} d^{mn}]$ is negative definite, positive definite, or null respectively [1,7].

The critical zone is described by

$$\phi^1 = \frac{1}{2\Pi_1^1 \Pi_{111}} (c^m c^n - \Pi_{111} d^{mn}) \phi^m \phi^n \tag{28}$$

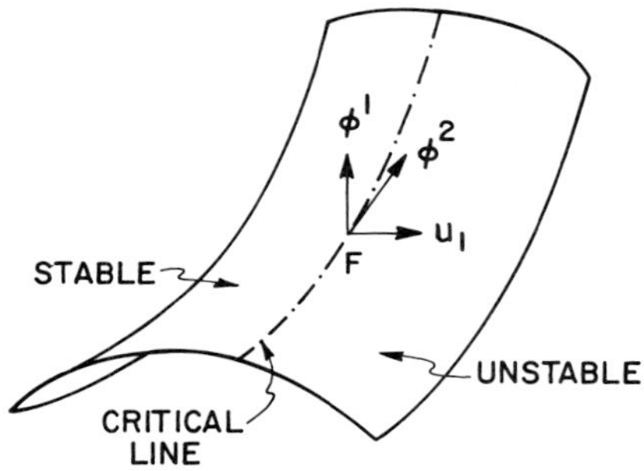

Fig. 1

and

$$u_1 = -\frac{1}{\Pi_{111}} c^m \phi^m , \tag{29}$$

the former equation giving the *stability boundary*. The convexity of this boundary, as well as the stability of the states in (27) were discussed in Refs. [1,2,7]. It was shown that if the system has only one behaviour variable the equilibrium surface is always anticlastic and the stability boundary cannot have convexity with respect to the region of stability. It is significant to note that this result cannot be generalized to systems having more than one behaviour variable and although (27) does not involve the non-critical coordinates, the influence of the latter is preserved through the coefficients c^m and d^{mn}.

ii) Singular general critical points (Riemann-Hugoniot catastrophe)

A distinct phenomenon arises when H_{111} (and consequently Π_{111}) vanishes at the general critical point. Using the transformation (24) and the resulting function (25), the multiple parameter perturbation technique yields the equilibrium surface

$$\Pi_1^1 \phi^1 + \frac{1}{3!} \overline{\Pi}_{1111} (u_1)^3 + c^m u_1 \phi^m + \frac{1}{2} d^{mn} \phi^m \phi^n = 0 \qquad (30)$$

This first order equation can further be simplified by introducing appropriate transformations. It is first noted that qualitatively the behaviour of the system will be similar for all control paths ξ in the ϕ^m space. Without loss of generality one can then, consider ϕ^2 as a representative parameter, thus reducing (30) to a two-parameter equation. It can be shown that if control is restricted to ϕ^2 (i.e., for $\phi^1=0$) the critical point F appears as a point of bifurcation. To this end, differentiate (30) with respect to u_1 twice and evaluate at F to obtain

$$\left.\frac{d\,\phi^2}{d\,u_1}\right|_F = 0 \quad \text{or} \quad \left.\frac{d\,\phi^2}{d\,u_1}\right|_F = -2\,\frac{c^2}{d^{22}} ,$$

indicating a symmetric point of bifurcation.

Based on this property, one can reformulate the problem by introducing the sliding and rotating coordinates $\overline{u}_j$ through

$$u_i = u_i^F (\phi^2) + \alpha_{ij} (\phi^2)\, \overline{u}_j \qquad (31)$$

which effectively refers the system to the single-valued path through F. Thus, a new potential function,

$$S(\overline{u}_i, \phi^2, \phi^1) \equiv \Pi(u_i^F + \alpha_{ij}\, \overline{u}_j, \phi^2, \phi^1) , \qquad (32)$$

is obtained with the properties

$$\left.\begin{aligned} S_i(0,\phi^2,0) &= S_i^2(0,\phi^2,0) = S_i^{22}(0,\phi^2,0) = \ldots = 0 \\ S_{ij}(0,\phi^2 0) &= S_i^2(0,\phi^2,0) = \ldots = 0 \quad \text{for } i \neq j \end{aligned}\right\} \qquad (33)$$

Applying the multiple-parameter perturbation technique and using these additional properties, the first order equation of the equilibrium surface can be constructed as

$$S_1^1 \phi^1 + \frac{1}{3!} \bar{S}_{1111} (\bar{u}_1)^3 + S_{11}^2 \bar{u}_1 \phi^2 = 0 \tag{34}$$

where

$$\bar{S}_{1111} = S_{1111} - 3 \sum_{s=2}^{N} \frac{(S_{s11})^2}{S_{ss}}$$

Equation (34) was obtained in Ref. [12] in a different way and can be recognized immediately as the Riemann-Hugoniot Catastrophe. The stability boundary (bifurcation set) can readily be determined as

$$\phi^2 = -\frac{1}{2} (\bar{S}_{1111})^{\frac{1}{3}} \frac{(3S_1^1)^{\frac{2}{3}}}{S_{11}^2} (\phi^1)^{\frac{2}{3}} \tag{35}$$

which describes a cusp in the control space. The equilibrium surface (34) and the stability boundary are shown in Figure 2. Obviously, the point F appears as a singular point in the overall picture, resulting in an abrupt change in the convexity of the boundary, and it was therefore called *singular*. It must be noted, however, that in fact (34) represents a general critical point of order three.

The transformed control parameters ϕ^1 and ϕ^2 can be identified as the shock and normal parameters of the catastrophe theory respectively, and the importance of the transformation (24) in identifying the significant parameters is observed. In the following analysis of the higher order surfaces a similar transformation will be introduced which will produce third, fourth, etc. order as well as the primary (ϕ^1) and secondary (ϕ^2) parameters.

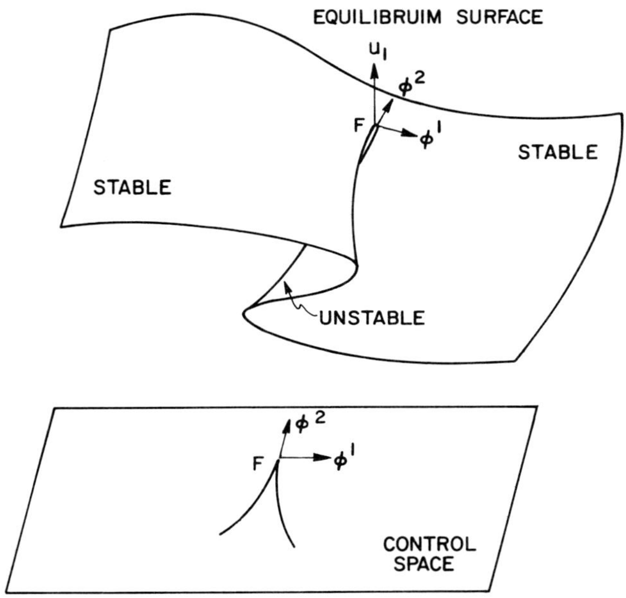
EQUILIBRUIM SURFACE
u_1
ϕ^2
F
ϕ^1
STABLE
STABLE
UNSTABLE
ϕ^2
F
ϕ^1
CONTROL
SPACE

Fig. 2

iii) General critical point of order four (swallow tail).

Similarly, the remaining two elementary catastrophes of dimension one, namely the swallow tail and butterfly can also be obtained. In fact, one can keep on generating additional higher order surfaces if no limit is imposed on the number of control parameters.

It is clear from the preceding analysis that the noncritical behaviour variables and higher order parameters can effectively be eliminated from the final result by suitable transformations. Consider, therefore, the simple case of one behaviour variable now and assume that the potential function is linear in control parameters. A fourth order surface arises when a further coefficient, H_{1111}, vanishes at the singular critical point.

In this case, introduce the linear, nonsingular transformation

$$\lambda^i = \beta^{ij}\,\phi^j \tag{36}$$

such that

$$\left.\begin{aligned} \Pi^1_1 &= 1, \quad \Pi^m_1 = 0 && \text{for } m\neq 1 \\ \Pi^2_{11} &= 1, \quad \Pi^r_{11} = 0 && \text{for } r\neq 2 \\ \Pi^3_{111} &= 1, \quad \Pi^t_{111} = 0 && \text{for } t\neq 3 \\ \Pi^4_{1111} &= 1, \quad \Pi^x_{111} = 0 && \text{for } x\neq 4 \end{aligned}\right\} \tag{37}$$

in the resulting function

$$\Pi(u_1,\phi^i) \equiv H(u_1,\beta^{ij}\,\phi^j)\ . \tag{38}$$

Following the familiar perturbation procedure, the equilibrium surface in the vicinity of the critical point F can be constructed as

$$\phi^1 + \frac{1}{4!}\overline{\Pi}_{11111}(u_1)^4 + \frac{1}{3!}(u_1)^3\ \phi^4 + \frac{1}{2!}(u_1)^2\ \phi^3 + u_1\ \phi^2 = 0 \tag{39}$$

On the other hand, any polynomial

$$u^m + \alpha_m\ u^{m-1} + \ldots + \alpha_1$$

may be brought to the form

$$\upsilon^m + \sum_{k=0}^{m-2} \beta_{k+1}\ \upsilon^k$$

by the transformation

$$\upsilon = u + \frac{1}{m}\alpha_m \tag{40}$$

Applying this to the polynomial in u_1, (39) can be expressed as

$$\phi^1 + \frac{1}{4!}\ a(u)^4 + \frac{1}{2!}\ b(u)^2\ \overline{\phi}^3 + cu\ \overline{\phi}^2 = 0 \tag{41}$$

where a, b, c are constants; $\overline{\phi}^2$, $\overline{\phi}^3$ are the new parameters and u is the new behaviour variable. Equation (41) is now recognized as the swallow tail of the catastrophe theory. It is interesting to note that the transformation (40) reduced the number of *significant* parameters to three.

The equilibrium surface in the vicinity of the singular critical point could also be obtained by this method and, in fact, one can continue to generate higher order surfaces by following the pattern. The Butterfly catastrophe, for example, is the next surface and can be obtained by assuming that the fifth order coefficient H_{11111} vanishes simultaneously with the previous coefficients.

4. APPLICATIONS

i) Mechanics: A two-degree-of-freedom model

Ref. [2] contains several applications and illustrative examples of the theory discussed in the preceding sections. The two-degree-of-freedom model shown in Figure 3 will be considered here as an example. The model under the axial force Λ and small horizontal force ε consists of two pin-jointed straight numbers. Each joint has a linear torsional spring of stiffness $k_1 = 2k$ and $k_2 = k$.

The total potential energy of the system may be expressed [2] as

$$\begin{aligned} V &= k\, q_1^2 + \frac{1}{2} k\, (q_2 - q_1)^2 \\ &\quad - \Lambda\, \ell\left[\frac{1}{2}\,(q_1^2 + q_3^2) - \frac{1}{24}\,(q_1^4 + q_2^4) + \ldots\right] \\ &\quad - \varepsilon\, \ell\left(q_1 + \frac{1}{3!}\, q_1^3 + \ldots\right) . \end{aligned} \tag{42}$$

Despite the presence of two behaviour variables, the system does not have the *dimension* two of the catastrophe theory, and

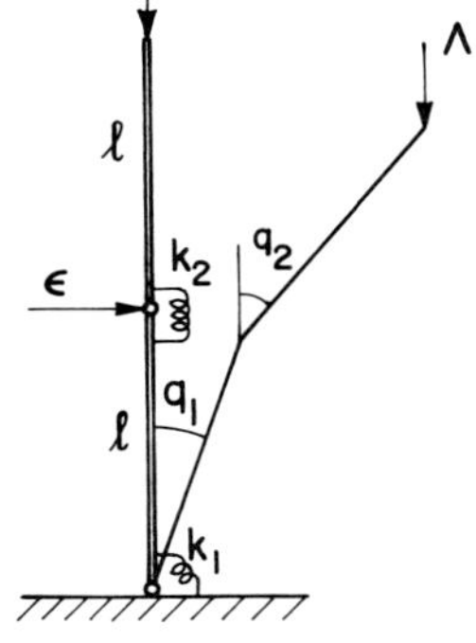

Fig. 3

none of the "universal unfoldings" is applicable. In order to elucidate the behaviour of the system, the potential function must first be transformed into appropriate canoncial forms discussed in the theory. Observing that $q_i = 0$ $(i=1,2)$ satisfies the equilibrium equations (3), the orthogonal transformation

$$q_i = \alpha_{ij} u_j \quad \text{where} \quad \alpha_{ij} = \begin{bmatrix} 0.3827 & 0.9239 \\ 0.9239 & -0.3827 \end{bmatrix} \tag{43}$$

may be introduced to obtain the function (32) directly,

$$\begin{aligned} S(u_i,\Lambda,\varepsilon) = \frac{1}{2} k\ (0.59\ u_1^2 + 3.41\ u_2^2) - \Lambda\ \ell \Big\{\frac{1}{2}\ (u_1^2 + u_2^2) \\ - \frac{1}{24}\ [(0.38\ u_1 + 0.92\ u_2)^4 + (0.92\ u_1 - 0.38\ u_2)^4]\Big\} \\ - \varepsilon\ \ell\ (0.38\ u_1 + 0.92\ u_2 + \frac{1}{3!} \ldots.) \ . \end{aligned} \tag{44}$$

One can then evaluate the derivatives

$$S_{11}(0,\Lambda,0) = 0.59k - \ell\ \Lambda$$

$$S_{22}(0,\Lambda,0) = 3.4k - \ell\ \Lambda$$

which yields the critical load

$$\Lambda^c = 0.59\ k/\ell \tag{45}$$

and indicates the critical coordinate u_1.

Following the theory of Section 3(ii) the first order equation of the equilibrium surface in the vicinity of the critical point (45) can now be obtained as

$$0.38\ell\ \varepsilon + \frac{0.44}{3!}\ k\ u_1^3 - \ell\ u_1\ \lambda = 0 \tag{46}$$

which is identical to (34) with $\phi^1 = \varepsilon$ and $\lambda = \phi^2$, indicating the Riemann-Hugoniot Catastrophe. It is noted that although the original system had two behaviour variables, the equilibrium surface (46) is in terms of u_1, the critical coordinate, only.

The stability boundary (bifurcation set) in $\lambda - \varepsilon$ parameter space can readily be obtained as

$$\Lambda = 0.59\,\frac{k}{\ell} + \frac{1}{2}\,(0.44k)^{\frac{1}{3}}\,\frac{(3 \times 0.38\ell\,\varepsilon)^{\frac{2}{3}}}{\ell} \tag{47}$$

by using (35).

ii) Thermodynamics: Phase transition

The phenomena of liquid/gas/liquid transitions can be explained within the context of stability theory. Consider the Van der Waals equation

$$\left[P + \frac{a}{V^2}\right](V-b) = c\,T \tag{48}$$

where P, V and T denote pressure, volume and temperature respectively and a, b, c are constants. Treating (48) as an equilibrium equation relating the *control* parameters P and T, and the *behaviour* variable V, one can proceed with higher order derivatives with respect to the latter to find the critical points and identify the system according to the general theory. To this end, write (48) in the form

$$P\,V^3 - (bP + cT)V^2 + aV - ab = 0\,, \tag{49}$$

and differentiate with respect to V twice to obtain

$$3P\,V^2 - 2(bP + cT)V + a = 0 \tag{50}$$

and

$$6P\,V - 2(bP + cT) = 0\,. \tag{51}$$

The simultaneous solution of these three equations yields the *singular* critical point F in the (V,P,T) space as

$$V_{cr} = 3b, \qquad P_{cr} = \frac{a}{27b^2}, \qquad T_{cr} = \frac{8a}{27bc}, \tag{52}$$

since all the first, second and third order derivatives of an assumed potential vanish simultaneously at this point as discussed in Section 3(ii).

The coordinates can be shifted to the cusp point (52) together with the transformation (24) (not orthogonal) by introducing

$$\begin{bmatrix} \frac{P}{P_{cr}} \\ \frac{T}{T_{cr}} \end{bmatrix} = \begin{bmatrix} 1 \\ 1 \end{bmatrix} + \frac{1}{3} \begin{bmatrix} -1 & 1 \\ \frac{1}{8} & \frac{1}{4} \end{bmatrix} \begin{bmatrix} \phi^1 \\ \phi^2 \end{bmatrix} \tag{53}$$

$$u = \frac{V_{cr}}{V} - 1$$

which transforms the Van der Waals equation in the vicinity of the cusp point into

$$\phi^1 + 3u^3 + u\phi^2 = 0 \tag{54}$$

Here the density u is used as the behaviour variable [13]. Equation (54) is identified again as the Riemann-Hugoniot Catastrophe as given by (34). Phase transition from liquid to gas and/or from gas to liquid occurs on the critical line given by (54) and

$$9u^2 + \phi^2 = 0. \tag{55}$$

The stability boundary in the parameter space can be obtained as before by eliminating u between (54) and (55). Physically, the jumps occurring on the critical line (Figure 4) represent boiling and condensation. In changing from liquid to gas, for example,

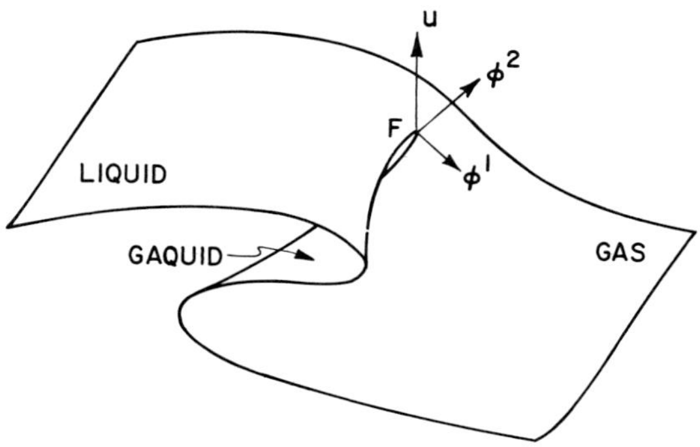

Fig. 4

the same state can be reached by a smooth path rather than jumps if the control parameters pressure and temperature are controlled appropriately. The interior sheet of the surface represents unstable states referred to as "gaquid".

iii) Urban systems

Amson discussed [14] various *Urbanitic Laws* concerning the behaviour of an abstract city. These laws relate the density τ, rental p and opulence Ω. For example, "evanescent city" is characterized by

$$p + \alpha \tau^2 = k \Omega \tau \tag{56}$$

while "saccadic city" is described by

$$p + \alpha\tau^2 = \frac{k\Omega\tau}{1-\beta\tau} \tag{57}$$

Treating τ as the behaviour variable and p and Ω as parameters, (56) is immediately recognized as the general critical point (fold catastrophe) of Section 3(i). It is further noted that the surface is anticlastic as shown in Figure 1. The city becomes evanescent at the fold line given by (56) and

$$2\,\alpha\,\tau = k\,\Omega. \tag{58}$$

On the other hand, (57) can be written as

$$(p + \alpha\,\tau^2)\left(\frac{1}{\tau} - \beta\right) = k\,\Omega$$

which resembles Van der Waals equation, and can be transformed into the cusp catastrophe. The stability boundary is shown in Figure 5. On this boundary the density undergoes a saccadic leap and the city is at catastrophic growth.

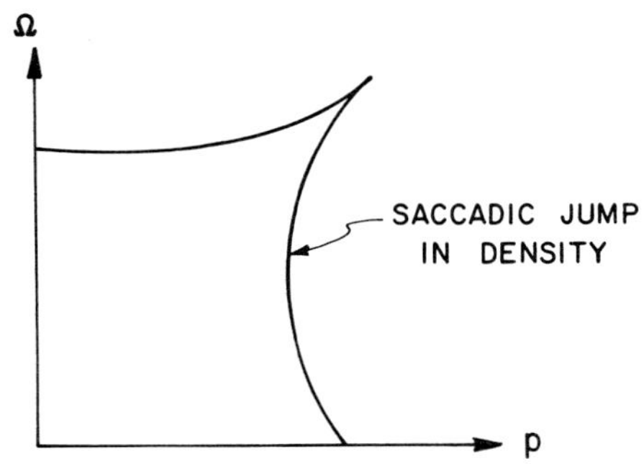

Fig. 5

iv) Zeeman's model for the heart beat

Zeeman discussed a mathematical model of the heart beat [15].

Treating the length (u) of the muscle fibre as the behaviour variable and the tension in the fibre (ϕ^2) together with the electrochemical control (ϕ^1) as parameters, the behaviour of the heart can be described by

$$\phi^1 + u^3 + \phi^2 u = 0 \tag{59}$$

which is shown in Figure 5. As the control parameters change, the length of the muscle fibre changes according to (59) and jumps occurring at the fold line correspond to heart beats.

A full discussion of this phenomenon can be found in Ref. [15].

v) Experimental results

A series of experiments on simply supported shallow arches made of high tensile steel strips was carried out [1] at University College. The arch shown in Figure 5 was subjected to two independent loads, P_1 and P_2. The stability boundary shown in Figure 4 clearly illustrates the familiar cusp. All points on the boundary except F are associated with *general* critical points. The point F is *singular* as discussed in Section 3(ii), and the behaviour of the system is governed by Riemann-Hugoniot Catastrophe.

The similarity between the stability boundary of this arch and that of the saccadic city is particularly striking. Applications of the theory to several other disciplines ranging from Engineering to cosmology, economics, meteorology, psychology, etc. reveal further analogies between the behaviour of systems from seemingly disconnected fields. It is now hoped that the application of the theory will spread throughout natural as well as social sciences.

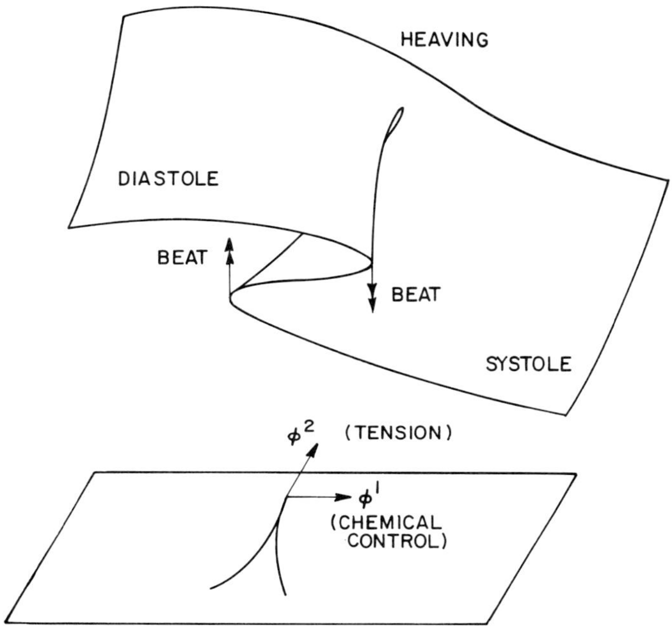
HEAVING
DIASTOLE
BEAT
BEAT
SYSTOLE
φ² (TENSION)
φ¹
(CHEMICAL CONTROL)

Fig. 6

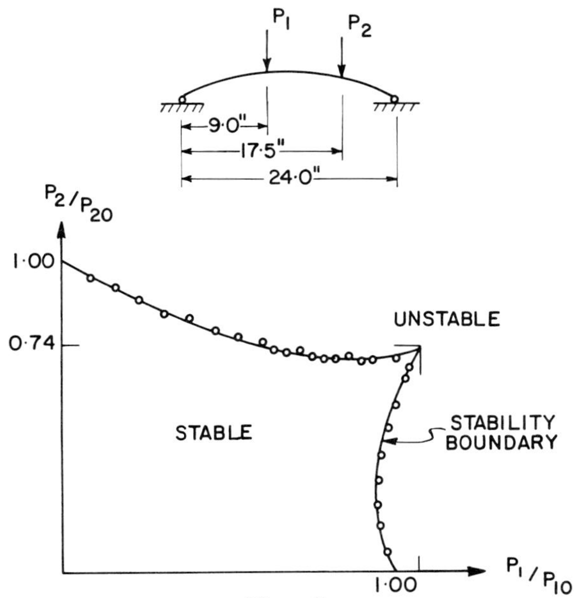
P1
P2
9·0"
17·5"
24·0"
P2/P20
1·00
0·74
UNSTABLE
STABLE
STABILITY BOUNDARY
1·00
P1/P10

Fig. 7

REFERENCES

1. K. Huseyin, "Elastic Stability of Structures Under Combined Loading", Ph.D. Thesis, University of London (1967).
2. K. Huseyin, *Nonlinear Theory of Elastic Stability*, Noordhoff International Publishing (1975), 220 pages.
3. W.J. Koiter, "On the Stability of Elastic Equilibrium", Dissertation, Holland (1945).
4. J.M. Thompson and G.W. Hunt, *A General Theory of Elastic Stability*, John Wiley & Sons (1973).
5. J. Roorda, "Concepts in Elastic Structural Stability", *Mechanics Today* (ed. by N. Nasser), Vol. 1, Pergamon (1972).
6. K. Huseyin, "Fundamental Principles in the Buckling of Structures Under Combined Loading", *Int. J. Solids Struct.*, 6, 479-487 (1970).
7. K. Huseyin, "The Elastic Stability of Structural Systems with Independent Loading Parameters", *Int. J. Solids Struct.*, 6, 677-691 (1970).
8. K. Huseyin, "The Post-Buckling Behaviour of Structures Under Combined Loading", *Z. Angew. Math. Mech.*, 51, 177-182 (1971)
9. R. Thom, "Topological Models in Biology", *Topology*, 8, 1969.
10. V. Mandadi and K. Huseyin, "On the General Theory of Stability", to be published.
11. K. Huseyin, "The Multiple-Parameter Perturbation Technique for the Analysis of Nonlinear Systems", *Int. J. of Non-Linear Mech.*, 8, 431-443 (1973).
12. K. Huseyin, "Singular Critical Points in the General Theory of Elastic Stability", *Meccanica*, 7, 58-68 (1972).
13. D.H. Fowler, "The Riemann-Hugoniot Catastrophe and Van der Waals' Equation", *Towards a Theoretical Biology* (ed. by C.H. Waddington), 4, 1-7 (1972).
14. J.C. Amson, "Equilibrium and Catastrophe Models of Urban Growth", *Space-Time Concepts in Urban and Regional Models*

(ed. by E.L. Cripps), 108-128 (1974).

15. E.C. Zeeman, "Differential Equations for the Heart Beat and Nerve Impulse", *Towards a Theoretical Biology* (ed. by C.H. Waddington), 4, 8-67 (1972).

TENSORS VERSUS MATRICES IN DISCRETE MECHANICS

HAYRETTIN KARDESTUNCER
University of Connecticut
Storrs, Connecticut

SUMMARY

In spite of being advocated by such great physisists and mathematicians as Einstein, Levi-Civita and Ricci, tensors have so far received little attention. Their use has seldom exceeded the formulation stages of the problems and their behaviour during the solution procedure has not been properly studied. Especially for problems with a large number of variables, such as those encountered in discrete mechanics, the tensor transformations have always been replaced by matrix manipulations. The study presented here, therefore, advocates the use of tensors not only for formulation but for the solution of problems through the search of principal directions. To accomplish this, the complete system is divided into sub-systems (elements) and reassembled through their stiffness centers. Much more work is needed to establish the practicality of the concept.

1. INTRODUCTION

The words 'tensor' and 'matrix' have often been employed interchangeably, yet although every second order tensor can be referred to as a matrix, the reverse is not necessarily true. Similar ambiguity exists between first order tensors and vectors. A physical vector is certainly a first order tensor but not all vectors are tensors. When an entity is proven to be a tensor, then it would be most appropriate to refer to it as a tensor instead of as a vector or a matrix which are nothing more than arrays of elements. Merely reaching proper results does not justify the use of general terms such as vector or matrix when such an entity follows well-defined rules during the coordinate transformations (linear or non-linear). The advocacy of the tensor concept, therefore, goes significantly beyond the terminology.

Although Gabriel Kron was one of the very first authors referring to tensors in structural analysis, even he avoided tensor notations in order to allow his readers to think of the more familiar term 'matrix'. He states in page 415 of Ref. (9), for instance, "the reader not acquainted with tensors may keep on calling the above symbols matrices instead of tensors". There seems to be nothing wrong in doing so. The reader should, however, be aware of the fact that this concept is not reversible. The familiar routes are not always the best. Leaving a familiar route and getting lost in the darkness is, in fact, the prime essence of scientific research.

The fact that we are most familiar with the Cartesian coordinates (in reference to which the distinction between the covariant and contravariant properties of tensors disappears) should not necessarily stop researchers from investigating the other coordinate systems and employing powerful rules of coordinate transformation between them. Since most physical laws are invariant during the coordinate transformation, they are in fact expressible by tensorial equations. The problems in elasticity, for instance, have been thoroughly treated by Brillouin [1], Sokolnikoff [2] and Michal [3]. More recent contributions to the field of continuum have been made by Eringen [4] and Fung [5] by using cartesian tensors.

Besides Kron a few others such as Soule [6], Borg [7] and the Author [8] have attempted to formulate problems in discrete systems by tensors. Still, very powerful properties of tensors, on the other hand, have not yet been utilized to the fullest extent. The majority of researchers seem to be content with or have restricted themselves to the tensor notations only. In doing so, the formulations have become condensed and precise. At the end of the formulation, one confronts the classical barrier, i.e., the solution of large simultaneous equations. The methodologies at this stage pay no attention to how the formulation has been done. Physics and geometry leave the stage to arithmetic in the form of number crunching. Undoubtedly, tensor notations are very useful at the formulation stages of the problems in eliminating the rigid rules and regulations of matrix algebra; nonetheless, one can not help to think in terms of tensors during the solution stage of the same problem. Should it be desirable once the formulation is complete to disassociate oneself from the physics and geometry of the problem? Instead of referring to some general techniques such as Gauss' elimination or Choleski's triangularization, which are equally applied to physical or nonphysical systems, could one proceed with physical reasoning and geometrical guidance toward the solution of a well-defined physical problem? Considering that the solution of a set of simultaneous equations is referred to as changing basis in mathematics, which can also be interpreted as a sort of coordinate transformation, would it then be possible to take advantage of the well-defined tensor transformations in effort to attain the solution for the problem? The principal directions for the stress or strain tensor (or any second order tensor, as a matter of fact) can be determined by elimination routines or by the tensor invariants given in Eq. (2.11). The latter may or may not necessarily be less time-consuming than the former, nevertheless it does not divorce itself from physics while changing the geometry in the form of observations. Granting that all physical phenomena can be observed from any angle (coordinate system) and from any location (origin of the coordinate system), yet the outcome (the result) may stand out more vividly for one observer than for another. To determine the location and the angle of such a particular observation, then, requires a most general coordinate transformation (rotational and translational). Although for tensors the rules of transformation are well-defined by geometry alone, the type of transformation is yet to be determined by physics. While it is rather obvious for some problems (symmetric problems), it may require tedious investigation for most others. Here we are, therefore, investigating the use of physical and geometrical reasoning toward the solution of problems in discrete mechanics. Emphasis, however, is placed upon those problems which are discrete in nature such as the skeletal systems.

2. TENSOR CONCEPT

As stated by Eisele and Mason [10], "The subject of tensor analysis possesses an undeserved aura of mystery, as intellectual fallout". The term 'tensor' is a highly ambiguous word being employed loosely to mean almost anything. According to Borg [11], for instance, Hamilton defined it as the positive square root of the sum of the squares of the four elements of the quaternion. Sokonikoff [2] refers to them as abstract objects whose properties are independent of the reference frames. While Kron [12] says "a tensor is a matrix with a definite law of transformation", according to Happ [13], Einstein's definition of tensors is as follows: "Let certain things ("tensors") be defined with respect to any system of coordinates by a number of functions of the coordinates, called the "components" of the tensor. There are then certain rules by which these components can be calculated for a new system of coordinates if they are known for the original system of coordinates, and if the transformation connecting the two systems is known. The things hereafter called tensors are further characterized by the fact that the equations of transformation for their components are linear and homogeneous. Accordingly, all the components in the new system vanish, if they all vanish in the original system." We refer to tensors as entities that follow one of the following rules under coordinate transformations.

Contravariant: $$T^{i'j'k'\ldots\ell'} = \frac{\partial x^{i'}}{\partial x^{i}} \frac{\partial x^{j'}}{\partial x^{j}} \frac{\partial x^{k'}}{\partial x^{k}} \cdots \frac{\partial x^{\ell'}}{\partial x^{\ell}} T^{ijk\ldots\ell}$$

Covariant: $$T_{i'j'k'\ldots\ell'} = \frac{\partial x^{i}}{\partial x^{i'}} \frac{\partial x^{j}}{\partial x^{j'}} \frac{\partial x^{k}}{\partial x^{k'}} \cdots \frac{\partial x^{\ell}}{\partial x^{\ell'}} T_{ijk\ldots\ell}$$

Mixed: $$T^{i'j'1'\ldots\ell'}_{p'\ldots t'} = \frac{\partial x^{i'}}{\partial x^{i}} \frac{\partial x^{i'}}{\partial x^{j}} \cdots \frac{\partial x^{\ell'}}{\partial x^{\ell}} \frac{\partial x^{p}}{\partial x^{p'}} \cdots \frac{\partial x^{t}}{\partial x^{t'}} T^{ijk\ldots\ell}_{p\ldots t}$$

2.1 Summation convention

Whenever the same lateral index occurs as a subscript, the summation is understood (convention is from Einstein) to be

$$u_m = v_{mn} w^n$$

$$u_m = v_{m1}w^1 + v_{m2}w^2 + v_{m3}w^3 + \ldots$$

The index m is called a free index and it takes all the values 1,2,3,... The index n is a dummy index since its replacement by another index does not alter the meaning of the equation. In order to avoid confusion, the same suffix must not be used more than twice in any one term.

2.2 Tensorial equations

If an equation consisting of various holors is valid in all coordinate systems, it is said to be a tensorial equation.

$$u_m^n = c_{rm} b^{nr} + e_m v^n \tag{2.1}$$

or

$$u_{m'}^{n'} = c_{r'm'} b^{n'r'} + e_{m'} v^{n'} \tag{2.2}$$

The invariance of tensorial equations during the coordinate transformations can be proven, for instance, by introducing the transformation rules to Eq. (2.2) which yields Eq. (2.1).

$$\frac{\partial x^m}{\partial x^{m'}} \frac{\partial x^{n'}}{\partial x^n} u_m^n = \frac{\partial x^r}{\partial x^{r'}} \frac{\partial x^m}{\partial x^{m'}} c_{rm} \frac{\partial x^{n'}}{\partial x^n} \frac{\partial x^{r'}}{\partial x^r} b^{nr} + \frac{\partial x^m}{\partial x^{m'}} \frac{\partial x^{n'}}{\partial x^n} e_m v^n$$

in which the term

$$\frac{\partial x^m}{\partial x^{m'}} \frac{\partial x^{n'}}{\partial x^n}$$

is common and

$$\frac{\partial x^r}{\partial x^{r'}} \frac{\partial x^{r'}}{\partial x^r} = I$$

In order to determine whether a given holor is a tensor or not, one must employ a valid method which is known as a 'tensor test'. For instance, let a univalent holor u^i be investigated. Depending upon the physical properties of such a holor, let us assume that there exists an equation such that

$$u^i v_i = s \tag{2.3}$$

in which v_i and s are known as univalent and nilvalent tensors. Since s = s' (being a scalar) and

$$v_i = \frac{\partial x^{i'}}{\partial x^i} v_{i'}$$

Eq. (2.3) can be written as

$$u^i \frac{\partial x^{i'}}{\partial x^i} v_{i'} = u^{i'} v_{i'} \tag{2.4}$$

or

$$[u^i \frac{\partial x^{i'}}{\partial x^i} - u^{i'}]\, v_{i'} = 0$$

If $v_{i'}$ is an arbitrary univalent tensor, then the bracket must be zero, which yields

$$u^{i'} = \frac{\partial x^{i'}}{\partial x^i} u^i$$

indicating that u^i is a contravarient univalent tensor.

Let us now consider the derivative of a scalar function ϕ

$$v_i = \frac{\partial \phi}{\partial x^i} \tag{2.5}$$

Since $\phi = \phi'$ (a nilvalent tensor), then

$$\frac{\partial \phi'}{\partial x^{i'}} = \frac{\partial \phi}{\partial x^1} \frac{\partial x^1}{\partial x^{i'}} + \frac{\partial \phi}{\partial x^2} \frac{\partial x^2}{\partial x^{i'}} + \ldots$$

or

$$\frac{\partial \phi'}{\partial x^{i'}} = \frac{\partial x^i}{\partial x^{i'}} \frac{\partial \phi}{\partial x^i} \tag{2.6}$$

from which one can conclude that: The derivative of a scalar function follows the covariant transformation rule.

In engineering sciences, Eq. (2.5) is frequently used to calculate the gradient of a potential function.

$$\text{grad } \phi \equiv \nabla\phi = g_i \frac{\partial \phi}{\partial x^i}$$

where g_i are the base vectors through which the metric tensors are defined.

$$g_{k\ell} = g_k \cdot g_\ell = \frac{\partial x^m}{\partial x^k} \frac{\partial x^n}{\partial x^\ell} \delta_{mn}$$

2.3 Non-tensorial equations

Consider the following equation

$$w^i = \frac{\partial v^i}{\partial u} \tag{2.7}$$

in which v^i is known to be a contravariant tensor and w^i is the derivative. Judging from the balance of this equation, the reader may assume that this equation is a tensorial equation, that is, w^i is also a tensor. However, if one substitutes

$$v^i = \frac{\partial x^i}{\partial x^{i'}} v^{i'} \tag{2.8}$$

and carries the chain rule of derivation, the resulting equation would be

$$w^i = \frac{\partial x^i}{\partial x^{i'}} \frac{\partial v^{i'}}{\partial u} + v^{i'} \frac{d}{du} \left(\frac{\partial x^i}{\partial x^{i'}}\right) \tag{2.9}$$

This, certainly, is not a tensorial transformation unless, of course, the second term vanishes. The vanishing second term, which implies that the Jacobian is not a function of position, indicates that both coordinate systems x^i and $x^{i'}$ are cartesian. Otherwise the derivative w^i is not a tensor. This is one of the reasons that most researchers attempt to use cartesian frames for their problems.

Similarly, if one employs components of v^i instead of its merates (elements), the tensorial transformation indicated by Eq. (2.8) will not hold. Although the user often has a choice between merates and components, Spencer and Moon [14] recommend employing merates even if a change to components is necessary immediately after the transformations.

2.4 Principal directions and invariants

Consider the following tensorial equation

$$a^i = b^i_j \, c^j \tag{2.10}$$

Because of its merit this equation is valid (invariant) in all coordinate systems.

$$\frac{\partial x^{i'}}{\partial x^i} a^i = \frac{\partial x^{i'}}{\partial x^i} \frac{\partial x^j}{\partial x^{j'}} b^i_j \frac{\partial x^{j'}}{\partial x^j} c^j$$

It is then conceivable that there must be a coordinate system according to which the expression of the second order tensors on the right possesses the following property:

$$b^{i'}_{j'} = \frac{\partial x^{i'}}{\partial x^i} \frac{\partial x^j}{\partial x^{j'}} b^i_j \quad \therefore \quad b^{i'}_{j'} = 0 \quad \text{for } i \neq j'$$

We shall refer to such a coordinate system as the 'principal coordinate system'. While in matrix terminology this is referred to as eigen-vectors and the corresponding components of b as eigen-values, in mathematics the above procedure is called the changing of basis.

If b is a second order tensor in three dimension, the following invariants hold during the transformation

$$
\begin{aligned}
I_1 &= \delta^i_i\, b^i_i \\
I_2 &= \frac{1}{2!}\, \delta^{ij}_{k\ell}\, b^k_i\, b^\ell_j \\
I_3 &= \frac{1}{3!}\, \delta^{ijk}_{\ell mn}\, b^\ell_i\, b^m_j\, b^n_k
\end{aligned}
\tag{2.11}
$$

These invariants (in the case of b representing stress or strain tensors) play an important role in establishing the yield criteria or theories of failure [the reader is referred to Chapter 6 of reference 5]. In the principal coordinate systems, the components of a tensor take extremal values and an equation such as Eq. (2.10) takes its simplest form. Consider, for instance, the equation of an ellipse expressed in the principal coordinate system.

$$\left(\frac{x^{1'}}{a}\right)^2 + \left(\frac{x^{2'}}{b}\right)^2 = 1$$

The equation of the same ellipse in reference to any other frame would be much lengthier. If the new system differs from the old + 30°, the above equation will take the following form

$$(0.75\, b^2 + 0.25\, a^2)(x^1)^2 + (0.25\, b^2 + 0.75\, a^2)(x^2)^2$$
$$+\, 2x^1x^2(-0.433\, b^2 + 0.433\, a^2) = a^2b^2$$

The lengthiness of this equation stems from the fact that the coordinate frame according to which the equation is written is not the principal frame of this configuration. If one, therefore, wishes to observe any phenomenon, he should rather observe it in the principal coordinate frame. The principal coordinate frames on the other hand are not always so obvious. They can be determined either by mathematical procedure, i.e., changing basis, determining eigen-values etc., or by physical reasoning.

The same thing can also be established by using tensor invariants shown in Eq. (2.11). The Mohr Circle, for instance, is a fine application of it from a geometrical point of view.

The beauty embedded in these invariants becomes extremely difficult to unearth when a tensor is expressed in n-dimensional mathematical space instead of in three-dimensional euclidean space, especially when the transformations are in a general sense instead of merely rotational. Nevertheless, once the equation is proven to be tensorial, the existance of the associated principal coordinates and the validity of the phenomenon in such a frame would be assured.

3. TENSORS IN CONTINUUM MECHANICS

Tensors in continuum mechanics have been known to researchers for a good many years. Brillouin [1], Sokolnikoof [2] and Eringen [4] presented thorough treatment of tensors in continuous media.

The very fundamental tensor in mechanics is 'energy'. The potential energy of a linear continuous medium, for instance, can be written as

$$\Pi = \frac{1}{2}\int_V \sigma^{iq}\varepsilon_{iq}\,dv - \int_V \sigma^{ij}_{,j}v_i\,dv - \int_S \sigma^{ij}\eta_j v_i\,ds \tag{3.1}$$

where σ and ε are the well-known stress and strain tensors, and v and η are the displacement and unit normal vectors (univalent tensors). The three conditions with which the solution of any problem in continuum must comply are given by the following tensorial equations.

(a) Equilibrium equations

$$\sigma^{ji}_{,j} + F^i = 0$$

(b) Compatibility equations

$$\varepsilon_{ij,k\ell} + \varepsilon_{k\ell,ij} - e_{ik,j\ell} - e_{j\ell,ik} = 0$$

(c) Boundary conditions

$$\sigma^{ij}\eta_j = \bar{t}^i \quad \text{on } s_t$$

$$v_i = \bar{v}_i \quad \text{on } s_v$$

The fundamental tensors in these equations are stress and strain tensors which in the cartesian coordinate system are given as

$$\sigma^{ij} = \frac{\partial}{\partial s^j}(F^i) \tag{3.2}$$

and

$$\varepsilon_{ij} = \frac{1}{2}(v_{i,j}+v_{j,i}) \tag{3.3}$$

The well-known relationship between these two tensors is

$$\sigma^{ij} = E^{ijk\ell}\,\varepsilon_{k\ell} \tag{3.4}$$

where $E^{ijk\ell}$ is known as the tensor of moduli. It contains the physical properties of the domain.

It can be shown that the stress tensor is the gradient of the strain energy (the first term in Eq. 3.1) with respect to the strain tensor

$$\sigma^{ij} = \frac{\partial U}{\partial \varepsilon_{ij}} \tag{3.5}$$

This relationship of continuum corresponds to Castigliano's first theorem in discrete mechanics [16]. Considering that the stress and strain tensors and the quadratic form of strain energy are all symmetric,

$$U = \frac{1}{2}E^{ijk\ell}\,\varepsilon_{ijk\ell}$$

the number of independent elements in E reduces to 21 which represents the most general form of anistropy.

Equation (3.4) in cartesian coordinates for isotropic elastic material takes the following form [5]

$$\sigma_{ij} = \lambda e \delta_{ij} + 2G\varepsilon_{ij} \tag{3.6}$$

where

$$\varepsilon_{ij} = \frac{1+\nu}{E}\sigma_{ij} - \frac{\nu}{E} e \, \delta_{ij}$$

The constants λ and G are called Lame's constants. If one substitutes Eq. (3.6) into the equilibrium equation

$$(\lambda e \delta_{ij} + 2G\varepsilon_{ij})_{,j} + F_i = 0$$

and refers to Eq. (3.3) for the strain tensor, this will yield

$$(\lambda + G) e_{,i} + G\nabla^2 v_i + F_i = 0 \tag{3.7}$$

which is the celebrated Navier equation in the theory of elasticity. The introduction of boundary conditions to this equation yields a unique solution for v_i. The reader is advised to refer to Sokolnikoff [2] for the use of this equation in elasticity.

Needless to say, Eq. (3.7) is a tensor equation and can easily be transformed into any other curvilinear coordinates by the transformation rules presented earlier. Such a transformation is surely recommended when the region of interest involves curved boundaries.

4. TENSORS IN DISCRETE MECHANICS

While tensors in continuum mechanics are localized at infinitesimally small elements, their counterparts in discrete mechanics are expressed at the nodal points of the domain. Regardless of the method employed for the solution of the problems in continuum mechanics, for instance, the results are expected to yield either one of the two tensorial fields namely, that of stress and strain tensors. The analogous tensorial fields in discrete mechanics are the force and displacement tensors with the latter two being the differential and integral values of the former two [15].

Let Ω be a continuous, simply connected and bounded domain with Γ being its boundary where P and V representing tractions and displacements are prescribed. Further assume that Ω is decomposed into a finite number of subdomains (the concept of discrete mechanics) with a finite number of nodal points where the compatibilities between the domains (elements) are maintained. The solution of this problem, then, reduces to the minimization of the following functional.

$$\Pi = \frac{1}{2} P^{iq} V_{iq} - P^{mq} V_{mq} - P^{sq} V_{sq} \tag{4.1}$$

While the first term in this equation represents the strain energy of the system, the second and third terms represent the body and surface force potentials. Considering that the total potential energy Π is a scalar quantity (nilvalent tensor), Eq. (4.1) is fundamentally a tensorial equation. The minimization of this equation yields

$$\frac{\partial \Pi}{\partial V_{jr}} = \frac{1}{2} \frac{\partial P^{iq}}{\partial V_{jr}} V_{iq} + \frac{1}{2} P^{iq} \frac{\partial V_{iq}}{\partial V_{jr}} - \frac{\partial P^{mq}}{\partial V_{jr}} V_{mq}$$

$$- P^{mq} \frac{\partial V_{mq}}{\partial V_{jr}} - \frac{\partial P^{sq}}{\partial V_{jr}} V_{sq} - P^{sq} \frac{\partial V_{sq}}{\partial V_{jr}} = 0$$

Since

$$\frac{\partial V_{iq}}{\partial V_{jr}} = 1 \quad \text{for} \quad i,q = j,r; \quad \text{zero otherwise}$$

the above equation reduces to

$$\frac{1}{2} P^{iq} + \frac{1}{2} K^{iqjr} V_{iq} - K^{mqjr} V_{mq} - K^{sqjr} V_{sq} = 0$$

where

$$K^{iqjr} = \frac{\partial P^{iq}}{\partial V_{jr}}$$

and similar substitutions for the other terms.

Considering, however, that m+s = i, then

$$P^{iq} = K^{iqjr} V_{jr} \tag{4.2}$$

This is the basic equation that is formed and solved in discrete mechanics.

Before going any further, it would be necessary for us to prove that K in this equation is also a tensor. So, let us write the external work of the system

$$W = K^{iqjr} V_{jr} V_{iq} \tag{4.3}$$

Since W = W' Eq. (4.3) is also valid in any other coordinate system, therefore,

$$K^{iqjr} V_{jr} V_{iq} = K^{iq'jr'} V_{jr'} V_{iq'} \tag{4.4}$$

Note that the transformation of coordinates is done over q and r not on i and j which take the numbering of nodal points. Considering that

$$V_{jr} = \frac{\partial x^{r'}}{\partial x^{r}} V_{jr'}$$

$$V_{iq} = \frac{\partial x^{q'}}{\partial x^{q}} V_{iq'}$$

Eq. (4.4) becomes

$$[K^{iqjr} \frac{\partial x^{r'}}{\partial x^{r}} \frac{\partial x^{q'}}{\partial x^{q}} - K^{iq'jr'}] V_{jr'} V_{iq'} = 0$$

Since this equation represents a summation q', r' = 1,2,... 6 it is difficult at this point to say that K is a tensor. We may, however, write

$$[\frac{\partial x^{1'}}{\partial x^{r}} \frac{\partial x^{2'}}{\partial x^{q}} K^{iqjr} - K^{i1'j2'}] V_{j1'} V_{i2'} +$$

$$+ [\frac{\partial x^{2'}}{\partial x^{r}} \frac{\partial x^{1'}}{\partial x^{q}} K^{iqjr} - K^{i2'j1'}] V_{j2'} V_{i1'} + \ldots = 0$$

Considering further that $V_{j1'} V_{i2'}$ is independent of $V_{j2'} V_{i1'}$ etc. and the interchange of dummy indices is allowed, then

$$[\frac{\partial x^{1'}}{\partial x^{r}} \frac{\partial x^{2'}}{\partial x^{q}} (K^{iqjr} + K^{irjq}) - (K^{i1'j2'} + K^{i2'j1'})] = 0$$

Or, in general, for any r' and q'

$$\frac{\partial x^{r'}}{\partial x^{r}} \frac{\partial x^{q'}}{\partial x^{q}} (K^{iqjr} + K^{irjq}) = (K^{ir'jq'} + K^{iq'jr'})$$

which says that the sum in parenthesis is a tensor. If, however, one remembers Maxwell's Law of Reciprocity

$$K^{iqjr} = K^{irjq}$$

it is then assured that K is a tensor. This we shall refer to as the 'stiffness tensor' instead of stiffness matrix.

At this point one should observe the similarities between Eq. (4.2) and its counterpart in continuum mechanics, Eq. (3.4). While the tensor of moduli strictly depends upon the physical characteristics of the domain, the stiffness tensors involve geometry as well.

Furthermore, if one wishes to reduce the valence of K in Eq. (4.2), he may do so by writing it in matrix form

$$\underset{\sim}{P} = \underset{\sim}{K} \underset{\sim}{V} \tag{4.5}$$

This equation, it may be noted, resembles the following equations in electric networks [20]

$$\underset{\sim}{I} = \underset{\sim}{Y} \underset{\sim}{E} , \quad \underset{\sim}{E} = \underset{\sim}{Z} \underset{\sim}{I} \tag{4.6}$$

where $\underset{\sim}{E}$, $\underset{\sim}{I}$, $\underset{\sim}{Y}$ and $\underset{\sim}{Z}$ represent voltage, current, admittance and impedance tensors respectively.

In spite of the striking similarities between these two equations [Eq. (4.5) and Eq. (4.6)], a closer examination of them indicates that while the elements of Eq. (4.6) are nilvalent tensors, those of Eq. (4.5) are bivalent tensors. This may be considered one of the main differences of the tensorial equations encountered in electric networks and in discrete mechanics. Such a difference between the two makes the former slightly more complicated than the latter [20].

When the continuum is divided into sub-domains (elements), the governing equation for the entire domain is certainly valid for sub-domains as well. Such elements could be one dimensional, as in the case of skeletal systems, or higher dimensional. Since the skeletal systems which are discrete by nature hold a unique place in mechanics (although theoretically not different than the higher dimensional continuum), they are treated separately in the next article.

4.1 One dimensional elements

The behaviour (load-deflection relationship) of a skeletal system is made up from the behaviour of its elements. Since a line element (whether prismatic or not) itself is the smallest skeletal system, Eq. (4.2) is also valid. In three dimensional space having three linear and three angular freedoms, this equation for member ab, for instance, becomes

$$p^{aq} = k^{aqbr} v_{br} \qquad q,r=1,2,\ldots,6 \quad \text{b not summed} \tag{4.7}$$

The elements of k which are in the form of

$$k^{aqbr} = \frac{\partial p^{aq}}{\partial v_{br}}$$

can be found in most texts on matrix analysis of structures [16]. Considering that q and r take a value from 1 to 6, there will at the most be 144 elements of which only 36 are independent. Since this is a tensorial equation, it can be established in reference to any coordinate system. The coordinate systems shown in Figs. (4.1.a and b) are those most often used by researchers. While the stiffness tensor is completely full in coordinate frame (a), 26 out of 36 elements are zero in coordinate frame (b). This is correct provided, of course, that both frames coincide with the principal direction of the cross-section.

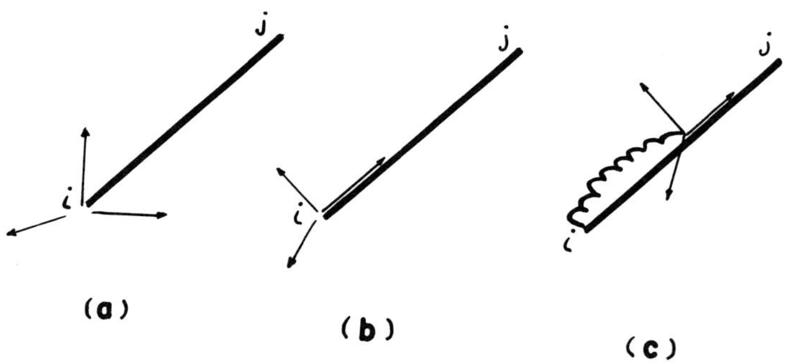

The principal directions of the cross-section are those in reference to which the inertia tensor I becomes diagonal.

$$I_{qr} = \int (x^m x_m \delta_{qr} - x_q x_r) dA \qquad q,r=1,2,3 \tag{4.8}$$

Although the principal directions for most cross-sections are known in advance, for complicated sections they can easily be determined by the tensor invariants given by Eq. (2.11). The inertia tensor in Eq. (4.8) which represents the area moments of inertia can be written in matrix form as

$$\underset{\sim}{I} = \begin{bmatrix} I_{xx} & -I_{xy} & -I_{xz} \\ -I_{yx} & I_{yy} & -I_{yz} \\ -I_{zx} & -I_{zy} & I_{zz} \end{bmatrix} \tag{4.9}$$

where the area product of inertia terms (the off-diagonal terms) do not vanish unless xyz is the principal coordinate frame of the cross-section. This entity is a bivalent tensor since it obeys the transformation rules cited earlier. If on the other hand one alters the sign of the off-diagonal terms from negative to positive, for instance, the new entity is still a matrix but would no longer be a tensor (neither the transformation rules nor the invariance will hold).

The orientation of the coordinate frame in the principal direction of the cross-section does not guarantee that the stiffness tensor in Eq. (4.7) will be diagonal (simplest form). The proper location of the coordinate axes in addition to their orientation do, on the other hand, accomplish such a diagonalization.

In order to illustrate this phenomenon, let us consider the force-displacement relationship for a prismatic element in 2-space. The force-displacement equation in the principal coordinate frame (Fig. 4.1.c) is

$$\begin{bmatrix} \Delta_o \\ \theta_o \end{bmatrix} = \begin{bmatrix} \frac{L^3}{12EI} & 0 \\ 0 & \frac{L}{EI} \end{bmatrix} \begin{bmatrix} V_o \\ M_o \end{bmatrix} \tag{4.10}$$

Now, assume that we wish to know the required forces to be applied at one end of this member in order to develop a certain displacement at that end while the other end remains fixed. The answer to this question requires nothing more than carrying out the following multiplication:

$$\begin{bmatrix} V \\ \\ M \end{bmatrix} = \begin{bmatrix} \frac{12EI}{L^3} & \frac{-6EI}{L^2} \\ \\ \frac{-6EI}{L^2} & \frac{4EI}{L} \end{bmatrix} \begin{bmatrix} \Delta \\ \\ \theta \end{bmatrix}$$

$$= \begin{bmatrix} \frac{12EI}{L^3}\Delta - \frac{6EI}{L^2}\theta \\ \\ \frac{-6EI}{L^2}\Delta + \frac{4EI}{L}\theta \end{bmatrix} \qquad (4.11)$$

provided, of course, that the phenomenon is observed in coordinate frames, similar to that of Fig. (4.1b), located at the end of the member under consideration. Instead, however, one may observe the phenomenon in the principal coordinate frames (Fig. 4.1c) and refer to the translational transformation of coordinates later on. Since such a coordinate transformation is well-defined, it introduces no difficulty and in most cases may not even be necessary. For this case it becomes

$$\begin{bmatrix} V \\ \\ M \end{bmatrix} = \underset{\sim}{T}^* \begin{bmatrix} \frac{12EI}{L^3} & 0 \\ \\ 0 & \frac{EI}{L} \end{bmatrix} \underset{\sim}{T} \begin{bmatrix} \Delta \\ \\ \theta \end{bmatrix} \qquad (4.12)$$

where

$$\underset{\sim}{T} = \begin{bmatrix} 1 & -\frac{L}{2} \\ 0 & 1 \end{bmatrix}$$

Note that Eqs. (4.11) and (4.12) are identical. In other words, dealing with Eq. (4.10) instead of (4.11) will yield identical results. The flexibility matrix in Eq. (4.10), however, can be evaluated more accurately than its counterpart in Eq. (4.11).

The comparison of Eqs. (4.11) and (4.12) indicates that there always exists the following relationship between the two stiffness tensors expressed in two different coordinate frames.

$$\begin{bmatrix} \frac{12EI}{L^3} & 0 \\ 0 & \frac{EI}{L} \end{bmatrix} = \underset{\sim}{T} \begin{bmatrix} \frac{12EI}{L^3} & -\frac{6EI}{L^2} \\ -\frac{6EI}{L^2} & \frac{4EI}{L} \end{bmatrix} \underset{\sim}{T}^* \tag{4.13}$$

The one on the left is certainly simpler than the one on the right. The point of observation of this phenomenon, which has been referred to by the author [17] as 'the stiffness center', has been known for a good many years as 'the elastic center' [18, 19]. While it is a mid-point for prismatic line elements, it can be determined for the other configurations by invariants of the inertia tensor in Eq. (4.8). In this case, however, dA = EI representing the stiffnesses of individual elements and the orientation of the principal direction is the one which makes the product inertia term zero. Consider, for instance, the system in Fig. (4.2a) which consists of members in series. We shall first disconnect the elements and spell out the stiffness tensor of each element in their own principal coordinate systems Fig. (b), then assemble these elements by infinitely rigid members through their stiffness centers Fig. (c).

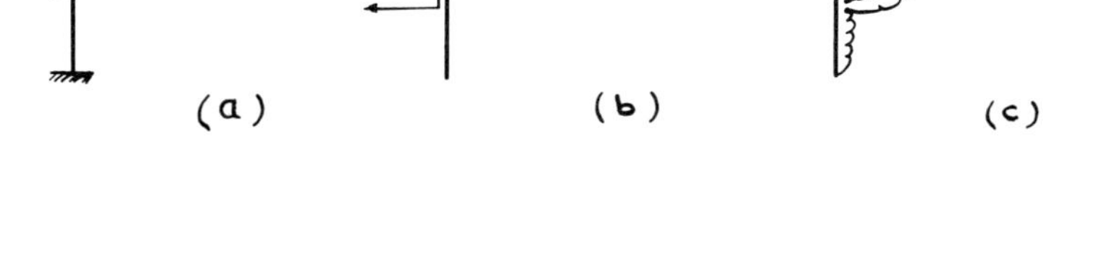

(a) (b) (c)

The final flexibility matrix of the system, which would also be in diagonal form, will be obtained by the translation (refer to reference 16) of the element flexibility matrices from their own center to the center of the system. Furthermore, the translation of displacements to the stiffness center results in forces developed at the center due to applied loads on the elements without any inversion procedure.

The concept of the 'stiffness center' from elements to sub-systems such as series, parallel and closed loops can also be extended to the entire system. In other words, the system can be divided into sub-systems and the sub-systems can be connected through their stiffness centers by infinitely rigid elements. By doing so, one will be able to reduce the dimension of the stiffness tensor (equivalent stiffness tensor) for the sub-systems which results in a considerably smaller size of stiffness tensor for the entire system. For complicated systems with multi-connected loops, however, this portion of the research has not yet been finalized.

4.2 Higher dimensional elements

The concept of diagonalization of stiffness tensors for elements with higher dimensions would also be valid. In this case, however, what is more important is not the simplicity (formwise) of the tensor but the ease of integration of the local functions within the domain of the element.

$$k^{ijqr} = \int_V b^q_r \, D^{ijqr} \, b^r_q \, dv \tag{4.14}$$

in which D^{ijqr} is the material tensor and

$$b^q_r = \frac{1}{2}\left(\frac{\partial}{\partial x_q} + \frac{\partial}{\partial x_r}\right) a(x_q, x_r)$$

The integration shown in Eq. (4.14) is often difficult. The geometry of the element and the assumed shape function $a(x_q, x_r)$

which are both expressed in a chosen local coordinate system play important roles in facilitating this integration. In the case of linear element boundaries, the choice of a 'natural coordinate' system $x^{i'}$ as a local coordinate system in the form of

$$x^i = b^i_{i'} x^{i'} \qquad \begin{array}{l} i = 0,1,\ldots,n \\ i' = 1,2,\ldots,n+1 \end{array} \tag{4.15}$$

where $x^o = b^o_{i'} = 1$ facilitates this integration drastically. The elements of matrix $\underset{\sim}{b}$ in this equation are the coordinates of the nodes in the global system. The integration formulas in natural coordinates can be found in most texts on the finite element method.

In the case of curved element boundaries, however, one may attempt to choose the local coordinate system such that in reference to it the geometry of the element becomes linear. Depending upon the desired continuity requirements between the element boundaries, the shape function can also be chosen as identical with the element geometry. Such an element is often referred to as an isoparametric element. If, on the other hand, the shape functions are chosen to be higher or lower order than the geometry of the element, such elements are called super-parametric and sub-parametric respectively. In all cases, however, once the integration in Eq. (4.14) is done, the transformation of the stiffness tensor from the chosen curvi-linear coordinates to the global coordinates (most often rectangular) becomes necessary. Such a transformation, which follows one of the standard tensor transformations mentioned in Art. 1, is always valid regardless of the chosen local or global coordinate systems.

Needless to say, the question of the most suitable coordinate system for a given element will always remain to be determined. A thorough knowledge of the physical problem and of tensor transformations is invaluable in finding the answers to this question.

5. CONCLUSION

There are always two distinct phases to every problem, namely, formulation and solution. If the problem is physical in nature, tensor notations are well-suited for the first phase. For the second phase, however, the nature of the problem and its formulation make no difference. Once the researcher reaches the second phase, he suddenly finds himself in an entirely non-physical world having lost contact with all he has done until then. For problems in discrete mechanics especially, he remains entirely in a numerical world until the solutions are attained. During such number crunching procedures, no reference is normally made to the physical characteristics of the problem. Considering that any linear operation which is done over an equality representing a physical phenomenon is nothing more than the observance of the same phenomenon in another coordinate frame, one may attempt to determine the best course of operation from the physical nature of the problem. The price for such an action might very well be as costly as any other non-discriminant action, nevertheless, it is bound to enlighten the sequence of observations while approaching the final results. It may even yield some intermediate results which may be found equally as important as the final results. Or, in some cases, it may increase the accuracy of the results if not necessarily reducing the labor. The search for principal directions in this study may be considered as an attempt in this line.

REFERENCES

1. Brillouin, L., *Les Tenseurs en Mechanique et en Elasticite*, Paris, 1938.
2. Sokolnikoff, I.S., *Tensor Analysis, Theory and Applications to Geometry and Mechanics of Continua*, John Wiley and Sons,

Inc., New York, 1964.

3. Michal, A.D., *Matrix and Tensor Calculus*, John Wiley and Sons, New York, 1960.
4. Eringen, A.C., *Mechanics of Continua*, John Wiley and Sons, New York, 1967.
5. Fung, Y.C., *Foundations of Solid Mechanics*, Prentice-Hall, Inc., New Jersey, 1965.
6. Soule, J.W., *The Solution of Multiple Branch Piping Flexibility Problems by Tensor Analysis*, Journal of Applied Mechanics, June, 1956 and March, 1958.
7. Borg, S.F., *The Tensor in Structural Theory*, Annals of Academy of Science, New York, 1962.
8. Kardestuncer, H., *Finite Element Methods via Tensors*, CISM, Springer-Verlag, Udine, 1972.
9. Kron, G., *Tensorial Analysis and Equivalent Circuits of Elastic Structures*, Journal of the Franklin Institute, Vol. 238, December, 1944.
10. Eisele, A.J. and Mason, M.R., *Applied Matrix and Tensor Analysis*, Wiley-Interscience, New York, 1970.
11. Borg, S.F., *Matrix-Tensor Methods in Continuum Mechanics*, D. van Nostrand Co., Inc., New Jersey, 1963.
12. Kron, G., *Tensor for Circuits*, Dover Publications, Inc., New York, 1959.
13. Happ, H.H., *Diakoptics and Networks*, Academic Press, New York, 1971.
14. Moon, P. and Spencer, D.E., *Vectors*, D. van Nostrand Co., Inc., New Jersey, 1965.
15. Kardestuncer, H., *Tensors in Discrete Elastic Systems*, The Matrix and Tensor Quarterly, Vol. 20, No. 1, Great Britain, September, 1969.
16. Kardestuncer, H., *Elementary Matrix Analysis of Structures*, McGraw-Hill Book Co., New York, 1974.
17. Kardestuncer, H., *Stiffness Matrix Method*, Proceedings of the ASCE, Vol. 100, No. ST7, July, 1974.

18. Parcel, J.I. and Moorman, R.B.B., *Analysis of Statically Indeterminate Structures*, John Wiley and Sons, Inc., New York, 1955.

19. Kinney, J.S., *Indeterminate Structural Analysis*, Addison-Wesley, Massachusetts, 1957.

20. Kardestuncer, H., *Flexibility Analysis of Closed Loops Subject to Temperature Variations*, VI IKM, Weimar, 1972.

ENERGY, ECOLOGY AND ECONOMICS: ELEMENTS OF A THERMODYNAMICALLY BASED ECONOMY

HERMAN E. KOENIG
Department of Electrical Engineering and Systems Science
Michigan State University
East Lansing, Michigan

THOMAS C. EDENS
Department of Agricultural Economics
Michigan State University
East Lansing, Michigan

1. INTRODUCTION

The brief segment of human history referred to as the "industrial revolution" is commonly taken as the norm upon which to base our economic expectations for the future. Key technological weapons for the triumphant allies of the Second World War were perfected and the western world achieved the highest material standard of living in history. The miracles of modern medicine altered the population dynamics of the world and the systematic exploration of the outer reaches of space was initiated. The profound successes of this super-industrialism led many to believe there are scientific and technological fixes for all of the nation's problems, including the most serious environmental and social ills as well as growing resource shortages. Within the past decade, for example, U. Thant, Secretary General of the United Nations, declared that "The central stupendous truth about developed economies today is that they can have (in anything but the shortest run) the kind of scale of resources they desire to have... It is no longer resources that limit decisions. It is the decision that makes the resources. This is a fundamental revolutionary change -- perhaps the most revolutionary man has ever known." [28]

It is very easy to see how "... the history of technological advance (over this period of time) suggests an optimistic outlook for future economic growth" [17]. Such optimism, however, reflects an overly narrow time perspective [14, 21]. It ignores the limitations of our material and energy reserves and overlooks certain thermodynamic and ecological constraints that limit the accessibility of these resources to our economy [6].

The aspects of the world economy of deepest possible concern are the rapid increase in population and nearly complete and (still) growing dependence on the terrestrial energy reserves which have been accumulating over millions of years [11, 12].

The rate of increase in energy consumption and human population over the past several centuries is illustrated in Fig. 1. Prior to 1800, population levels were held by starvation and pestilence. The technology for releasing energy from fossil reserves had not yet been developed. In the brief period between 1800 and 1975, the rate of energy use and population growth increased geometrically. High speed energy driven transport and communication systems drastically reduced economic isolation and modern medical advances eliminated pestilence and extended life expectancies.

There is growing concern that the world birth rate cannot be controlled in time by the use of contraceptives, control clinics, increased living standards, changes in attitude toward family size,

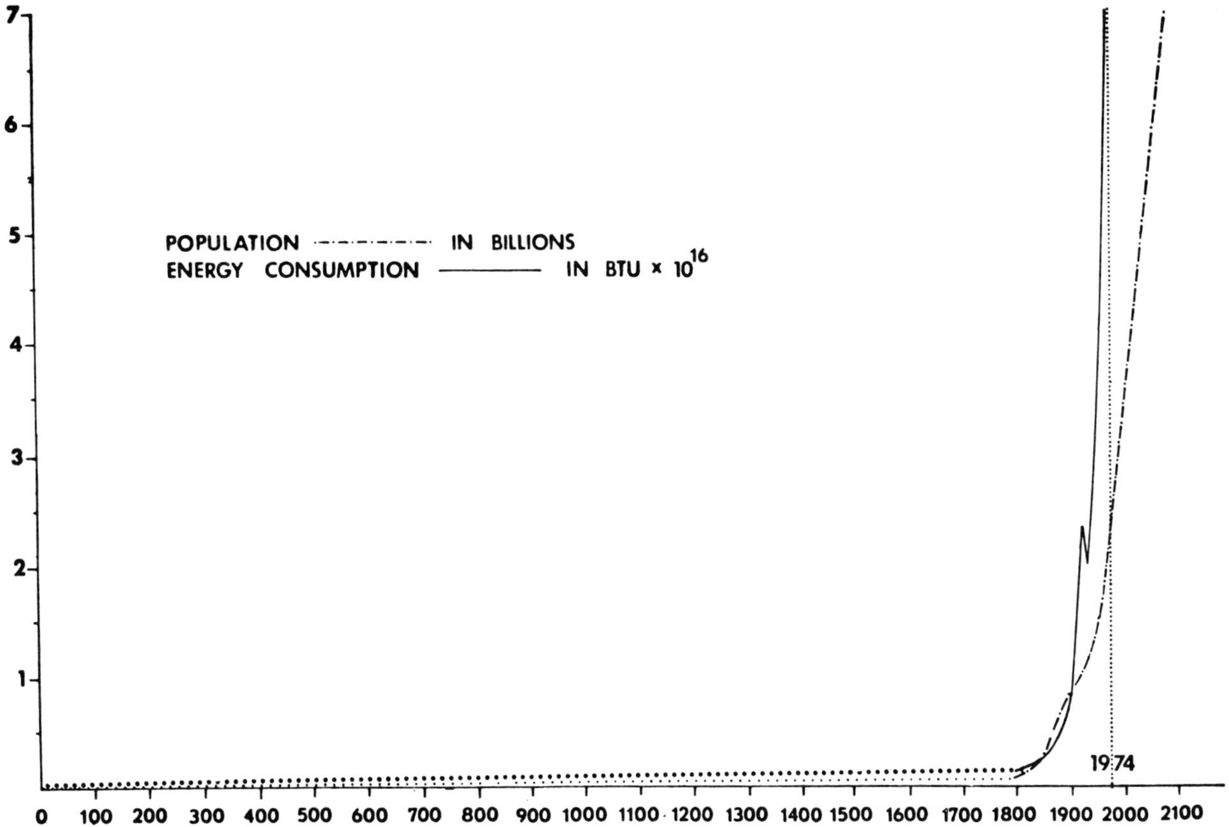

Fig. 1. Energy consumption [BTU's] and population trends from 0 A.D. to present. Compiled from data presented in [4]; reprinted in [19].

improving the status of women, and the legalization of abortion. We seem to be faced more and more with the unpleasant fact that death rates will go up. There are three ways this can happen: famine, pestilence and war [2,27]. The rate at which death rates go up will most surely depend upon the availability of energy and how efficiently and effectively it is used the world over in support of human populations. In view of the finiteness of virtually all raw material stocks and other limitations of our natural environment, the development of new and exotic energy sources will not in itself alleviate the need for long-range resource allocation procedures. It can only change the outcome of the allocation process and perhaps the planning horizon. The laws of thermodynamics that govern the use of finite stocks of materials and energy must eventually be reconciled with growing levels of human population and the limitations of the natural environment to accept material and thermal residuals.

The first six sections of this paper focus specifically on the thermodynamic and ecological constraints that circumscribe our systems of production and consumption with specific reference to long-term inconsistencies that now exist. The remaining sections develop basic economic policies and pricing mechanisms which might be used to redirect the trajectory of economic activity toward long-term thermodynamic and ecological consistency.

2. THERMODYNAMICS OF RESOURCE UTILIZATION

It is helpful to the understanding of the thermodynamic constraints that circumscribe economic activity to conceptually separate the human systems of production and human consumption from their supporting environment -- the biological processes of organic subsystems in the natural environment.* For further clarification, the physics and biochemistry of production and consumption systems are conceptually separated from the financial and other institutional aspects of these systems (see Fig. 2). The conceptual separation or dichotomy between human societies and the environment does not imply that the natural environment and human societies are mutually independent. Rather, it provides a conceptual framework for looking at the interdependence between these two aspects of human existence from a thermodynamic perspective. Likewise, the conceptual separation of the physical and biochemical aspects of production and consumption from the institutional aspects provides a framework for identifying the thermodynamic constraints the former will ultimately impose on the latter.

To understand the interdependences between human societies and their natural environment, the "quality" of potential energy sources must be defined and characterized. In addition, it must be determined what quantities might be *accessible* in the decades ahead.

Energy Stocks

One of the most highly respected assessments of the remaining world reserves of natural gas, petroleum and coal stocks is presented by Hubbert [16]. His analysis, reproduced in part in Figures 3, 4, and 5, presents optimistic and pessimistic estimates of these stocks as areas under "bell" shaped curves bounded by temporal changes in use rates. In each case the lower part of the bell represents the historical use rates from 1900 to the present.

---

* The natural environment as used here can be considered essentially as the geophysical, meteorological and biological processes on the surface of the earth as they have developed without a dominant influence by the human species. The systems of production and consumption represent essentially the man-made and managed systems of production and consumption.

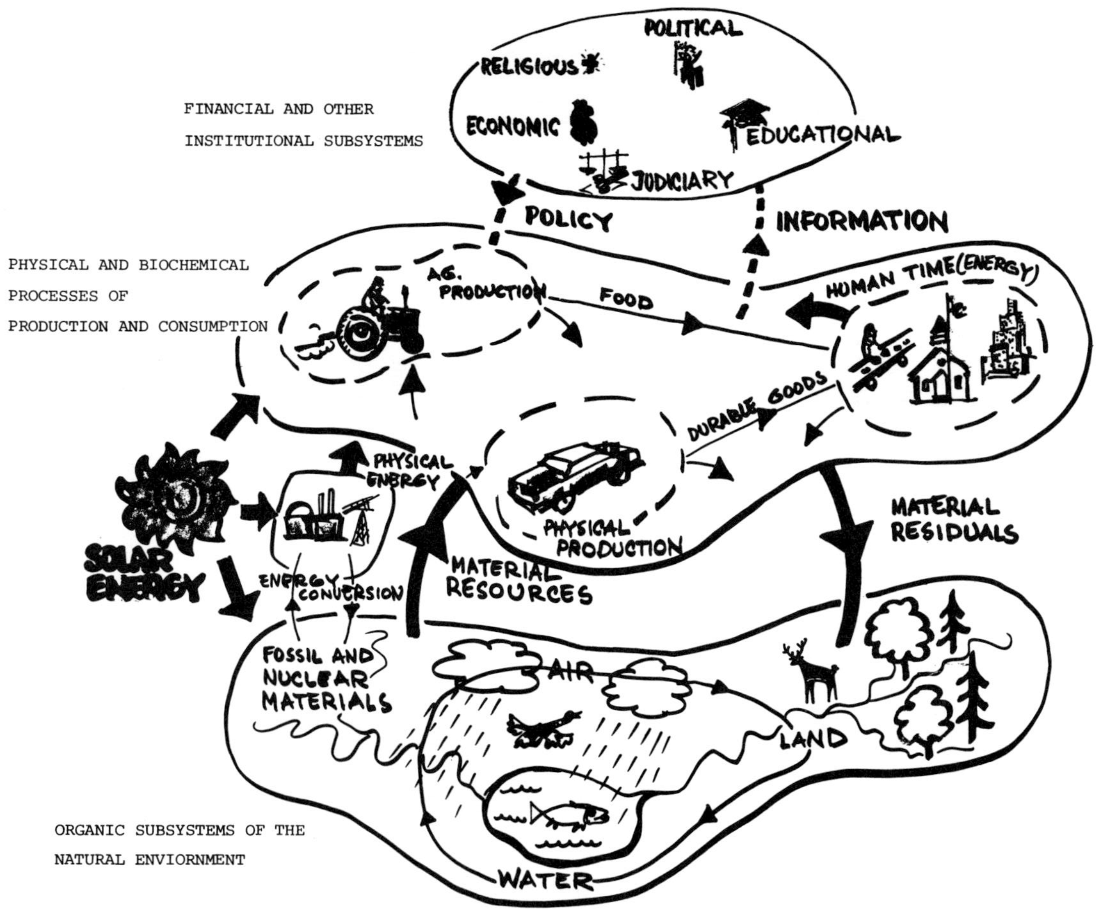

Figure 2. Conceptual elements of the relationship of human societies and the natural environment.

Chart 5

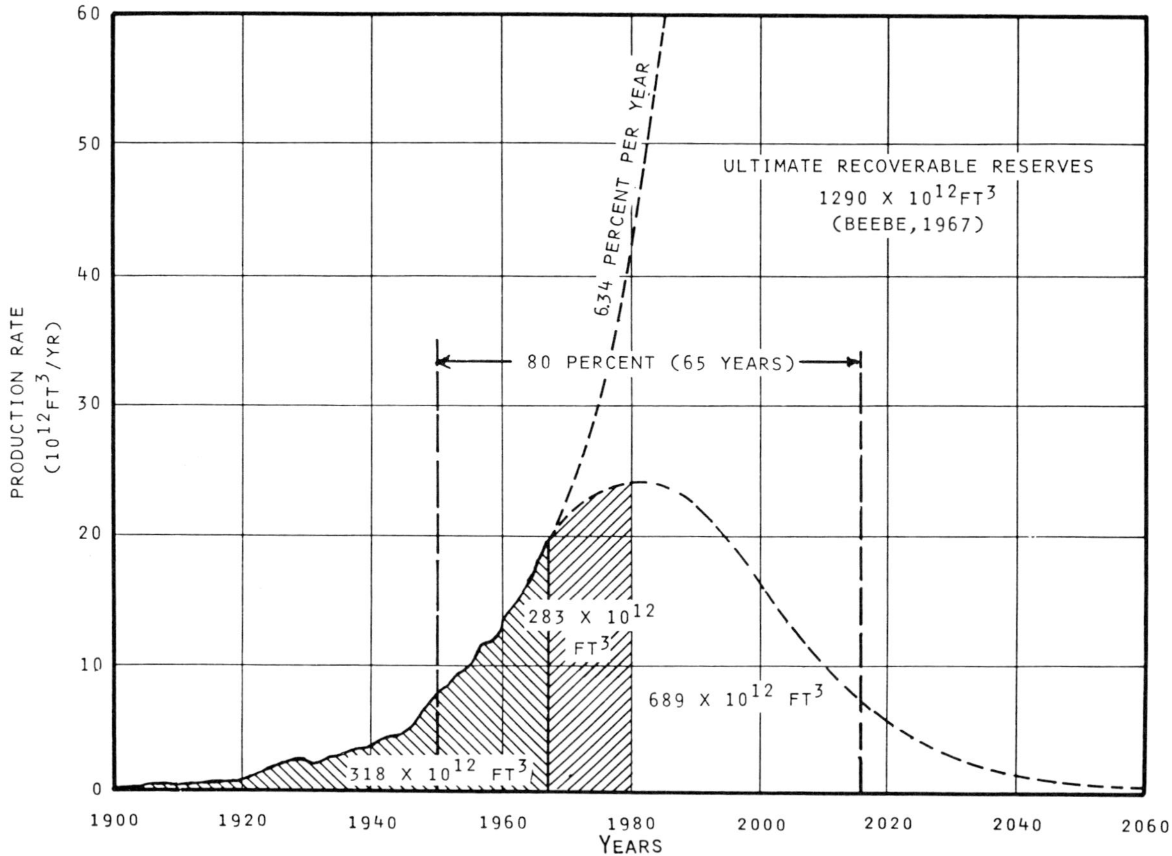

Fig. 3. Complete cycle of natural gas production in the United States. [23, p. 10]

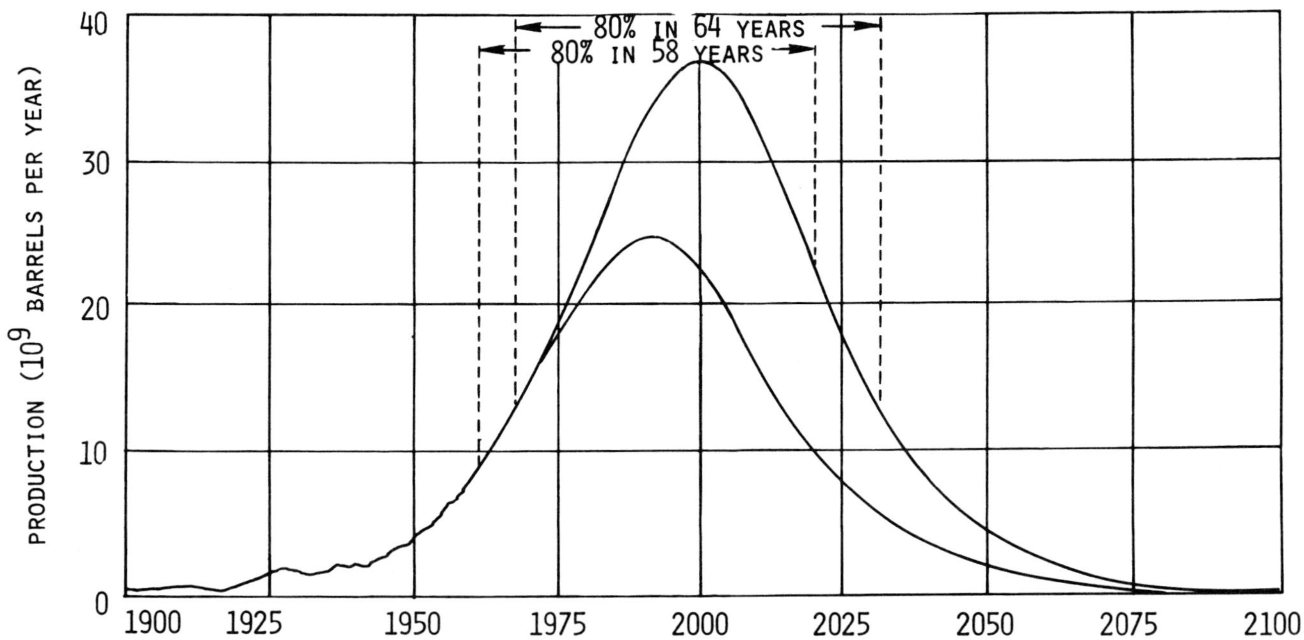

Fig. 4. Cycle of world oil production is plotted on the basis of two estimates of the amount of oil that will ultimately be produced. [16, p. 9]

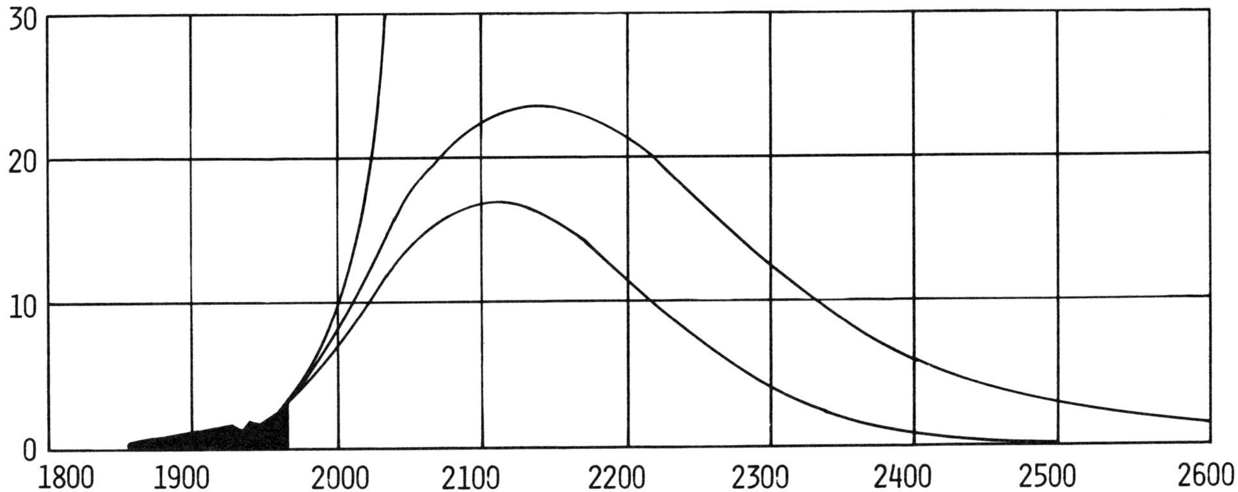

Fig. 5. Cycle of world coal production is plotted on the basis of estimated supplies and rates of production. [16, p. 9]

Use rates beyond the present are strictly hypothetical (but logical) scenarios based on the fact that, even under the best of conditions, substantial time will be required for technological and economic adjustments (however motivated) to first reduce the *rate of growth* in usage of each energy form to zero with a subsequent reduction in absolute magnitude to "phase it out". It is totally illogical to assume, as some do, that present growth or even present use rates of natural gas and petroleum will continue until we "suddenly run out".

Hubbert's scenarios vividly portray the importance of intrinsic technological and socioeconomic delays associated with shifting dependence from one form of energy to another. It suggests that within twenty or thirty years the "phase out" of petroleum as a basic source of energy will have begun and that in a period of some 58 to 64 years, 80% of the world's petroleum stocks will have been exhausted.* This period of time is relatively short compared to the lead time that will be required to alter production technologies, transport facilities, demographic distribution, and other features of the economy to accommodate alternative energy sources. Even if the impact of fossil fuel shortages is reduced through the implementation of new technologies (e.g., solar, fission, and/or fusion), the lead times required to shift the technical and socioeconomic features of our economy consistent with these new energy sources must be taken into consideration.

---

* As a world aggregate, Hubbert's analysis does not attempt to relate to the political and economic problems of world distribution of these stocks, nor does it attempt to relate to the technical problems and growing capital facilities required to make the energy potentially available from these stocks accessible to the production and consumption activities of our economy. These additional technical and political factors will almost surely reduce significantly the effective time period over which petroleum resources are available to any given economic system.

· Accessible Energy

A sharp distinction must be made between the *stocks* of fossil and atomic fuels stored in the geosphere on one hand, and the energy that is ultimately *accessible* to the human species from these stocks on the other. It takes energy to get energy. If, for example, a process requires an input of 100 billion BTU's per day to produce 300 billion BTU's of energy, the net energy is computed as 200 BTU's per day. Such "net energy"* calculations should also take into consideration the qualitative distinction between energy sources. If the energy inputs consist solely of fossil fuels, for example, the policy implications of the calculation are quite different from a situation in which the inputs are exclusively solar and human forms of energy. The expansion of current energy supplies and the development of new energy sources unfortunately require large amounts of terrestrial energy inputs.

The usable energy derived from a conversion process may be less than the energy required to build the capital facilities and carry out the extraction, processing and transportation operations. Such seems to be the situation with regard to many of the oil shale deposits in the western sections in the United States. On an aggregate energy input basis, some experts compute that even in the case of nuclear fission, the net return of energy may be as low as 3 to 12 times that required to develop and maintain a nuclear power plant over its expected life of 30 years. [1] Thus, it can be expected that the net energy accessible for production and consumption will decline even more rapidly than the stocks. When the growing costs (measured in real energy terms) of recovering remaining reserves are considered in relationship to the historical geometric increase in energy consumption, the lead time available for remedial action may be even shorter than the analy-

---

* "That energy (regardless of form) remaining for use by society after all energy uses chargeable to producing that energy (in its final form) are subtracted." [26]

sis presented in Fig. 6 implies.

The decreasing accessibility of energy supplies is formally embodied in the second law of thermodynamics which states that all forms of energy are ultimately transformed into heat and that eventually the heat energy becomes so diluted (dispersed) that it can no longer be converted to mechanical and other forms. Thus, energy derived, for example, from the combustion of fossil fuels is "dissipated" and cannot be recycled in the same sense as the materials (the chemically bound elements of carbon, oxygen, etc.) in which the energy was originally bound or stored.

The concept of entropy is sometimes used as an index of the amount of unavailable energy in a thermodynamic system at any given time. Thus, the entropy index of fossil fuel is said to be very low, thereby signifying that much of the bound thermal energy in fossil fuels is available for useful work. After combustion, the entropy of the same thermodynamic system is regarded as very high to indicate that most of the energy is no longer available in a usable form.

In view of the second law of thermodynamics, efficient use of low grade (high entropy) solar energy sources may require major adjustments in land use patterns and the physical, technological, and socioeconomic organization of the processes of production and consumption as well as adjustments in cultural values and life styles. Effective use of solar energy sources may require demographic and technological decentralization. It would be very difficult (if not totally impractical), for example, to "collect" or concentrate solar energy at a level required to support many of our larger urban and industrial centers.

Terrestrial energy sources in contrast to solar (extra terrestrial) forms are said to be of a very high quality (low entropy) since the accessible energy density is very high. This is particularly true for petroleum, natural gas and nuclear sources. Thermodynamically efficient use of these sources permits a greater variety of land use patterns, technologies, and levels of demo-

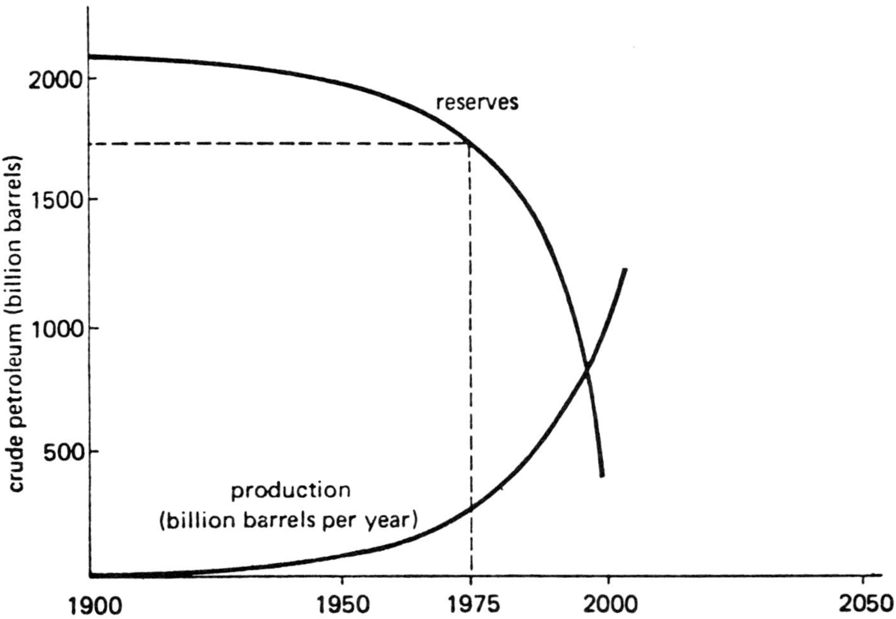

Fig. 6. World reserves of crude petroleum at exponential rate of consumption. [13, p. 7]

graphic concentrations. Hydrogen fusion (if it develops) perhaps will require even greater demographic and technological centralization than we now have. The institutional problems of developing, managing and maintaining high density energy sources in themselves may prove to be socially and culturally intractable.

The important thermodynamic principle of central concern to economic policy is that the quality of our energy sources as defined above (solar on one extreme and fusion on the other) is a major determinant of the basic structure of our production economy. Since we cannot know the ultimate outcome of fusion technology and the social reaction to high density nuclear sources of energy, one of the most critical scientific "boundary conditions" to long-range economic policy must remain at least somewhat open. Yet the geometric reductions in the energy that is ultimately accessible (by any technology) from our terrestrial energy resources implies that the time horizons for making the adjustments (one way or another) may be much shorter than normally perceived, particularly when viewed against the several decades or more required to affect the associated technological changes and to appreciably modify land-use patterns.

· Accessible Materials

Because the chemical elements of the fossil fuels are much more organized and "concentrated" prior to combustion, entropy has come to be regarded as an index of *disorder of matter* as well as a measure of unavailable energy. Thus, inorganic phosphorus as found in the phosphate mines of Florida is characterized as having a state of low material entropy, whereas the same material is in a state of high entropy when scattered in the form of fertilizer (at a very low concentration) on agricultural lands. The entropy of phosphorus may again be reduced as it is bound up in the biomass produced by photosynthesis (based upon continual solar energy input) and is physically collected and transported (based on the input of physical energy) to an urban community for human ingestion. If these materials, now contained as an integral part of the urban wastes, are redistributed on the agricultural land, the entropy is again increased at an energy cost but the material cycle is closed. Closure of the material cycle takes place at an energy cost which is a direct function of the accumulated changes (both positive and negative) in the entropy of the material as it completes the loop. On the other hand, if these urban wastes are discharged into a lake or stream, the phosphorus becomes so disorganized or scattered (increasing entropy) that it is virtually impossible to reorganize and make it accessible to the food chain. Thus, matter also is subject to irrecoverable dissipation and ultimately becomes inaccessible because the energy required to reorganize it is prohibitive in any practical sense.

It is possible to partially recycle some of our critical materials and in this sense make them continuously accessible to the economy, but the recycling can never be complete. The long-term implication is that the systems of production and consumption in agriculture and industry ultimately depend upon finite stocks for material resources. Because the concentrated stocks are extracted first, the entropy of the material stocks increases with time as they are drawn upon and the system of production and consumption, as a thermodynamic process, is moving progressively to a state of increased resource entropy. This process implies that unchanging flow rates of economic substances* ultimately will require expanding energy budgets.

In principle, material and energy equilibrium (a steady-state production economy) is an illusionary concept. This is not to conclude that the goal of moving *toward* a material balance is an empty gesture. On the contrary, efforts in this direction are critical if the rate of entropy increase is to be minimized, thereby extending material and energy stocks. Recycling of materials not only conserves material and energy in the popular sense, but has the compounding effect of greatly reducing the rate at which resources move to a state of increased entropy. In economic terms, it reduces the rate at which society moves toward the inevitable position of diminishing returns in the utilization of remaining material stocks.** Improved thermodynamic efficiency

---

* The term "economic substance" is employed in this and subsequent passages to represent the "real" basis of all human activity. These substances include materials and energy as well as residuals and thermal emissions -- all of which carry economic as well as physical impacts.

**Diminishing returns in this context can include increased capital facilities per unit of output as well as reduced output per unit of labor and energy input.

therefore has the basic elements of compounded savings rather than simple savings and should be given much more careful consideration in economic policy than it has thus far received.

3. ECOLOGICAL CONSTRAINTS

All materials (whether ultimately used for energy sources or structural artifacts) extracted from the geosphere must be accounted for somewhere in the system. They are either retained as material and structural stocks within the human system of production and consumption or they are returned in one form or another to the natural environment (geosphere, biosphere, or atmosphere). All thermal energy is eventually re-radiated to outer space. The ability of the natural environment to "accept" residual materials and energy is therefore of fundamental concern.

The capacity of the natural environment to assimilate material and energy residuals is a complex subject from both a scientific and aesthetic point of view. From a scientific point of view the physical and biological *states* of a lake, a stream, an estuary, or a regional "airshed" depend upon the temporal rates at which man-made materials of a particular chemical composition (or thermal quality) are injected into these environmental components. On the other hand, the aesthetic, recreational and other cultural factors of environmental "quality" are directly or indirectly related to the physical and biological states.

Both of these considerations lead ultimately to the concept of a set of regionally specific constraints on the quantity and quality of emissions that can be discharged into any given region of the natural environment -- regionally specific constraints that are partially a "technical" question and partially a "social choice" issue relating to the "kind of environment we wish to live in". Just as the regional environments have a natural capacity to support certain types of production more effectively than others (given that a particular biological state is to be preserved), they also have varying tolerances for accepting the residuals of production and consumption activities. A lake in southern climates, for example, may have a higher tolerance for nitrogen and phosphorus than northern lakes and the "airshed" in the windy city of Chicago has a higher tolerance for oxides of nitrogen than the Los Angeles basin.

The central significance of these constraints from an economic point of view is that they impose discernable restrictions on the technical structures of the production and consumption processes -- including product mix, technology employed, scale of operation, and spatial organization. Without such constraints the concentrations of the externalities inevitably associated with the process become excessive. The work of Meadows, *et al.* [22,8], points out the temporal limitations imposed on exponential growth in material and energy by the natural environment. But it is not generally accepted that the same limits constrain the spatial and technical organization of the means of production and consumption. Many of the most serious environmental problems are appropriately characterized as environmental "hot spots" -- too many people and/or processes in one physical location.

Pollution abatement processes are costly (both in terms of energy and human resources) and attempts to rid the environment of material residuals by technical processing ultimately leads to frustration. It is now being generally recognized that the natural (and "free") processing capabilities of regional environments are an economically valuable *resource* of the region, and that these tolerances must be treated with the same respect as regional material producing capabilities.

4. THERMODYNAMIC ASPECTS OF PRODUCTION AND CONSUMPTION

The energy required to produce a unit of a given economic substance is affected simultaneously by the thermodynamic characteristics of the material and energy resources *and* the technical *structure* of the economy. The technical structure as used here refers specifically to the technologies of production, degree of material recycling, scale of operation, and spatial organization of the components of production and consumption. The difference in the relative "efficiencies" of alternative systems of production and consumption in converting heat energy in the form of solar radiation and terrestrial energy (of all forms) into economic substances must be more fully understood.

It is scientifically correct (though economically unusual) to view the systems of production and consumption as "heat engines". In the final analysis, these systems literally convert thermal forms of energy into economic substances, which can in reality be viewed as energy derivatives. The field of thermodynamics (defined narrowly) is a discipline for analyzing, designing and controlling heat engines -- a heat engine being defined as any device or system used to convert heat (thermal energy) into mechanical energy. Thermodynamics is used in a broader sense here to refer to the concepts, principles and procedures for understanding, designing and managing systems of production and consumption (viewed in total) to effectively utilize energy resources (including both solar and terrestrial forms) in the production of economic substances. In this larger context, the relative *thermodynamic efficiencies* (economic substances per unit of thermal energy) of alternative systems in carrying out these transformations, their relative compatibilities with the natural environment, their economic parameters, and their temporal stability -- the ability of the system to sustain itself over a long period of time -- must be examined.

An understanding of the relationship between the thermodynamic characteristics and economic parameters is an absolute prerequisite to any comprehensive effort to manage and to conserve (make effective use of) natural resources. These include arable land -- through which solar energy is converted to economic substances

-- as well as the finite stocks of terrestrial energy. Some of the greatest conservation gains are surely to be found in the analysis of how arable land resources (solar energy) are combined with terrestrial energy resources, rather than in how they are managed individually. Thus, arable land -- including in particular prime agricultural land -- is regarded as a solar energy input to economic substances other than food.

The material flows through the systems of production and consumption are sustained under the influence of three *energetic* factors: *solar energy* (measured in land area allocated to manage photosynthesis), *human energy* or time (measured in man hours), and physical energy (conventionally measured in BTU's or kilowatt-hours).*

During the past few decades "technology substitution" (primarily capital for labor and land) has taken place at unprecedented rates. In addition, many new technologies serve a complementary role vis-a-vis existing factor inputs. Through the use of more synthetic fertilizers, for example, physical energy is substituted for land. With energy driven machines, physical energy replaces human time and effort. And through communication and information systems, human time is, to a degree, substituted for physical energy. However, there do exist limits to the technical substitution of one energetic factor for another -- limits due to resource depletion (fossil based energy inputs) and to purely technical constraints.

Within the sphere of what is technically feasible, the degree to which systems of production and consumption are designed to make a partial substitution of one energetic factor for another certainly should depend upon the availability of the three "ener-

* Note that in this conceptualization, labor is explicitly referred to as an energetic variable rather than a material flow in the usual economic sense. A day of human time, like a kilowatt of energy or solar radiation, is dissipated, never to be recaptured or recycled. For a more detailed discussion, see [18,20].

getic resources" -- the balance between population, arable land, and accessible physical energy. The central point is that neither maximum terrestrial energy efficiency, manpower efficiency, nor land efficiency (however defined) in itself is necessarily a valid criteria for the design of the physical and technical organization of the means of production and consumption. The objective, rather, is to identify economic and institutional policies which collectively will maintain a *dynamic balance* between these basic energetic factors over time. Balance is used here in a generic sense with no implications as to what is "optimal" or desirable in any specific socioeconomic context. Regardless of what this balance is, there is, or should be, a concern for effectively using the energetic factors of production.

· Technical Fixes

Freeman, *et al.* [9], make specific reference to a class of improved thermodynamic efficiencies in production and consumption called "technical fixes", which could easily result in an annual savings of from 30 to 40% of the present U.S. energy budget. When these savings are viewed against the inevitable increase in material entropy, the implied savings are even more significant than the raw figures would indicate.

Technical fixes as discussed by Freeman relate to:

(1) improved thermodynamic efficiencies at the *final point* of energy use (improved insulation, i.e., improved efficiencies in automobiles, household appliances, heating systems, etc.), and

(2) improved thermo-efficiencies in the *energy processing sector* (power plants, petroleum refineries, uranium enrichment processes, coal mining and processing techniques, etc.).

The important feature of these adjustments, from an economic point of view, is that they can take place within the basic framework of the present system of production and consumption without major changes in the structure itself. Further, there is no implied reduction in the "standard of living" or changes in life style. It is simply a matter of providing the economic motivation to adopt techniques which are already known.

- Product Durability

In retrospect, it is now clear that the economic policies of much of the industrialized world have given undue emphasis to labor productivity (material flow rates per man hour) and aggregate leisure in a rather narrow sense without adequate consideration to thermodynamic efficiency. For example, a labor force A that produces 50 refrigerators per week with an average life of 25 years provides more leisure time than labor force B (of the same size) producing 100 refrigerators per week with an average life of 10 years. Yet, by conventional economic accounting procedures, the productivity of labor force B may be higher than A, even though labor force B may have a much lower thermodynamic efficiency than A. Conventional economic concepts would regard this as "good", since it contributes more to Gross National Product (GNP) and creates additional jobs. Well-being in terms of nutrition and diet is perhaps appropriately measured by the production rates of specific classes (whole grains, meats, etc.) of food. But the material well-being of the human species in matters of "durable" products such as clothing, automobiles, housing and household appliances should be measured by the standing stock (states) and not solely by the production rates (flows). Unfortunately, GNP, as currently conceived and computed, makes no such distinction. Major improvements, in terms of thermodynamic efficiency are clearly possible by increasing the technical life expectancy of durable products. Given the fact that in excess of 40% of the energy consumed in the U.S. goes to industrial production (excluding transportation) [9], the potential savings in the national energy budget are self-evident.

Although much of the requisite technology is available, it is clear that a transition to more durable products may require changes in the basic structure of the economy to deal with the potential unemployment and other socioeconomic factors associated with such a change. Throughout most of the period of industrialization, the ability to economically motivate increases in the flow rates of materials in the system (much of which occurs through planned obsolescence and consumerism) according to unemployment levels has been a prerequisite to maintaining economic stability.

One potential ramification of increasing product durability is the effect on employment levels. Increased durability, broadly defined, can be accomplished in two general ways:*

(1) by increasing the physical or technical useful lifetime of a product, and
(2) by reducing or eliminating cosmetic changes (e.g., style changes in automobiles).

The employment effect, in the case of technical refinements, is uncertain. However, the elimination of purely cosmetic alterations in products would lead to increased unemployment. In a time when general levels of unemployment are high, the capacity of the economic system to absorb more workers is limited. To what extent is it possible, for example, to implement land use policies which would facilitate the absorption of displaced workers? Would the maintenance or re-establishment of the small family farm encourage urban-rural migration? These and other selected questions should be addressed by land use planners and policy makers at all levels of policy formulation.

· Spatial Organization

The spatial organization of production and consumption systems has a major influence on thermodynamic efficiency in two important respects. The first is in reference to spatial organization and

---

* Product durability and perceived standard of living have sociological and behavioral aspects as well as technical aspects. This is discussed in the next section.

the *scale* of production and consumption processes vis-a-vis the capabilities and tolerances of the natural environment to accept the residuals of production and consumption *without* energy intensive "pollution abatement" technologies. The second involves the logistics of material and human transport.* Constraints on regional product specialization (with the implied limitations on scale) also have the potential of greatly reducing transportation requirements without a significant sacrifice in material standard of living.

Given the fact that roughly 60% of U.S. gasoline consumption is allocated to the transport sector, the potential reduction in energy requirements is very significant. But a shift from automotive to rail, trolley systems, or other forms of transit requires, in many cases, major changes in the physical layout of the various components of urban and rural communities. Reductions in logistics of transport place many additional constraints on the size of the urban community and the levels of regional specialization, and these constraints are always specific to the tolerances of the landscape to accept (with limited abatement technologies) the inevitable residuals.

---

* The trade-off between the three energetic factors of production in relationship to the scale (size) of feed cattle operations is dramatically portrayed in the work of Connor, *et al.* [3,29]. These results were developed for the ecological conditions that prevail in the North Central Region of the U.S. They show for virtually all feasible technologies and all operations larger than 300 or 300 head, the physical energy requirements per hundred weight of animal gain increase monotonically while land and labor requirements per hundred weight gain remain essentially independent of scale. The monotonic increases in physical energy requirements are attributed to transportation costs.

Land-use patterns with their environmental and transport implications (in the broad sense) are of increasing import as attempts are made to raise thermodynamic efficiencies through recycling. Spray irrigation of urban wastes from New York, Detroit or Los Angeles to close the phosphorus cycle in the food chain, for example, is virtually impossible in the context of present land-use patterns (the energy requirements would be prohibitive).

Thermodynamically efficient and environmentally sound land-use patterns will require basic structural changes in the economy -- decentralization of many existing technologies (industries), limitation of the size and location of urban communities, and an increase in regional economic self-sufficiency. These adjustments in physical structure are certain to have major impacts on labor requirements and many other socioeconomic factors. Increased thermodynamic efficiency through both product durability and decentralization *can*, in principle, result in increased environmental quality.

5. BEHAVIORAL ASPECTS OF PRODUCTION AND CONSUMPTION

How the production processes are utilized, who they serve, as well as the output levels are determined in part by the patterns of ownership and of control, and in part by patterns of consumption. [10] That there is a relationship between consumption behavior and production is indisputable, and the sequence of causation may be as much from the latter to the former as vice versa. The relationship is undoubtedly *bilateral*. But, deleterious environmental and thermodynamic consequences of their joint actions are *external* to the loop.

Production

Production influences consumption patterns at basically two levels:

(1) at a superficial or *functional* level, and

(2) at a deep rooted or *structural* level.

At the functional level there are readily identifiable control elements which exert a strong influence on what is available for consumption and the quantities in which it is made available. Cosmetic changes in automotive and other product designs, for example, do not emanate entirely from individual consumer demands. They are to a great extent promoted by the producers through a variety of advertising media and marketing techniques.

At the structural level, a similar market power exists, but is somewhat more subtle. The social and physical infrastructure which lends itself to the intensive utilization of the automobile, for example, was generated over time, in part, by the market power of a few large producers. According to Sawyers: [24]

> "In the 1930's, Los Angeles had the largest trolley system in the world; it was the trolley car and not the auto which had given the metropolis its sprawling, suburban form. The 700 mile trolley network reached out to a distance of 75 miles from downtown Los Angeles and served 80 million passengers a year. Between 1938 and 1944, GM subsidiaries scrapped trolley cars, ripped up track, tore down power lines, and replaced them with costly, lumbering, odorous buses, thereby driving Los Angelines to the automobile."

And in 1949, General Motors was convicted

> "... of having criminally conspired with Standard Oil of California, Firestone Tire and others to replace electric transportation with gas -- or diesel -- power buses and to monopolize the sale of buses and related products to local transportation companies throughout the country." [25]

This is not to imply that General Motors was alone responsible for the subsequent developments in the transportation sector. It does, however, provide one specific example of how agents in control of production were able to influence basic structural parameters, including migration and settlement patterns of both firms and individuals, thereby altering both land use patterns and the basic transportation systems associated with them. This example also illustrates how the technical structure of the transportation system is an inextricable component of land-use patterns. On one hand, it is a major determinant of land use patterns, yet on the other hand, many aspects of its evolving structure can be said to be responsive to changing land use patterns.

· Consumption

Prior to the early part of the 19th century, individual rights and wants were generally conceived of as secondary to community goals and constrained by the natural environment. In a sense, economic activities took their place among other social institutions and were not a dominating factor. In the process of industrialization, the justification of the processes of production became inextricably linked to the satisfaction of individual wants and desires. Perhaps the greatest ramification of the end results of this process was the elevation of the individual and of his expressed desires to the highest point on the social structure of hierarchical constraints. Natural environmental and ecosystemic limitations which were once viewed as constraints on individual behavior appeared more and more to be overridden by technology -- a technology which is critically dependent upon terrestrial material and energy stocks with a rapidly increasing entropy.

In the case of public goods, where the products of the system are indivisible, the demands are determined in large part through political processes through which authority is delegated to individual agencies (e.g., highway commissions; city, state, and local planning; etc.) responsible for the aggregation of societal needs. Even at the public level, individual desires and expectations are elevated to the highest point on the social structure of hierarchical constraints. Roads, for example, are created on the basis of the projected "needs" of various communities and power generating facilities are planned on the basis of historical growth expectations.

Individual and public expectations and preferences are influenced by a number of factors in the cultural milieu -- peer groups, status symbols, income levels -- in addition to the promotional influences of the producer. [10] The point is that individual preferences are *not* literally individual and that a variety of elements of the cultural matrix can be used to influence individual and public expectations and preferences. It is not entirely a question of interference versus non-interference with individual preferences but a question of substituting one influence for another.

The physical and technical design of the production and consumption processes and land-use patterns serves as a "superstructure" for social, cultural, and economic activities. Regulation of the thermodynamic characteristics (broadly defined) will therefore indirectly influence demographic factors, job specialization, recreational opportunities and many other aspects of the social and cultural "fabric". Land tenure rules, in particular, have always been a major element of the social and cultural forms, and it is not uncommon to use land tenure as a mechanism for employment opportunities.

Highly specialized, centralized and energy intensive production systems, for example, provide a favorable atmosphere for high *labor* productivity (high output flow rates per unit of human time) and it is precisely this productivity that provides the opportunity to pursue cultural goals. But the highly specialized nature of employment opportunities in such systems tends also to isolate the individual from the overall thermodynamic and environmental consequences of his or her actions. The "tragedy of the commons", referred to by Garrett Hardin in a slightly different context, is a pervasive element of a culture conditioned by specialization and centralization. [14] Deleterious environmental and thermodynamic effects, stemming from the aggregate of individual actions, do not serve to limit or constrain individual behavior. The feedback loops between the individual and the environment are too long and too weak to be functional. Everyone -- and therefore no one -- is responsible for the effects of their individual actions. The system, therefore, tends to perpetuate myths, misunderstandings, and thermodynamically absurd consumption patterns and life styles.

Increasing entropy of materials requires a high degree of centralization and specialization -- it takes increasing commitments of capital and energy to maintain extraction and production rates. Accordingly, the influence of large firms in determining what will be available for use by individual consumers can be expected to increase.

De-emphasis on individual and public consumption for thermodynamic and ecological reasons (however motivated) will undoubtedly result in a reduction of the flow rates of redundant goods and services through the economy. This in turn will alter employment patterns (and levels) and almost surely will place additional requirements on land use for recreational purposes and for employment opportunities. These and other social, cultural, and behavioral considerations may impose restrictions on future alternatives that are more confining than thermodynamic and ecological considerations. On the other hand, physical, ecological and sociological constraints may be largely congruent and overlapping. In the long term, the precepts of social and political freedom must be consistent with the laws of thermodynamics and ecology -- *all political decisions pertaining to resource management must ultimately fall within the sphere of what is thermodynamically and ecologically feasible*. For this reason, discussions in this and subsequent chapters relate primarily to control instruments that might be used to constrain economic activity within the sphere of thermodynamic and ecological feasibility.

6. CONTROL INSTRUMENTS FOR RESOURCE MANAGEMENT

Efforts to render the system of production and consumption -- as it exists today -- compatible with ecological and thermodynamic constraints may necessitate very fundamental changes in the economy and in the institutional interactions which give direction to the entire process. The concept of control must be enlarged to include the management of the physical and technical organization or *structure* of the system of production and consumption, as well as regulation of the material, energy, and monetary flows -- the usual *functional* aspects of a given structure.

The structure of the production and consumption system changes with time as new technologies are developed and implemented and as new resources are uncovered and old ones are depleted. These structural changes must be viewed in the context of an "evolutionary scale" (decades or centuries) in comparison to temporal changes in the functional aspects (days, weeks or months). The central point is that control of the functional aspects in and of themselves is not sufficient unless the system structure has few deficiencies and is on an "acceptable" trajectory of evolutionary change.

It is *not* necessary to perceive an optimum system of production and consumption, optimum land-use patterns, or optimal environmental quality. Nor is it rational to perceive of moving the existing economy to a particular static state. This is thermodynamically impossible. The best that apparently can be achieved is an alteration of the direction of changes in system structure -- increased technical efficiencies, improved land-use patterns, increased product durability, and changes in life style and values. The particular land-use patterns through which such goals can be achieved depend in part on the quality and technical properties of energy systems developed in the future. Since the outcome of some technological developments in energy systems is uncertain, many detailed parameters of economic policy cannot be determined *a priori*. It would appear prudent, however, in the light of present knowledge to move deliberately and with dispatch in the general direction of structures that will promote major improvements in material recycling and thermal efficiencies (broadly defined) and greater reliance on continuous solar forms of energy during the decades immediately ahead. Such is the objective of the control policies discussed in this paper. The intent is to identify alternative control mechanisms *only to the extent of avoiding certain classes of technologies, scales of operation, and land-use patterns.* The objective is to constrain the decision processes of market mechanisms and other land-use determinants within the sphere of long-term ecological and thermodynamic consistency.

In retrospect, it is clear that market systems, as we now know them, developed largely outside the influence of absolute resource shortages or the limitations of the natural systems in our environment to accept the inevitable residuals of production and consumption. The monetary costs now assigned to material and energy residuals injected into the natural environment too often are taken as zero. Likewise, the monetary costs assigned to materials and energy extracted from the environment are heavily weighted in favor of short-term or even immediate opportunities rather than long-term thermodynamic and ecological considerations. In this sense, the market system, even if it is functioning, is misdirecting land and other resource allocations because some of the primary input signals to the system are improperly weighted or do not include all the relevant information.

Many physical resources such as water, forest products, and solar energy (broadly defined to include wind and hydro forms as well as direct solar flux) can be extracted from the natural environment at continuous rates for indefinite periods of time if properly managed. They are, of course, referred to as the *renewable resources.* Likewise, certain classes of physical residuals such as compounds of carbon and nitrogen and other organic compounds found in the natural systems can be injected into the environment at continuous rates for indefinite periods of time if properly managed. In a context parallel to that of renewable resources, it is perhaps appropriate to refer to these as *renewable residuals.*

The *non-renewable resources* -- fossil fuels, uranium, steel, etc. -- are drawn from finite stocks with an essentially monotonic increase in entropy (discussed in earlier sections). Material residuals that accumulate in the environment -- such as DDT and inorganic solid wastes -- are referred to in subsequent paragraphs as *non-renewable residuals*. The entropy of these residuals typically will decrease with time as they accumulate in the environment.*

· Control of Renewable Residuals

Control over the rate at which renewable residuals are injected into the environment is most readily affected in terms of properly identified homogeneous ecological regions -- wet lands, biomes, or specific components of the environment such as lakes, streams and water sheds. Regions can be identified in such a way that a given control action(s) has the same qualitative (directional) effect on the ecological state of the region.

Control of the quality and rate at which residuals from the man-made systems of production and consumption enter these regions or components of the environment can be exercised either through direct institutional control of physical rates or indirect institutional control through the use of economic pricing policies. Either procedure requires theoretical and/or empirical understanding of the temporal rates at which residuals of various qualities can be "processed" by regional environments without excessive degradation of the quality of the natural processes. Given this information, the alternative forms of direct and indirect control can be viewed as illustrated in Fig. 7. Direct control is equivalent to assigning a zero price to the residual until such time that the threshold level of inputs to the environment is reached. After this point, a monetary cost of indefinite magnitude is indirectly imposed because of the restriction that no additional

---

* Holdren refers to these residuals as quantitative and qualitative, respectively. [15]

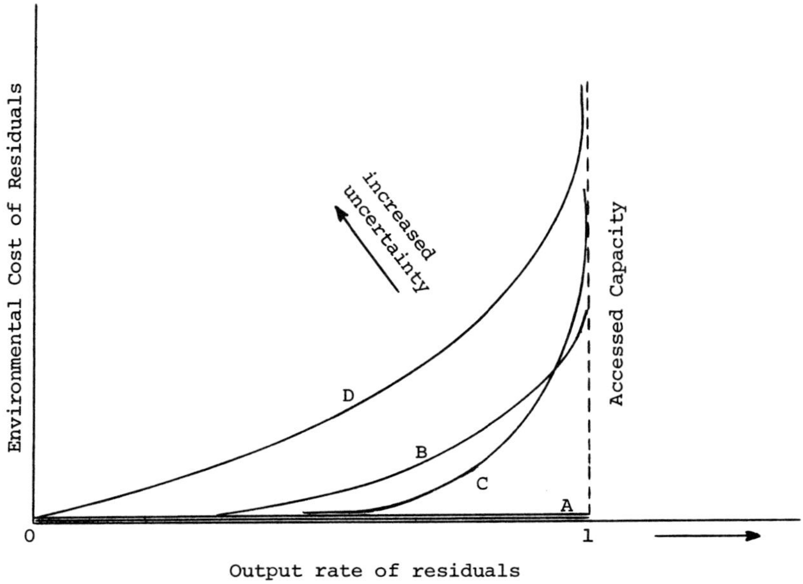

Fig. 7. Hypothetical environmental costs as a function of regional output rate of residual, normalized with respect to designated capacity [1]. Schedule of costs correspond to alternative regulatory policies as follows:

A - Direct control
B - Market determined price (sale of discharge rights)
C - Tax schedule for low uncertainty
D - Tax schedule for high uncertainty

discharge is permitted. This form of direct control is illustrated in Fig. 7 by price schedule A.

Indirect control through pricing policies may be achieved through market processes such as the sale of discharge rights (discussed by Ayres and Kneese [1] and Dales [5] or through taxes imposed on the residuals discharged into the environment. Whether a market process or tax procedure is used in any particular situation depends upon many physical, environmental and economic factors including the nature of the production and consumption processes, the type of residuals, and the nature of the recipient environmental components. In either case, an essential feature of the control instrument is that the monetary price per unit of discharge increases as the designated capacity of the region is approached. If the price is determined through market processes such as the sale of discharge rights, then the price will progress perhaps as indicated by schedule B in Fig. 7.

If the price is institutionally imposed as a tax, it is essential that the price at the threshold level be sufficiently high to ensure that the injection rate does not exceed the designated value. The rate at which the price progresses from zero at very low volume rates to the "cut-off levels" can be based on a variety of considerations, including in particular the degree of uncertainty associated with the assessed capacity. If the acceptable capacity is known with a high degree of certainty, a price schedule such as C may be used -- i.e., a *liberal* control. However, if there is considerable uncertainty as to what constitutes an acceptable discharge rate and/or the cost of being wrong (ecologically) is very high, the pricing policy may progress as indicated by schedule D -- i.e., a *conservative* control. Thus, in principle, at least, it is possible to internalize the cost of not knowing exactly what the environmental limits are by the shape of the price schedule curve. The point to be emphasized is that the rationale for the indicated tax schedule is based on control considerations in the presence of uncertainty rather than on pollution damage as such.

Each of the above control mechanisms will clearly have differing effects on land-use patterns. The general scope and nature of these differing effects are apparent when the pricing schedules in Fig. 7 are viewed in relationship to the manner in which the monetary costs of production vary with production rate and technological innovation (Fig. 8). Economists traditionally have emphasized the eventual diseconomies of scale at any given stage of technological development, as indicated by the U-shaped broken curves. Diminishing returns alter scale relationships over time

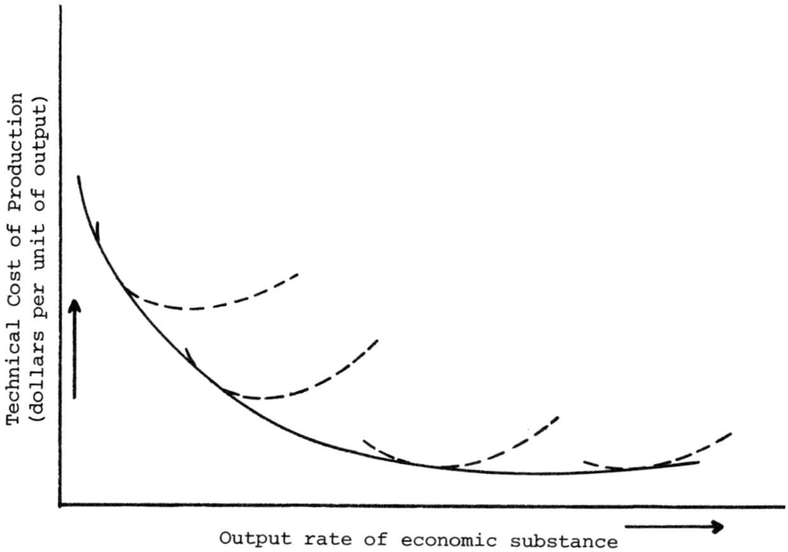

Fig. 8. Technical cost (hypothetical) of producing an economic substance as a function of scale of operation and temporal advances in technology; broken curves represent short-term economies for a given technology.

and technological advances in automation, scale of mechanization, and physical and managerial technologies produce a series of such U-shaped curves, the envelope of which typically would be of the form indicated by the solid lines in this figure.* The costs of production illustrated in Fig. 8 support the empirical observation that in many industries, the law of economic survival requires both technological innovation and increased scale of operation -- a

* The envelope is indicated as an abstraction of numerical results obtained by Connor, *et al.* [3], and apparently is characteristic of the technical costs of production found in many areas of physical and agricultural production.

"competitive" pattern that every "successful" firm understands well. When applied to the aggregate of human use of economic substance the technical cost curves in Fig. 8 would suggest that shorter term economic motivations in the direction of both technological innovation and increased consumption rates increases aggregate total consumption and reduces the monetary cost.

It is usually the case that at any level of technology, the rate at which residuals are produced by a given production process is directly proportional to the rate of production of economic substances. When this is the case, the technical costs of production in Fig. 8 and the environmental costs represented by the varying price schedules illustrated in Fig. 7 are simply added to establish the total cost of production. In the case of physical regulation (schedule A), there are no environmental costs added to the technical cost of production, except perhaps indirectly as cost opportunities beyond the cut-off point. All other pricing schedules tend to introduce diseconomies of scale and are said to internalize to the economic system the environmental costs of production. When the resulting cost schedule (technical costs plus environmental costs) is applied at the level of interfirm competition, it suggests that progressive environmental taxes will have a stabilizing effect on the size of the firm if, in fact, the technical costs of production display the characteristics hypothesized here. Likewise, when applied at the aggregate level of human consumption, it suggests that the utilization of the economic substance associated with the residual will also be stabilized -- the cost of production can no longer be decreased monotonically through technological innovation and expanded scale of operation.

· Economic Implications of Increasing Entropy

The relationship between production costs, the rate of production and increasing entropy is shown in Fig. 9 for a hypothetical extraction firm. The curve AB represents the long run average cost of extraction for low resource entropy. It is usually as-

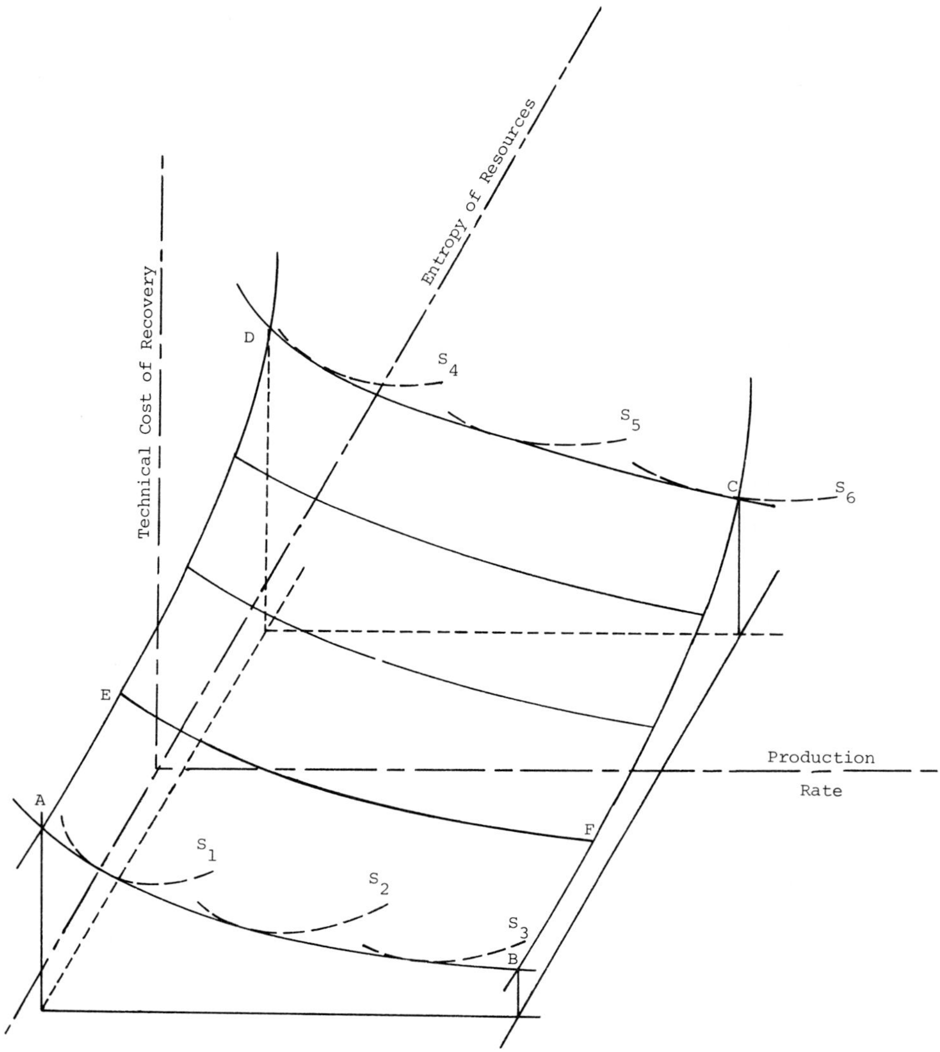

Fig. 9. Technical cost of recovery as a function of production rate and changing entropy of resource.

sumed that the average cost is inversely related to the rate of production except possibly at the extremely high level. The broken lines (S_n) illustrate various plant sizes at different production rates. The "U" shape of these curves is conventionally explained by the law of diminishing returns in production theory.* [7, p. 118]

As the entropy of the resources increases (AD and BC), the average cost of extraction tends also to increase for each rate of production. Over time, plant size is increased in an attempt to cope with the higher entropy levels of the material(s). As extraction proves more difficult technically, the cost of capital equipment required in the process increases (witness, for example, the capital costs of offshore drilling compared to the earlier wildcat rigs required to recover shallow inland deposits). The long-term average cost curve AB eventually shifts to new levels (EF and DC, in Fig. 9) reflecting the progressively high technical costs of extracting higher entropy resources. Note that as the entropy of the resource increases, would-be entrants into the extraction industry are faced with higher initial costs. This is true at both the domestic and international levels and implies that increased entropy itself will lead eventually to greater firm size and market power.

Progressive increases in entropy also eventually lead to price increases that cannot be countered by increased scale of operation and technological innovation. The point at which this occurs

---

* Average Product (AP) is the ratio of output to the number of units of the variable input [O/V]. And average variable cost (AVC) is the ratio of total variable cost to output [TVC/O]. Total variable cost is equal to price times the number of variable input units (PV). Therefore, AVC = TVC/O = PV/O, and AVC = P(V/O), AVC = P(1/AP). The law of diminishing returns shows average product increasing at first and then decreasing. The inverse of this is the "U" shaped AVC curve. [7, pp. 191-2]

depends upon the quantitative parameters of the particular resource. In the case of some extraction industries, there undoubtedly has been a movement "downhill", so to speak, on the three-dimensional surface in Fig. 9. In all likelihood, the bottom of the hill has been reached for some industries and they are on the way up the entropy-scale incline. The implications to the economy are far reaching indeed.

· Control of Non-Renewable Resources

The inescapable consequence of non-renewable resource extraction is an eventual increase in entropy with (more often than not) a concommitant increase in the size of the firm required to carry out the extraction process, and eventually, higher recovery costs. These inevitabilities stem from the critical role assigned to certain non-renewable resources (fossil fuels, for example), and by the *structure* of the economy which has evolved over decades of changing technologies, land-use patterns, cultural values and life styles.

Redirection of the trajectory of the structure away from critical dependence on finite stocks, can, at least in principle, be achieved through extraction regulation, calculated to extend the availability of the resource over a given planning horizon. The temporal rates of production for petroleum, coal and natural gas shown in Figs. 2, 3, and 4, for example, represent logical regulatory targets for phasing out these forms of energy. As pointed out by Hubbert, these bell shaped curves provide time for making the technological, economic and land use adjustments necessary to reduce energy consumption and/or shift to other forms of energy.

The indicated annual changes in rate of production might be controlled by institutionally regulating the physical production. Alternatively they might be controlled indirectly by prices that are established through appropriate market mechanisms (such as extraction rights) or by institutionally establishing a new price schedule. The procedures are essentially similar to those already discussed for renewable residuals except that in this case, the allowable rates vary from year to year as illustrated in Fig. 10. The imputed price of the resource as established by market mechanisms would undoubtedly increase from year to year (e.g., BD to FH in Fig. 10) as the hypothetical price schedules indicate. This would be particularly true if energy of all forms becomes relatively less available and more expensive in the future. If the imputed prices are institutionally imposed through taxes, the progressive increase in price with production rate during any given year (e.g., EG to FH in Fig. 10), as well as the progression from year to year, can be based on a variety of other considerations including the uncertainties surrounding the planning horizon and the estimated resource stocks.

The annual production rates indicated in Fig. 10 can be represented as a trajectory on the three-dimensional technical cost surface shown in Fig. 9. The results are shown in Fig. 11 along with the trajectory of price (technical cost of recovery plus imputed price). On these trajectories, time is an independent parameter.

The hypothetical relationships assumed for illustrative purposes in Figs. 9 and 11 would suggest that as a nonrenewable resource such as petroleum is phased out, the price of the resource increases rather precipitously for three reasons:

(1) the increasing entropy of the remaining resource,
(2) reduced cost advantages imposed by monotonic reduction in production rates, and
(3) increasing imputed administered prices that are reflective of the future values to the economy of a vanishing resource.

The past two hundred years represent a unique era in the history of human societies -- depicted as the "down-hill" region of the cost surface illustrated in Fig. 11 for at least some of the critical resources. During this time, control policies have been relatively successful in regulating the functional aspects of the

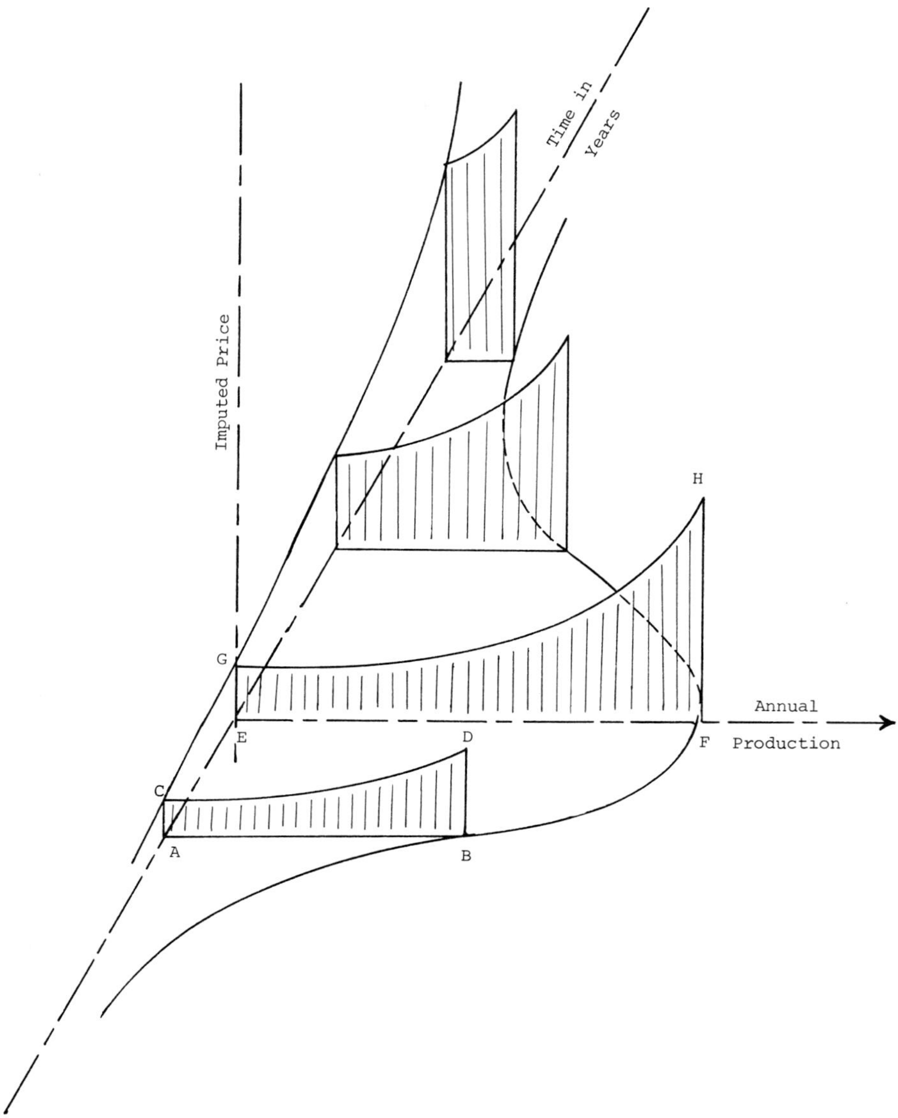

Fig. 10. Administered prices of finite resource as a function of use rate.

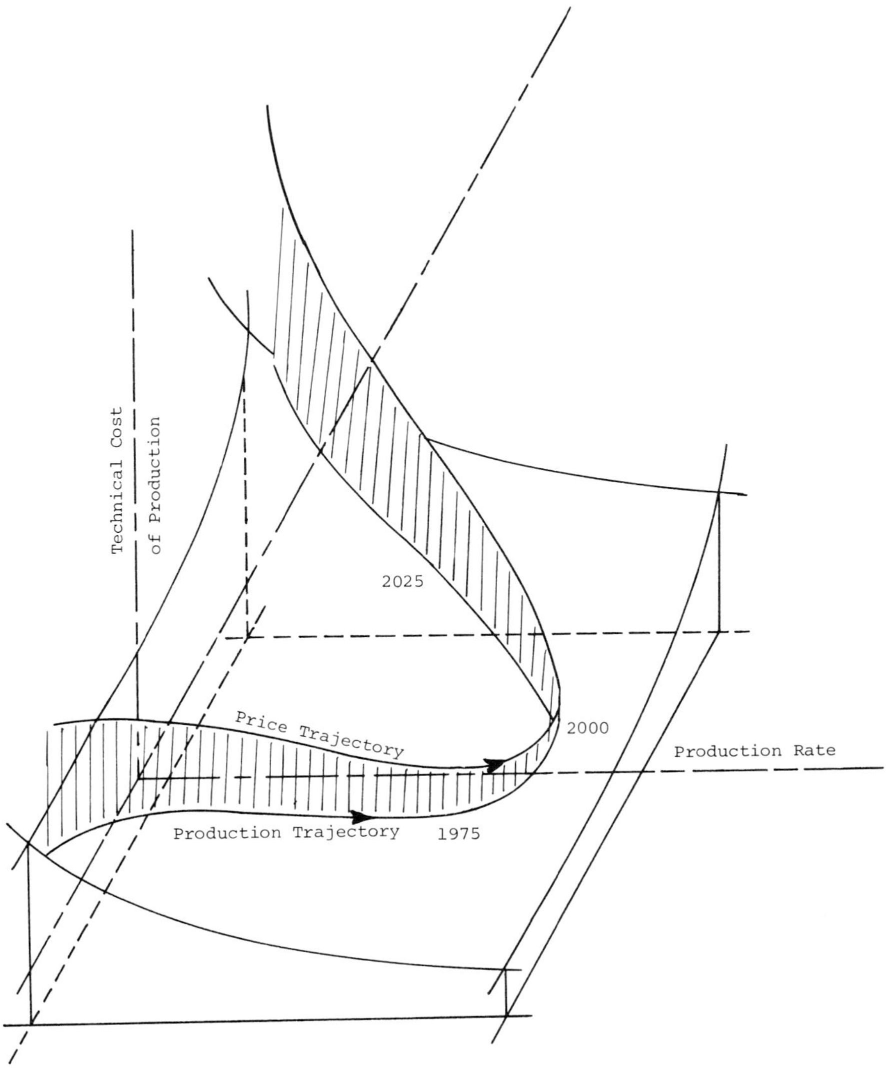

Fig. 11. Administered price as a function of production rate and changing entropy for a finite resource.

economy about an implied direction of evolutionary changes in structure. Present directions of change cannot be sustained indefinitely because of thermodynamic and environmental constraints -- eventually western industrial societies will enter the "uphill" region of Fig. 11. Alternate control and/or structural regulation policies which individually and collectively redirect the path of evolutionary changes in the structure of our economy must be developed.

- Control of Irreversible Land Allocations

From a thermodynamic point of view, arable lands -- including in particular agricultural land and managed forests -- are, in the final analysis, capital stocks for biologically converting a continuous source of thermal energy into economic substances. The thermodynamic efficiency of these biological conversions are potentially as high as many physical processes (photovoltaic cells, solar panels, etc.). Draft animals, for example (with their associated food chain) are perhaps among the most efficient "solar cells" yet developed for converting solar energy into mechanical energy.

Through technological developments in genetic breeding, expanded use of synthetic nitrogen and pesticides, and more effective tillage and irrigation, the per capita arable land allocated to agricultural production in the U.S. has steadily decreased over the past half century, as indicated in Fig. 12. So far, these reductions in per capita arable land requirements more than compensated for increased U.S. population. But as this nation and the world enter a period of reduced growth rates and the eventual phase out of terrestrial energy stocks, demand for arable land will almost surely increase again.

During the past half century when low entropy terrestrial stocks of energy were available as substitutes for arable land, land market processes assigned a relatively low economic value to arable land with concomitant allocations to other use functions --

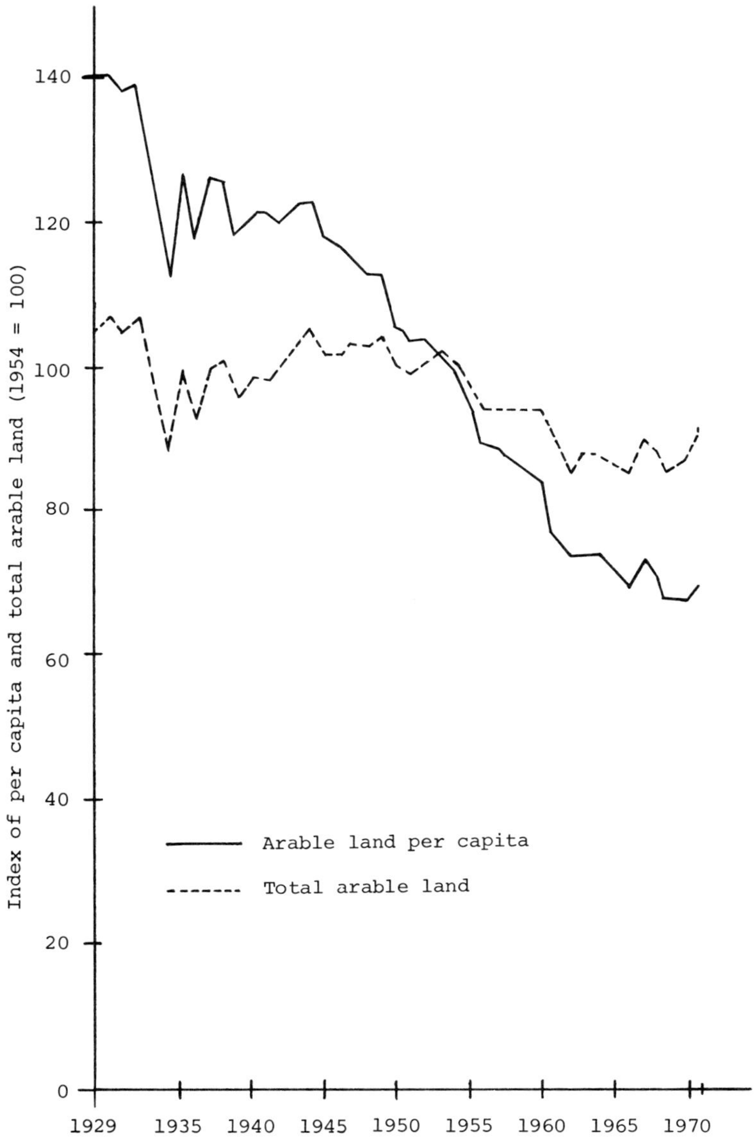

Fig. 12. Arable land per capita and total acreage -- 1929-1971 (1954 = 100). Source [30].

uses which, all too often exploit only the physical properties of the land and in no way utilize its energy-conversion characteristics. The problems of preserving the "surplus" arable land and simultaneously maintaining viable agricultural price levels in the U.S. during this period are well documented by federal and state conservation programs, agricultural price supports, the Soil Bank Act, and many other institutional processes attempting to deal, in one way or another, with an aspect of arable land conservation and price support. In retrospect, preservation of such temporary surplus production stocks within the framework of open market mechanisms has proven to be a very difficult problem for which a totally adequate solution was never found.

In view of the irreversible* land allocations that already have taken place the still rising world populations and the rapidly increasing entropy of the national and world stocks of materials and energy, the problems of land management in the decades ahead extend considerably beyond that of simply holding excess stocks in reserve. How and at what rate should any remaining reserves be brought back into production? What technologies and land-use patterns should be promoted over the next quarter century in rural regions to accommodate and promote the requisite shifts in demographic distributions, employment opportunities, and other factors of urban-rural "balance"?

Existing market allocation processes have been seriously questioned on the basis of tax equity for agricultural production.**

---

* A tract of land, once paved or converted to urban uses, for example, is no longer available for agricultural purposes, at least not within the time frame of rational planning. Such land uses are therefore referred to as irreversible allocations.

**See House Bill 4244, State of Michigan, 1974, for an example of legislation reflecting this concern.

The point of concern, from a thermodynamic and ecological point of view, however, is that they are fundamentally incapable of dealing with irreversibilities. Only when the initial allocations are reversible can normal market mechanisms (perhaps with appropriately managed price structures) reallocate these resources to photosynthesis as the economy becomes increasingly dependent upon solar energy flux. The market structure is essentially a *feedback* process in which correction signals (in the form of price changes) are generated *after the fact.*

Managing (controlling) irreversible allocations of arable land is confounded by the fact that there is no sharp distinction between arable and nonarable land. Further, to a limited extent, irreversible conversions are both necessary and desirable. Indeed, when taken to the extreme, prohibition of irreversible allocations of arable land in the corn belt even now would leave this vast region devoid of further developments in markets, food processing, transportation, and other elements of the infrastructure necessary to support agricultural production. Moreover, it would force the human populations which ultimately consume the agricultural products to settle elsewhere (perhaps in the desert) with implied increases in energy intensiveness (due to the need, for example, of long-distance transportation of food and water, and air conditioning).

The irreversible conversions of arable land can be constrained to a predetermined *rate* schedule by imposing a *conversion cost* on the individual or corporation making the conversion, according to the (hypothetical) schedule indicated by the three-dimensional surface in Fig. 13. The administratively determined conversion rate schedule (in acres per year) for the various quality grades of arable land is shown by the broken line A-B. The "shape" of this curve is determined by the relative availability of the various grades in the geographic region under consideration, and the changes in industrial and demographic features to be anticipated or promoted during the selected planning horizon. Typically,

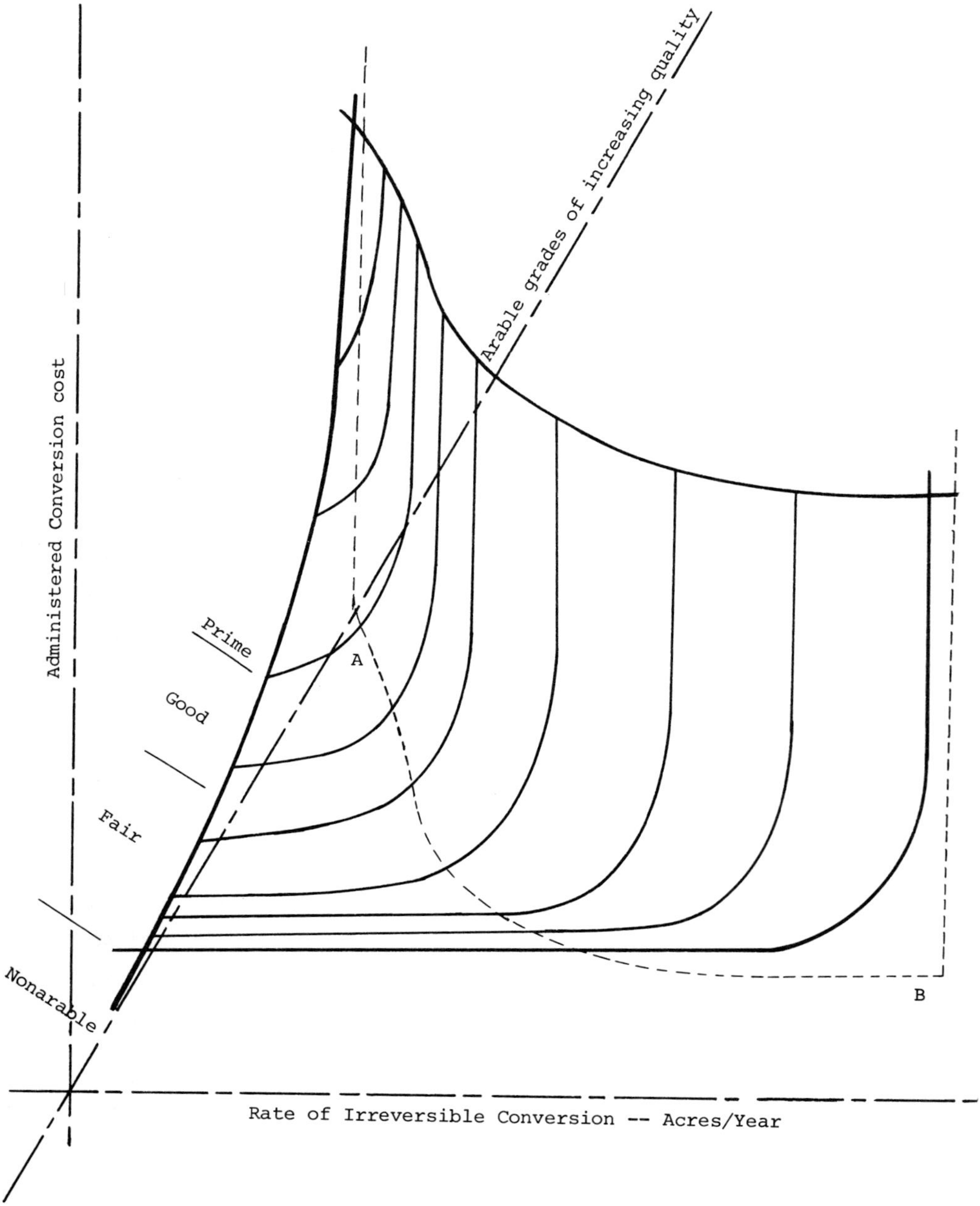

Fig. 13. Administered cost of converting arable land to nonarable land as a function of quality and conversion rate.

the allowable irreversible conversion rates would be less for high grades than the lower grades of arable land with no limit on the nonarable lands as the hypothetical curve indicates.* The administered conversion costs are progressive with respect to quality grade and with respect to the selected conversion schedule as indicated by the three-dimensional surface. In principle, at least, the rate of progression can be determined by open market sale of a given number of conversion rights per year of each grade of land. Alternatively, conversion can be controlled by a tax with the rate of progression determined by the uncertainties associated with the entire planning process, as in the case of residuals and nonrenewable resources.

The pricing instruments illustrated in Fig. 13 internalize into the market processes, the *future* costs of current irreversible allocations, vis-a-vis uncertain growth in world populations, increasing entropy of geophysical stocks and the unknowable outcomes of future technological developments. In this sense, the economic costs associated with maintaining flexibility (keeping the options of the future open) are included as economic weights or signals to the decision processes.

The basic principles, outlined in this section for managing the prices of renewable and nonrenewable resources and residuals and the irreversible conversions on arable land, are formulated to constrain economic activity and land-use patterns within a finite sphere -- a sphere of constraints that is based on a growing knowledge of the laws of thermodynamics and ecology that bound human activities over a given planning horizon. Within this sphere of constraints, market mechanisms, even as we now know them, continue to operate as allocation processes in determining what specific

---

* In principle, it is possible to set the schedule so that the *net* loss in solar energy conversion capability per year is the same at all grade levels, but other considerations would usually (but not always) put priority on preserving the high quality grades.

uses, when, and by whom the resources are utilized. In this sense, they represent *minimum* constraints, as determined by the geophysical and ecological features of the landscape, the political boundaries of the planning region, and the selected planning horizon.

Yet, as with any set of principles, implementation, at best, will involve a variety of embellishments to gain political and institutional acceptance, particularly in face of the inevitable socioeconomic dislocations and adjustments in perceived rights associated with the transition from a system that is essentially unplanned (very short planning horizons) to a system that explicitly takes into consideration the thermodynamic and ecological constraints that delimit our long-term existence. Such a transition is not here perceived as the responsibility of any existing or new central "planning agency" but largely the result of a gradual infusion of the above principles into the management policies of existing agencies within our institutional system -- a process of institutionalizing the control principles.

7. INSTITUTIONAL ASPECTS OF PRODUCTION AND CONSUMPTION

There are perhaps a variety of perspectives from which a working definition of institutions may be developed. From the systems perspective used in this paper (illustrated in Fig. 2), the *institutional system* consists of the entire collection of individuals, groups, agencies and organizations which collectively facilitate, constrain or otherwise give direction to the economic, thermodynamic and ecological characteristics of the systems of production and consumption. It encompasses the laws and legislative actions of the federal, state, regional, and local executive units, as well as the judicial branches of government that adjudicate and enforce legislative action. It includes also the complex of public, private, and voluntary organizations; the market processes;

financial arrangements; and customs and myths. Thus, literally all elements in the universe of concern to resource management, *exclusive* of the purely thermodynamic and ecological aspects, are included in the institutional system.

The various components of the institutional system, each of which is referred to as an *institution* (an individual, a group, an agency, an organization, etc.), ultimately affect land-use patterns -- directly or indirectly -- through the formulation and/or implementation of specific *policy instruments*. The policy instruments may be designed to deal with a broad range of policy objectives.* They may be designed to deal with short-term functional aspects of the economy -- e.g., fiscal and monetary policies -- or they may be focused on a particular aspect of resource management. Thus, the management of the resources and residuals and the irreversible conversions of arable land, like all other policy objectives, are ultimately the responsibility of *individuals* (or groups of individuals) operating within the framework of the institutional system.

In principle, each institution, through its members and representatives, interacts with and is constrained by its *institutional environment* (the other institutions taken collectively) *and* it relates to restricted aspects of the overall production and consumption system. That is, each institution of necessity must focus on a particular aspect of the overall system and formulate policy instruments which can be evaluated only in the context of particular objectives.

---

* Policy *objectives* can be either broad or specific. The Employment Act of 1946 and the National Environmental Policy Act (NEPA) exemplify the former. The goal of preserving agricultural land is an example of a specific objective. Policy instruments (also referred to as control instruments) consist of specific actions (e.g., land use legislation, the application of a specific tax, etc.) which will lead to the fulfillment of the stated objectives.

In the institutional environment, individuals within each institution must perform these inter-institutional functions:

(1) The development and implementation of policy instruments which relate to their sphere of direct responsibility.

(2) The coordination of the sphere of direct responsibility (including formulation and implementation of policy instruments) with other institutions.

(3) The evaluation of the efficacy of current policies -- both in terms of the objectives sought and the effectiveness of the means used in achieving them.

The collective goals and control strategies that emerge from these essential functions constitute what may be called the "intelligence" of the institutional system. The intelligence is that *intrinsic* characteristic of the institutional system that emerges from the interactions between the institutions and which ultimately gives direction to the economic, thermodynamic and ecological characteristics of the systems of production and consumption.

Formulation of policy objectives and their implementation (the intelligence) must be supported by timely flow of critical information describing the physical, ecological and economic behavior of the system. In practice, this information is provided in part through an array of statistical data gathering and processing agencies and organizations and in part by inquiries carried out by the individual institutions.

From a systems theoretical perspective, the institutional system is literally the "control element" that facilitates and/or constrains production and consumption activities -- including physical and biological processes. In principle, this element functions through its capacity to sense "information" about the current state of the system, "process" this information and apply "control signals" to the production and consumption systems that modify behavior and/or structure in subsequent time periods. *Unlike* the mechanical control devices in physical processes, the control logic is not established exogenously, but emerges from the actions *and* interactions of the individual units -- i.e., governmental, judicial, political, economic, educational, etc. -- as an intrinsic characteristic of the overall institutional structure. In this sense, the institutional system is said to have *self-regulating, goal-seeking,* features -- features that are usually attributed to a class of control systems referred to as cybernetic processes.

The self-regulatory features of the institutional system derive primarily from four basic activities carried out continuously or periodically and in sequence as "control cycles".

(1) Assessment of the current state of land use patterns and potential future options,

(2) Establishment and modification of current policy objectives and targets,

(3) Implementation of new objectives and targets by way of actions or functions performed by the responsible institutions, and

(4) Evaluation of the relative success of the outcomes in relation to the policy objective and the institution's specific role in achieving the objective.

In practice, this last process in the sequence is all too often overlooked, and the others are partial rather than complete. Evaluation is an essential element in accounting for the actions (or lack thereof) of particular individuals, agencies, or groups involved in the development and implementation of policy objectives. Moreover, follow-up evaluations of achievements and errors made in any given control cycle provides very valuable information for improving policies, strategies, and tactics. The process of learning from past mistakes provides "feedback" to the decision processes, facilitating the process of learning from past performance. This feedback is a critical aspect in the ongoing control process. The fund of knowledge and "experience" which is built up as a result of this aspect of the control process is a critical element of the intelligence of the institutional system.

In practice, control cycles also involve substantial time lags -- both in terms of the intra-institutional interactions and in the response of the physical and technical features of the landscape to control actions. These time delays are important considerations in formulating and implementing control policies and instruments in the sense that action may be too late to avoid undesirable conditions -- some of which may be irreversible. The components of the institutional system must ultimately develop at least a limited responsibility to predict how alternative control instruments (including the absence of cognizant controls) will in fact affect the behavior and structure of the production and consumption systems and to approximate how long the adjustments will take vis-a-vis the time delays associated with the control cycle of the institutional system. This predictive capability is said to provide "feed-forward" information (predictive analysis) and *components* the feedback information (error analysis) rather than replacing it.

The goal-seeking features of the institutional system derive from the interaction that takes place within and between specific institutions in the form of political haggling, debates, town meetings, scientific investigations, and other forms of communication which result in the formulation of specific goals and actions. Without this active interaction between the components of the institutional system, the institutional system can be little more than an aggregate of individuals and agencies, each selfishly and incongruently pursuing its own goals and expanding its own bureaucratic empire. The active mode of behavior is *the* element of institutional performance upon which democratic institutional systems are most critically dependent. *But* this performance must be guided, sobered and depoliticized by an understanding of the short-term and long-term thermodynamic, ecological and economic characteristics of the overall system, and the manner in which the particular sphere of responsibilities and authorities of the individual relate to these overall characteristics. The *genius* of the institutional system as a self-regulating, goal-seeking allocator of resources (broadly defined) is critically dependent upon this active mode of behavior. The alternative is an institutional structure which inculcates thermodynamic and ecological constraints through a centrally controlled hierarchy of authority.

From a pragmatic point of view, the market system is an important component of the self-regulating, goal-seeking process. It makes detailed allocations of resources within the overall framework of the managed prices (or costs), property rights and other relationships established by various components of the institutional system. For example, the monetary prices of *all* economic substances in the exchange system are determined by the managed costs of the material resources and residuals and the irreversible conversions of arable land discussed in the previous section. Adjustments in these basic resource costs will result in changes in relative prices of virtually all exchange transactions in the economy, with concomitant changes in industrial structure and eventual shifts in resource allocations and land-use patterns.

In reality, all prices are managed directly or indirectly. The question is not one of management or no management, but the criteria upon which the managed prices are based and the policy instruments and control strategies employed. The price of money (the interest rate, for example) is set indirectly by the Federal Reserve System and the price of utilities is set directly by the representative federal utility commissions. In most cases, the present criteria for administering prices has not included consideration of thermodynamic and ecological constraints. Managing the costs of resources, residuals, and irreversible conversions, according to the principles discussed in the previous section, simply extends the criteria of pricing to include these latter factors, and does not negate the importance of existing criteria.

If the prices of resources, residuals and irreversible arable land allocations are managed in accordance with thermodynamic and ecological principles, it is known only that the resource and residual rates and land-use patterns will ultimately fall within the sphere of what is thermodynamically and ecologically feasible over a given planning horizon. Policy instruments based on these principles would not selectively favor or penalize any specific industry or activity. Rather, they would constrain and bound general economic activity. In this sense, they might be viewed as *general* control instruments. This is not to imply that constraining production and consumption activities in the context of thermodynamically and ecologically sound principles will result in neutral impacts. On the contrary, the burdens of adjustment will undoubtedly be distributed unevenly. More selective control instruments -- instruments specific to particular industries, processes, regions, or socioeconomic groups -- must be developed to distribute the transitional costs equitably. Indeed, the socio-political acceptance of general control instruments will depend upon the perceived effectiveness of specific instruments in minimizing socioeconomic dislocations.

8. SUMMARY

Economic development in the industrialized world during the bicentennial period now being celebrated by the United States is commonly taken as the norm upon which essentially the entire world bases its expectations for the future. Yet, it is now understood with increasing clarity that the superindustrialism that developed during this period is based on *finite* stocks of fossil energy and certain raw materials that are critical to an industrialized economy. The finite nature of virtually all geophysical stocks and the laws of thermodynamics that constrain technologies and govern their use ultimately will lead to a rapid reduction in the standard of living as now perceived and measured *unless* there are major adjustments in the basic structure of the economy -- adjustments that improve the thermodynamic efficiency and make more effective use of continuous energy sources. These adjustments cannot change the fundamental principles of thermodynamics that govern the use of finite terrestrial stocks, however, they can greatly reduce the rate at which the entropy of material and energy stocks increases with time. Increases in the rate of extraction of natural gas, petroleum, coal, fissionable uranium, and other terrestrial stocks can only extend the time for making these inevitable adjustments in efficiency at the expense of a rapid increase in the entropy of the remaining stocks and an enormous potential detriment to future generations.

There are three basic ways in which major reductions in per capita energy requirements can be achieved without significantly reducing living standards:

(1) Technological *efficiencies* -- more efficient automobiles, refrigerators, home heating systems, etc.
(2) Increase the life expectancy (durability) of products -- automobiles, refrigerators, homes, clothing -- by capitalizing on technology that already exists and by reductions in style changes.
(3) More energy efficient *land-use patterns* -- more effectively manage the size and location of urban communities and industrial centers in relationship to agricultural production and recreation regions.

Adjustments in the first category can take place *within* the general framework of our present economy and life styles. Adjustments in the latter two categories require alterations in the basic *structure* of the economy.

Throughout most of the period of industrialization, expanded production (much of which occurred through planned obsolescence and induced consumerism) has been used as the primary means for providing employment opportunities and maintaining economic stability. Transition to more durable products and/or a reduction in the *rate of growth* of energy consumption (however motivated) may require fundamentally new instruments for providing employment opportunities. Market incentives, which promote thermodynamically efficient land-use patterns, can be achieved, at least in principle, by administering the prices of resources and residuals. Similarly, irreversible conversions of arable land can be managed by progressive taxation or the sale of conversion rights. Although such adjustments may be perceived as a contraction of "freedoms", they are inescapable elements of a thermodynamically based economy.

REFERENCES

1. Ayres, R.V., and Kneese, A.V., "Production, Consumption and Externalities", *American Economic Review*, Vol. 59 (1969).
2. Borgstrom, Georg, *The Food and People Dilemma*, Duxbury Press, North Scituate, Mass. (1973).
3. Connor, L.J., *et al.*, "Beef Feedlot Design and Management -- A Progress Report", Michigan State University, E. Lansing, Michigan, Agricultural Economics Staff Paper (73-74) (Sept. 1973).
4. Cook, E., "The Flow of Energy in an Industrial Society", *Scientific American*, Vol. 224, No. 3 (September 1971), pp. 134-144.
5. Dales, J.H., *Pollution, Property, and Prices*. U. of Toronto Press, Toronto (1968).
6. Edens, T.C., "Economic Growth", Letters: *Science*, AAAS, Vol. 189, No. 4201 (August 8, 1975), pp. 410-11.
7. Fergeson, C.E., *Microeconomic Theory*, Rev. Ed., Irwin Co. (1969).
8. Forrester, J., *World Dynamics*. Wright Allen Press, Cambridge, Mass. (1971).
9. Freeman, S.C., *et al.*, "Exploring Energy Choices", Energy Policy Project of the Ford Foundation, Washington, D.C. (1974).

10. Galbraith, Kenneth, *The New Industrial State*. Houghton Mifflin Co., Boston (1967), pp. 212-215.
11. Georgescu-Roegen, N., *The Entropy Law and the Economic Process*. Howard University Press, Cambridge, Mass. (1971).
12. Georgescu-Roegen, N., "Energy and Economic Myths", *Southern Economic Journal*, Vol. 41, No. 3 (January 1975), pp. 347-381.
13. Goldsmith, E., *et al.*, *Blueprint for Survival*. Houghton Mifflin Co., Boston, Mass. (1972), Figure 1, p. 7.
14. Hardin, Garrett, "The Tragedy of the Commons", *Science*, Vol. 62 (Dec. 1968), pp. 1243-1248.
15. Holdren, J.P., "Man as a Global Ecological Force", Cal. Tech. Popula. Progr., Occasional Papers, Series 1, No. 2, Cal. Inst. of Tech., Pasadena (1973).
16. Hubbert, M. King, "The Energy Resources of the Earth", *Scientific American* (Sept. 1971), pp. 61-70.
17. Hueckel, Glenn, "A Historical Approach to Future Economic Growth", *Science*, AAAS, Vol. 187, No. 4180 (March 14, 1975), pp. 925-931.
18. Koenig, H., Cooper, W., and Falvey, J., "Engineering for Ecological, Sociological, and Economic Compatibility", *IEEE Transactions on Systems, Man, and Cybernetics*, Vol. SMC-2, No. 3 (July 1972), pp. 319-331.
19. Koenig, H., Edens, T., Cooper, W., "Ecology, Engineering, and Economics", *Proceedings of the IEEE*, Vol. 63, No. 3 (March 1975), pp. 501-511.
20. Koenig, H., and Tummala, R., "Principles of Ecosystem Design and Management", *IEEE Transactions on Systems, Man, and Cybernetics*, Vol. SMC-2 (Sept. 1972), pp. 449-459.
21. Krauzbert, Melvin, "Can Technological Progress Continue to Provide for the Future?", in Weintraub, A., Schwartz, E., and Aronson, J. (Editors), *The Economic Growth Controversy*, I.A.S.P., New York (1973), pp. 62-81.
22. Meadows, D., *et al.*, *The Limits to Growth*. Universe Books, New York (1972).

23. The Office of Energy Research and Planning, Office of the Governor, "A Report of the Oregon Energy Council" (January 1, 1975).

24. Sawyers, Larry, "Urban Form and the Mode of Production", *URPE*, Vol. 7, No. 1 (Spring 1975), p. 55.

25. Snell, Bradford C., "American Ground Transport", Presented to the Sub-committee on Antitrust and Monopoly of the Committee on the Judiciary, U.S. Senate (February 26, 1974), p. 32.

26. *Synthetic Fuels*, "Energy Forecasts" (December 1974), p. 1-1.

27. T.I.E. (The Institute of Ecology), *T.I.E. Report*, Vol. 4, No. 4 (May 20, 1975).

28. Toffler, A., *Future Shock*. Random House, New York (1970), p. 16.

29. Tummala, R., and Connor, L., "Mass-Energy Based Economic Models", *IEEE Transactions on Systems, Man, and Cybernetics*, Vol. SMC-3 (Nov. 1973), pp. 548-555.

30. U.S.D.A., *Agricultural Statistics* (1972).

ECONOMIC ACTIVITIES UNDER PUBLIC REGULATION —A THERMODYNAMIC ANALOGY

NIELS C. LIND
Solid Mechanics Division
University of Waterloo
Waterloo, Ontario, Canada

SUMMARY

The optimum state of control of a system of economic activities subject to public control in the interest of maintaining safety, environmental cleanliness or other common objectives, is expressed in two postulates formally similar to the axioms of thermodynamics of general physical systems. Several extremum principles are derived.

1. INTRODUCTION

Risk is an important element in most human activity. The calculus of probability has been developed as a tool for making rational decisions in situations involving risks when all outcomes can be assigned (a) a probability and (b) a value on a common scale. In situations that involve uncertainty it may be possible only to assign a "subjective" probability measure; decisions in such contexts are "rational" in the sense of being consistent with the state of belief expressed in the probability measure. All practical decisions (with the possible exception of certain parlor games) involve subjective probability, and can be rational only in this narrow sense.

In many decision problems the outcomes cannot easily be assigned values on a common scale, as when loss of life or cultural values, or the demands of a clean or beautiful environment, are involved. In such cases decisions are often made in an intuitive fashion; if a sufficient number of such decisions are available it may be possible to uncover the scale of values they reflect. This scale may or may not be constant in time, so that it may be difficult to ascertain whether or not they are "rational".

When public safety (rather than the individual decision maker's safety) is involved, the element of judgement in the decision is usually constrained by safety regulations imposed by public authority. An example is a building code regulating some of the activities in the construction sector of the economy. Safety regulations are legal provisions and must, as such, reflect a stable set of values acceptable to the majority of the citizens. Similar statements can be made about regulations aiming to protect the environment for the sake of cleanliness, health or aesthetics.

This paper is concerned with the question faced by the regulating authority: How should the safety provisions (for example) be written in order that they impose a safety level appropriate to society as a whole? Some guiding principles are proposed and formalized into two postulates. These postulates constitute an ideal system with the same mathematical structure as the basic postulates of equilibrium thermodynamics of general physical systems in the form proposed by Callen [2], so that there is a far reaching analogy between elementary thermodynamic equilibrium of a physical system and the optimal state of equilibrium of regulated activities. With the addition of a third postulate, the "Axiom of Archimedes" [1] expressing a common value scale ("Everything has its price") for all accessible states of the system, the problem would reduce to a classical microeconomic problem that can be solved by optimization. Without this axiom, however, each regulated activity can be internally optimized, and the basic safety (or cleanliness or aesthetic) level decision for the activity is then made separately on a political plane.

2. REGULATED ACTIVITIES

Consider an autonomous society with a classification imposed on all transactions of the economy, defining a set of *activities* i {1,2, ...} (or *sectors*, or *industries*) producing and consuming goods and services. Some activities produce outputs that may perform at different levels of quality, or to different degrees of satisfaction depending on the amount of resources spent in the production; some of these, the *regulated activities* (i = 1, ...,n), are moreover, subject to regulation in the public interest. The regulating authority may be an arm of government, or a manufacturer's association, etc., that prescribes a standard norm, or code, or regulations for production. It is assumed that the economy as a whole is in static equilibrium and sufficiently large, such that the prices of input (goods and services) required by the regulated activities can be taken as independent of time and demand with acceptable accuracy. In other words, the *free activities* i = n + 1, n + 2, ..., constitute a constant price reservoir. Attention in the following is concerned mainly with the sub-economy of regulated activities subject to requirements of safety.

Each regulated activity produces q commodities in quantities N_1, N_2, ..., N_q per unit of time, called *production rates*. These variables are assumed known, i.e., measurable and controllable. The total monetary value consumed in the creation of these commodities per unit of time is called the *investment rate* U. It is similarly assumed measurable. It is, moreover, subject to the conservation principle called Walras' Law [3].

Parameters that have values in a composite system equal to the sum of the values in each of the subsystems are called *extensive* parameters. Investment rate U and production rates N_i are extensive parameters. The regulations specify certain numerical values (e.g., tolerances, safety factors, allowable response or design loads) called the *regulation parameters* P_0, P_1, ..., P_m. The price of each commodity depends on the unit price and the amount of the commodities consumed from the free activities in the production, assumed constant. The amount of these commodities necessary in the production depends on the regulation parameters and the technology. Technological improvements will reduce the amount consumed and hence reduce the price of the regulated activities. If, as an idealization, the technology is assumed static, the investment rate is a function only of the production rates and the regulation parameters:

$$U = U(P_0, \ldots, P_m, N_1, \ldots, N_q). \tag{2.1}$$

It is convenient to introduce the investment rate as a state parameter. Assuming that $\partial U/\partial P_0$ is nonzero, (2.1) may be inverted and written

$$P_0 = P_0(U, P_1 \ldots, P_m, N_1, \ldots, N_q) \tag{2.2}$$

A real economic system changes in time through a succession of states in a very complicated fashion. But with the assumption of static technology and free activities in equilibrium, it is assumed as a simplifying abstraction (ultimately to be verified by comparison with the performance of real systems) that the system of regulated activities possesses certain static states:

Postulate 1. There exist particular states (called equilibrium states) of the system of regulated activities that macroscopically are characterised completely by the investment rate U, the regulating parameters P_1, ..., P_m, and the production rates N_1, ..., N_q.

The production rates are, of course, required to be nonnegative. The regulating parameters may, similarly, be subject to constraints. The states that satisfy all these constraints are called *feasible equilibrium states*.

We consider in the following the general problem of how the regulated activities should be governed for best performance. First, criteria are needed to evaluate performance and the cost of implementing it. These criteria, reflecting the values of society are discussed below in the section on performance criteria. Performance in practice depends in a random and unknown fashion on input, so that the problem has two parts: identifying the ideal state of control, and finding the best way to approach this state from another state. It is therefore a dynamic problem in optimum control of an adaptive system. We shall here merely consider the static problem of identifying the optimum state of control.

3. THE LOSS FUNCTIONAL

3.1 Performance criteria

In order to characterize the optimum state of control of the system, the outcomes of the control decisions must be rated according to performance. When a ranking of the states is possible, for example in terms of utilities, the rating problem is trivial. Assume that this ranking is not possible. Then, it may still be possible to develop a ranking merely of the possible losses that are envisaged for the system. The elements of such a rating are developed in considerable detail in this section and applied within a theoretical framework in subsequent sections.

The outcomes are aggregates of goods or services produced, with satisfactory service performance realized to a degree that varies in time, varies with the individual product and has a random component due to uncontrollable factors. It is reasonable to assume that "satisfactory performance" is a well defined concept for each product, and it is convenient to characterize performance of a unit in terms of a loss relative to satisfactory performance.

It is convenient to discretize the time domain by division into periods of constant length. Consider a unit completed in time period t_0 with a given design life. When the design life is completed, the unit must be taken out of servie or replaced; this terminates the life history of the unit. We seek to quantify the loss incurred due to malperformance during any period $t > t_0$ of this history. Loss events may be instantaneous (e.g., structural collapse) or distributed in time (e.g., pavement failure); to avoid the tedium of maintaining this distinction, distributed losses are discretized into lumped variables assigned to the end of each time period. Malperformance is often considered as the result of a gradual deterioration which, however, in itself does not constitute a loss. For example, cumulative damage is conceived as the process leading to metal fatigue fracture. Such accumulated damage would not be counted as a cumulative loss. It represents in effect only a loss potential.

3.2 Loss

There are many difficulties associated with the quantization of loss. First, there may be uncertainty in the notion "satisfactory performance", as in the case of loss due to excessive wind sway of a tall building. In some cases, such uncertainty has been resolved by recourse to a panel of experts assessing performance (quality of food products or highway pavements, for example). These are difficulties of more practical statistics; in principle there is no difficulty in defining "satisfactory performance" with adequate precision. This is obvious for systems subject to formal performance specification. Systems for which the definition is impossible would not seem amenable to rational analysis. Moreover, their alleged malfunction would often lead to inefficient legal disputes, so such systems are generally avoided in a well-organized society and they are left out of the discussion here.

Second, the spectrum of possible malperformance is infinite. For example, no two accidents are alike in human, material or social impact. Superficially we may characterize an accident by a few statistics: number of fatalities, number of injuries, amount of material loss and so on. However, we recognize that the number of modalities of loss is actually infinite. For example, the loss of an eye is not in any sense commensurate with the loss of a limb. Furthermore, when it comes to characterizing the performance of an entire activity, or system of activities, a description of each loss in complete detail would provide far too much system state information to be manageable. For practical reasons microscopic detail is dispensed with and the macroscopic performance is expressed in terms of a few selected *loss variables*. Loss descriptors that survive in the transition from the microscopic part to macroscopic system are called the *socio-economic loss variables* of the regulated activities. An example of this process is evidenced in the American National Standard for measuring and recording work injury experience [10].

We therefore assume that it is possible to identify a finite number M of relevant *modalities* (components of socio-economic loss) of malperformance for a given system of regulated activities. For example, these might be: number of fatalities, number of severe injuries, number of minor injuries, replacement cost, and time out of service, all per period.

Loss events are usually quite rare, so that the question of what constitutes a single loss event can be answered with precision. For example, five air traffic accidents, each involving 100 casualties, would be perceived as separate loss events, whereas the same total number of casualties in automobile traffic in a single weekend is perceived (by the general public) as a single loss event - of course with a different value scale. Following Mauch and Schneider [7] it is presumed that the notion of "a single loss event" is well defined.

This establishes the loss statistic in time $\underset{\sim}{\ell}_j(t)$ of event j in time period t as an ordered set of M numbers

$$\underset{\sim}{\ell}_j(t) = \{\ell_1, \ell_2, \ldots, \ell_k, \ldots, \ell_m\}j . \qquad (3.1)$$

It may be assumed that such losses can be translated in time. Strictly economic components can simply be capitalized to a common point in time and added. Cultural or human losses occurring at different times would likely be assigned a constant value, reflecting a more permanent basis of value. The convention is adopted that each component of $\underset{\sim}{\ell}_j$ is thus capitalized to the same point in time, and the explicit reference to the time of the loss is dropped.

It is further assumed that there exists a common ratio scale of values for all loss events within each modality. This scale can be represented in the form of a loss function f, such that $L_{kj} = f_k(\ell_{kj})$ is the component in modality k of loss event j with loss statistic $\underset{\sim}{\ell}_j$. Loss functions are generally nonlinear. For example, one major accident involving 10 fatalities is perceived as more serious than 10 single-fatality accidents. The assumption means, however, that it is meaningful to compare losses within each modality. It means, moreover, that such losses can be discussed out of context. Thus, there is no induction of loss from one modality to another (readily acceptable), and the loss is dependent only on the loss statistics and not on the particular activity in which it was incurred. This establishes *loss* $\underset{\sim}{L}$ as an M-dimensional vector over the field of real numbers

$$\underset{\sim}{L} \equiv (L_1, L_2, \ldots, L_M) = (f_1(\ell_1), f_2(\ell_2), \ldots, f_M(\ell_M)). \qquad (3.2)$$

Decisions are made on the basis of hypothetical (possible) losses rather than real losses after the fact. With each loss $\underset{\sim}{L}_j$ is associated a probability p_j and an expected value $p_j\underset{\sim}{L}_j$. Risk affinity or risk aversion may well occur [1], and may be different in different modalities. In each modality, however, it is a function only of the loss statistic and may be considered as incorporated in the loss function f. Decisions can therefore be made on the basis of *expected loss* R_i (an M-vector) in each activity i. Since loss events are mutually exclusive, we have

$$\underset{\sim}{R}_i = \sum_j p_j L_j, \qquad (3.3)$$

in which the summation extends over all conceivable loss events associated with activity i.

3.3 Loss functional

The next step is to define a scalar measure F of loss for each activity. This scalar is a function of the expected loss and is therefore a functional on the linear vector space E^M. To show that the measure of loss is a linear functional, observe that the activity may conceptually be divided arbitrarily into two new activities. The envisaged losses may be assigned arbitrarily to the new activities, each governed by the same functional, without affecting the total loss from the viewpoint of society. Thus, an arbitrary partitioning of R into $\alpha R_1 + \beta R_2$ gives the requirement

$$F(\alpha R_1 + \beta R_2) = \alpha F(R_1) + \beta F(R_2) \qquad (3.4)$$

which establishes that F must be a linear functional. It is assumed continuous at a point in the range of loss for the activity, and is therefore continuous everywhere.

The scalar loss functional F can be represented by a vector of the specific values attached to the loss per unit in each modality. It amounts to a utility ordering of the loss spectrum for the activity. In practice, the range of losses within an activity is limited. Moreover, loss statistics within an activity tend to be proportional. That is, material losses are correlated with the number of fatalities, as are the numbers of casualties. The existence of a loss functional for an activity means merely that those responsible for its regulation have a self-consistent view of the severity of its possible malfunctions.

On the other hand, there seems to be a conflict in the value of the loss functional in between activities. A particular loss (for example 100 dead with $\$10^6$ material loss) receives different attention in different activities (e.g., in defense combat, in a commercial scheduled flight, or in the collapse of a terminal building). On closer analysis it seems, though, that the more "preventable" an accident is, the rarer it is, the higher its news content, and the higher the response if it occurs. This is in qualitative accord with the law of equal marginal returns: It is so relatively cheap to make the terminal building "perfectly" safe, that it is expected to be safer. The conflict in the loss functionals is probably more apparent than real. In a just society [8] the value placed on human life should not depend on the activity or any other loss attribute of the individual. The value of the economic components of loss are automatically equilibrated in a competitive economy. This leads to the assertion (perhaps more normative than descriptive) that the same loss functional F operates in all activities.

4. ECONOMIC EQUIVALENT OF LIFE

The question of "the value of a human life" cannot be avoided in the present formulation, although it is relegated to a relatively minor importance in situations such as the present context, where a direct exchange of lives for dollars is not in question.

There has been much effort expended in recent years to establish a measure of this equivalent. In some contexts, the value of a life is reflected by a negative price (war, family planning). The concern here is only with the hypothetical question of: "What is the amount we are willing to forego at a time in the future to prevent the loss of one life at the same time?" Compare Thoreau [9]: "The cost of a thing is the amount of what I will call life which is required to be exchanged for it ...".

Experience shows that the value of a dollar in these terms is finite and, depending on the context is between 2×10^{-5} and 2×10^{-6} lives. This value is to be interpreted in the sense that the decision maker will take equal preventative measures, in the interest of those exposed, against the loss of one life or of \$50,000 to \$500,000. The value is clearly not today's economic equivalent of that loss (except perhaps in a completely Marxistic society) since the loss will in both cases be private. While the question if far from resolved, it will be sufficient for the purpose of this paper to accept the existence of the economic equivalent on the basis of these bounds.

5. SAFETY AND SECURITY

We define the *safety* S^* of a regulated activity as the negative of the loss functional F. Safety is an extensive function of state. Evidently, if the states of a compound activity represent the same total production at the same total value input, but have different levels of safety, the state with the higher safety S^* would be preferred because the expectation of the loss is less. One state could be obtained from the other by relaxation of one or several of the control parameters, with a redistribution of investment rate or productions within the compound systems, that is by relaxation of an internal constraint of the state variables. All the information necessary to find an optimum state for a system with controlled rates of investment and production for a compound system of activities is contained in the safety function. The optimum state may not be unique. For example, there may be some of the parameters P_i that do not influence the safety; they may be redundant or irrelevant. We leave such parameters out of the discussion. Formalizing, this leads to the following postulate of rational preference.

Postulate 2. There exists a function (called the Safety S)* *of the investment rate U, the control parameters P_1, ..., P_m, and the production rates N_1, ..., N_q, defined for all feasible equilibrium states, with the property that the value it assumes in the optimum state in the absence of an internal constraint is maximum over the manifold of constrained feasible equilibrium states. Safety is additive over the constituent sub-activities, continuous and differentiable in all its arguments, and monotonically increasing with investment rate U.*

The safety function may be expressed as:

$$S^* = S^*(U, P_1, \ldots, P_m, N_1, \ldots, N_q). \tag{5.1}$$

The postulate of a maximum implies that $d^2S^* < 0$, which implies in particular that the determinant $|\partial^2 S^*/\partial P_i \partial P_j|$ is nonzero. Thus it is possible to perform a Legendre transformation on S* with respect to the variables P_i. This introduces the new variables $V_i \equiv \partial S^*/\partial P_i$, which are extensive. The partial Legendre transform of safety with respect to the parameters P_1, ..., P_m, is called the *security* S:

$$S = S^* - \sum_1^r V_i P_i = S(U, V_1, \ldots, V_m, N_1, \ldots, N_q). \tag{5.2}$$

This expression is called the *security fundamental equation* of the system, involving only extensive variables. An equivalent *investment rate fundamental equation* can be derived by solving (5.2) for U:

$$U = U(S, V_1, \ldots, V_m, N_1, \ldots, N_q). \tag{5.3}$$

This formulation is suitable for some contexts. The additivity property when applied to conceptually distinct subsystems implies that security is a homogeneous first-order function of the extensive parameters:

$$S(\lambda U, \lambda V_1, \ldots, \lambda N_q) = \lambda S(U, V_1, \ldots, N_q). \tag{5.4}$$

A similar relation holds for the investment rate, (5.3).

6. MAXIMUM SECURITY PRINCIPLE

Another implication of the postulate of maximum safety is that the security is also a maximum over the manifold of constrained values of its arguments. To see this, it is enough to observe that to an arbitrary line element *a* through state characterised by $(U,V_1,V_2, \ldots, V_n, N_1, \ldots, N_q)$ there corresponds a unique line element *b* through the corresponding point in $(U,P_1,P_2, \ldots, P_m, N_1, \ldots, N_q)$ because $|\partial V_i/\partial P_j|$ is nonzero. Over any line element b, S* is maximum at optimum. But S* and S are equal at corresponding points of the two line elements. It follows that S is maximum at optimum on line element *a*.

7. MINIMUM INVESTMENT RATE PRINCIPLE

From the maximum security principle, an extremum principle for the investment rate can be stated as follows. The optimum value of any of the parameters $V_1, \ldots, V_m, N_1, \ldots, N_q$ is such as to minimize the investment rate for a given value of the security. To prove this, assume for the sake of argument that the investment rate is not a minimum consistent with the security. The investment rate could then be lowered at constant security. Now, let the investment rate be restored to its original value, necessarily increasing the security since $\partial S/\partial U$ is positive. This has led the system to a neighbouring state of higher security inconsistent with the maximum security principle.

8. ALTERNATIVE REPRESENTATIONS

If a system of regulated activities is represented in the analysis by (5.2), the analysis is said to be in the *security repre-*

sentation, while it is said to be in the *investment rate representation* if (5.3) is used. Each can be considered as a fundamental representation from which other representations are derivable by partial Legendre transformations. By arguments similar to that employed in proving the security maximum principle from that of the safety, it can be shown that any partial Lengendre transform of the security is subject to a maximum principle and any partial Lengendre transform of the investment rate is subject to a minimum principle.

9. METHOD OF ANALYSIS

The problems of regulated activities can often be cast into the following general form. Let s activities be specified by their fundamental equation in the form

$$Y_i = Y(X_1, \ldots, X_r), \quad i = 1, \ldots, s \tag{9.1}$$

In the presence of a set of constraints, let the activities be in equilibrium and optimized in the state characterised by the values X_i^1, X_i^2, ..., X_i^r, $i = 1, \ldots, s$. Now, let some of the internal constraints be relaxed, and find the new state variables of the compound system that minimizes or maximizes the state function according to the governing principle.

For example, let two activities a and b produce q commodities at fixed rates. Let it be desired to change the investment rates U^a and U^b (by adjusting the parameters $P_o{}^a$ and $P_o{}^b$), keeping the overall investment rate unchanged, in order to maximize the safety. The activities are governed by

$$S^{*a} = S^{*a}(U^a, P_1^a, \ldots, P_m^a, N_1^a, \ldots, N_q^a) \tag{9.2}$$

$$S^{*b} = S^{*b}(U^b, P_1^b, \ldots, P_m^b, N_1^b, \ldots, N_q^b) \ . \tag{9.3}$$

Since safety and investment rate are extensive, they are for the compound system

$$S = S^{*a} + S^{*b}, \tag{9.4}$$

$$U = U^a + U^b. \tag{9.5}$$

By (9.5), $dU^a = - dU^b$. Differentiating (9.4) and inserting, gives the optimality condition $dS^* = 0$ in the form

$$dS^* = \left(\frac{\partial S^{*a}}{\partial U^a} - \frac{\partial S^{*b}}{\partial U^b}\right) dU^a, \tag{9.6}$$

yielding

$$\frac{\partial S^{*a}}{\partial U^a} = \frac{\partial S^{*b}}{\partial U^b}. \tag{9.7}$$

This expression shows that the marginal cost of safety must be the same for the two systems at optimum. For fixed parameters P_i and production rates N_i, (9.7) gives a relation between U^a and U^b which can be solved in conjunction with the conservation condition $U^a + U^b = U = \text{const}$.

As another example, the two activities may be governed by a common parameter $P_1^a = P_1^b$ and operate at prescribed safety levels S^{*a} and S^{*b}. How should the parameter be adjusted to minimize the total investment rate $U^a + U^b$? In the investment rate formulation

$$U^a = U^a(S^{*a}, P_1, \ldots, N_q^a) \tag{9.8}$$

$$U^b = U^b(S^{*b}, P_1, \ldots, N_q^b). \tag{9.9}$$

The optimality condition for $dP_1^a = dP_1^b$ gives $dU = 0$ as

$$\frac{dU^a}{\partial P_1} + \frac{\partial U^b}{\partial P_1} = 0. \tag{9.10}$$

The unknown, P_1, can be determined from this equation. If, on the other hand, the prescribed safety levels S^{*a} and S^{*b} are also relaxed subject to the total safety constraint

$$S^{*a} + S^{*b} = S^{*0}, \tag{9.11}$$

(9.10) would be supplemented with

$$\frac{\partial U^a}{\partial S^{*a}} - \frac{\partial U^b}{\partial S^{*b}} = 0. \tag{9.12}$$

The three unknowns S^{*a}, S^{*b} and P_1 are now to be determined from (9.10) - (9.12).

Other types of problem are readily identified and solved in the same general way. For example, maximizing the safety for a prescribed total investment rate by adjusting allocations of production between activities ($N_1^a + N_1^b + \ldots = N_1$), or minimizing the required investment rate for an activity with prescribed safety through adjustment of the regulation parameters, and so on.

10. THE THERMODYNAMICS ANALOGY

An axiomatic presentation of equilibrium thermodynamics of general physical systems has been given by Callen [2]. His postulates for a "simple system" reduce to:

I. There exist particular states (called equilibrium states) of simple systems that macroscopically are characterised completely by the internal energy U, the volume V, and the mole numbers N_1, N_2, ..., N_q of the chemical components.

II. There exists a function (called the entropy S) of the extensive parameters of any composite system, defined for all equilibrium states and having the following property. The values assumed by the extensive parameters in the absence of an internal constraint are those that maximize the entropy over the manifold of constrained equilibrium states.

III. The entropy of a composite system is additive over the constituent sub-systems. The entropy is continuous and differentiable and is a monotonically increasing function of the energy.

IV. The entropy of any system vanishes in the state for which $\partial U/\partial S = 0$.

Postulate I corresponds to the first postulate in this paper, if "energy" is taken for "investment rate", and "mole numbers" for "production rates". The "volume", an extensive quantity, is in the thermodynamics of solids replaced by (six) geometric extensive parameters V_1, V_2, ..., (elongations and angular dilatations), each analogous to volume. The analogy is therefore complete if a state specification in terms of $(U, V_1, \ldots, V_m, N_1, \ldots, N_q)$ is equivalent to one in terms of $(U, P_1, \ldots, P_m, N_1, \ldots, N_q)$. This is, however, assured because $|\partial V_i/\partial P_j|$ is nonzero.

Postulates II and III specify properties of the entropy that are analogous to postulate 2. It corresponds to similar statements in terms of security that are equivalent to the safety postulate 2.

Postulate IV has no counterpart in the present development. It plays no central role in thermodynamics and its counterpart seems unnecessary here.

The idea of an analogy between economics and a physical system is not new. 50 years ago, P.O. Pedersen presented an electrical network model of an economy in which commodities are manufactured in an activity (represented by a voltage source in series with a self-induction) for either consumption (a resistor) or inventory (a capacitance) [4]. Franksen [5] has presented a competitive market economy in the general terms of a system of two-terminal components and has also earlier [4] presented a thermodynamical analogy of this economy. His analysis is in terms of the first and second laws of thermodynamics, providing an interesting insight into the underlying analogy on a microscopic scale, but it could be recast in axiomatic form analogous to the present paper.

The existence of such analogies can be exploited in two different ways. Going from the general to the particular, specific results (such as Maxwell's relations) from one system can immediately be developed for application in the other field, or analog computers can be built to solve specific problems. Going the other way, a more concise and manageable formulation of systems become possible. Thus, it is possible that an abstract general thermodynamics theory could be developed with many essentially different realizations in economics and other fields, including the realm of classical thermodynamics. Both of these developments are outside the scope of the present paper.

11. CONCLUSIONS

1. The objective of regulating a system of economic activities in the public interest of safety is expressed in terms of two postulates. The first postulate defines the state variables for a subset of states, called equilibrium states, to which the attention is restricted. The second postulate expresses the principle of maximum safety, a function of (equilibrium) state.

2. The principle of maximum safety is shown to have many equivalents, for example principles of minimum investment rate or maximum security.

3. The two postulates are formally equivalent to a set of postulates for thermodynamic equilibrium of physical systems, in the form presented by Callen [2].

4. The safety functional is derived from the risk of losses by the introduction of the loss statistic (an ordered set of loss descriptors); a loss function for each element of the loss statistic (generally nonlinear, expressing aversion towards more serious losses); and a value vector, ordering the set of expected losses. Derivation of this functional in practice is illustrated by an example involving fire prevention and structural fire protection in an earlier version of this paper [6].

5. In contrast with classical presentations of thermodynamics, the presentation in Callen's postulates makes no references to processes. It is, however, a condition of application that the

states are approached quasistatically. The duration of real processes must be long in comparison with the relaxation time. For systems of regulated activities, the relaxation time is in the order of a few years, namely the time of implementation of changes in the regulations.

12. ACKNOWLEDGEMENT

This study was made in the course of a project sponsored by the National Research Council of Canada.

REFERENCES

1. Borch, K.H., *The Economics of Uncertainty*, Princeton University Press, New Jersey, 1968.
2. Callen, H.B., *Thermodynamics*, John Wiley and Sons, Inc., New York, N.Y., 1960.
3. Dorfman, R., Samuelson, P., and Solow, R., *Linear Programming and Economic Analysis*, McGraw-Hill Book Co., New York, N.Y., 1958.
4. Franksen, O.I., "Mathematical Programming in Economics by Physical Analogies", *Simulation*; Part I, No. 6, June 1969, pp. 297-314; Part II, No. 1, July 1969, pp. 25-42; Part III, No. 2, August 1969, pp. 63-87.
5. Franksen, O.I., "Basic Concepts in Engineering and Economics", paper presented at the Meeting of the Heads of Control Laboratories in North-Western Europe, Copenhagen, May 21, 1970.
6. Lind, N.C., "Optimization of Economic Activities under Public Regulation", University of Waterloo *Solid Mechanics Division Report No. 27*, Waterloo, Ont., April, 1974.

7. Mauch, S.P., and Schneider, T., "Die unmittelbare Gefährdung unseres Lebensraum" , *Schweitzer Archiv*, 1971.

8. Rawls, J., *A Theory of Justice*, Harvard University Press, Cambridge, Mass., 1972.

9. Thoreau, H.D., *Walden*, Boston, 1854.

10. USAS Z16.1, American National Standards Institute: *Method of Recording and Measuring Work Injury Experience*, New York, 1967.

A SPARSITY-ORIENTED APPROACH TO THE DESIGN OF MECHANICAL SYSTEMS*

N. ORLANDEA
Department of Mechanical Engineering
Iowa State University

D. A. CALAHAN
Department of Electrical and Computer Engineering
The University of Michigan
Ann Arbor, Michigan

* This work was supported by Grant AFOSR 75-2818 from the Directorate of Mathematical and Information Sciences, Air Force Office of Scientific Research.

SUMMARY

In this paper, the systematic formulation of the equations of motion for three-dimensional mechanical systems (mechanisms) is considered in view of sparsity requirements for large simulation problems. Topological approaches of previous authors are extended to include the practical case of joint-constraints, and the resultant formulation is also related to Lagrange multiplier methods. Finally, a simulation example is studied several times larger than heretofore state-of-the-art problems and the growth rate of computation time with problem size is noted.

1. INTRODUCTION

Computer-aided design of mechanical systems has recently been undergoing an evolution due to the highly efficient methods of solving systems of linear equations where the matrix of coefficients is sparse. These new techniques, when used in a computer-aided design program, affect even the initial formulation of the problem. This paper describes the ADAMS (*A*utomatic *D*ynamic *A*nalysis of *M*echanical *S*ystems) computer program whose formulation and implementation incorporates sparse matrix techniques and stiff integration algorithms.

To begin, let us consider a vector of generalized position functions $\underline{f} = [f_1\ f_2\ \ldots\ f_n]^T$. These functions depend on the generalized coordinates $\underline{q} = [q_1\ q_2\ \ldots\ q_k]^T$ that define uniquely the positions of a mechanical system. The velocities of the system are then given by the equation

$$\frac{\partial \underline{f}}{\partial \underline{q}} \dot{\underline{q}} = \dot{\underline{f}} \tag{1.1}$$

where $\dot{\underline{q}}$ and $\dot{\underline{f}}$ are the time derivatives of generalized coordinates and generalized functions. In general, the vector $\underline{f}$ is a nonlinear function of $\underline{q}$.

By definition, the vector of generalized constraint functions $\underline{f}_c$ of a constrained mechanical system satisfies the following conditions

$$\underline{f}_c \equiv \underline{0} \tag{1.2}$$

and

$$\frac{\partial \underline{f}_c}{\partial \underline{q}} \dot{\underline{q}} \equiv 0, \tag{1.3}$$

where in general, $\underline{f}_c$ are also nonlinear functions of $\underline{q}$.

To solve for the positions of a constrained mechanical system as functions of the generalized coordinates, it is customary to use the Newton iteration method, which can be expressed by

$$\frac{\partial \underline{f}_c}{\partial \underline{q}} \Delta \underline{q} = - \underline{f}_c \tag{1.4}$$

where $\Delta \underline{q} = [\Delta q_1 \ldots \Delta q_k]^T$ represents the Newton differences and $\underline{f}_c$ is the vector of errors.

By comparing relations (1.3) and (1.4), one can see that the matrix $\frac{\partial \underline{f}_c}{\partial \underline{q}}$ defined to be the Jacobian matrix for positions is used in both equations. This result will be used later in developing the sparse matrix formulation.

The sparse matrix formulation involves geometrical constraints, reaction and applied forces, and inertial forces. Let us define for a free body belonging to a constrained mechanical system

$$\underline{R} = [R_1 \ldots R_n]^T, \quad \underline{F} = [F_1 \ldots F_m]^T, \quad \text{and} \quad \underline{P} = [P_1 \; P_2 \ldots P_\ell]^T$$

as being the vectors of reaction forces, applied forces and inertial forces. The corresponding position functions for these vectors are

$$\underline{f}_r = [f_{r_1} \ldots f_{r_n}]^T, \; \underline{f}_a = [f_{a_1} \ldots f_{a_m}]^T, \; \text{and} \; \underline{f}_i = [f_{i_1} \ldots f_{i_\ell}]^T$$

The physical interpretation of the vectors $\underline{f}_r$, $\underline{f}_a$, and $\underline{f}_i$ is that these vectors define the application points of the vectors $\underline{R}$, $\underline{F}$, and $\underline{P}$.

On the basis of equation (1.1) we can write

$$\frac{\partial \underline{f}_r}{\partial \underline{q}} \dot{\underline{q}} = \underline{v}_r$$

$$\frac{\partial \underline{f}_a}{\partial \underline{q}} \dot{\underline{q}} = \underline{v}_a$$

and

$$\frac{\partial \underline{f}_i}{\partial \underline{q}} \dot{q} = \underline{v}_i$$

or by introducing the notations

$$\underset{\sim}{\alpha}_r = \frac{\partial \underline{f}_r}{\partial \underline{q}}, \quad \underset{\sim}{\alpha}_a = \frac{\partial \underline{f}_a}{\partial \underline{c}}, \quad \text{and } \underset{\sim}{\alpha}_i = \frac{\partial \underline{f}_i}{\partial \underline{q}}$$

we can write

$$\underset{\sim}{\alpha}_r \dot{\underline{q}} = \underline{v}_r \quad \underset{\sim}{\alpha}_a \dot{\underline{q}} = \underline{v}_a \quad \underset{\sim}{\alpha}_i \dot{\underline{q}} = \underline{v}_i \tag{1.5}$$

In the following, we take into consideration that the components of vectors $\underline{R}$, $\underline{F}$, and $\underline{P}$ are not usually expressed directly as generalized forces. A transformation of $\underline{R}$, $\underline{F}$, and $\underline{P}$ to a vector $\underline{A} = [A_1 \ \dots \ A_k]^T$ whose components are the generalized forces corresponding to generalized coordinates $\underline{q}$ is required.

The transformation is given by the mechanical power identity of vectors $\underline{R}$, $\underline{F}$, and $\underline{P}$ on one side and vector $\underline{A}$ on the other side. Therefore, we can write

$$[\underline{R}^T \ \underline{F}^T \ \underline{P}^T] \begin{vmatrix} \underline{v}_r \\ \underline{v}_a \\ \underline{v}_i \end{vmatrix} = \underset{\sim}{A}^T \dot{\underline{q}}$$

or by considering Equations (1.5)

$$[\underline{R}^T \ \underline{F}^T \ \underline{P}^T] \begin{vmatrix} \underset{\sim}{\alpha}_r \\ \underset{\sim}{\alpha}_a \\ \underset{\sim}{\alpha}_i \end{vmatrix} |\dot{\underline{q}}| = \underset{\sim}{A}^T \dot{\underline{q}} \tag{1.6}$$

or in transposed form

$$|\dot{\underline{q}}|^T [\underset{\sim}{\alpha}_r^T \ \underset{\sim}{\alpha}_a^T \ \underset{\sim}{\alpha}_i^T] \begin{vmatrix} \underline{R} \\ \underline{F} \\ \underline{P} \end{vmatrix} = \dot{\underline{q}}^T \underline{A}$$

From this by identifying the coefficients of q_j, $j = 1,\dots,k$, we obtain

$$[\underset{\sim}{\alpha}_r^T \; \underset{\sim}{\alpha}_a^T \; \underset{\sim}{\alpha}_i^T] \begin{vmatrix} \underline{R} \\ \underline{F} \\ \underline{P} \end{vmatrix} = \underline{A} \tag{1.7}$$

For a free body motion D'Alembert's principle tells that

$$\underline{A} = \underline{0}$$

By analogy with analysis of electrical circuits, this expression is similar to Kirchhoff's law for currents for a node when the forces are considered through variables and the free bodies as nodes.

For a mechanical system we can apply the above relation for each body if we consider that for two neighboring bodies s and s+1, (Fig. 1 and Table 1),

$$\underline{R}_s = - \underline{R}_{s+1}, \tag{1.8}$$

and

$$\underline{f}_{c_{s,s+1}} = \underset{\sim}{\alpha}_{r_s} \dot{\underline{q}}_s + \underset{\sim}{\alpha}_{r_{s+1}} \dot{\underline{q}}_{s+1} \tag{1.9}$$

where

$$\underline{f}_{c_{s,s+1}} = \underline{0}$$

On the basis of equation (1.4) we can write for two reciprocally constrained bodies that

$$|\underset{\sim}{\alpha}_{r_s} \; -\underset{\sim}{\alpha}_{r_{s+1}}| \begin{vmatrix} \Delta\underline{q}_s \\ \Delta\underline{q}_{s+1} \end{vmatrix} = - \underline{f}_{c_{s,s+1}}$$

or

$$|\underset{\sim}{\alpha}_{r_s} \; -\underset{\sim}{\alpha}_{r_{s+1}}| \; |\Delta\underline{q}| = - \underline{f}_{c_{s,s+1}} \tag{1.10}$$

where

$$\Delta\underline{q} = [\Delta\underline{q}_s^T \; \Delta\underline{q}_{s+1}^T]^T$$

The geometrical variables corresponding to the last two equa-

$\underline{F}_s$, $\underline{F}_{s+1}$	- The vectors of applied forces which act over bodies s and s+1.
$\underline{R}_s$, $\underline{R}_{s+1}$	- The reaction forces between reciprocally constrained bodies s and s+1.
$\underline{P}_s$, $\underline{P}_{s+1}$	- The inertial forces of bodies s and s+1. (The inertia forces and torque were considered to act in the same plane.)
$\underline{f}_{c_{s,s+1}}$	- The vector of constraint function between bodies s and s+1.
$\underline{r}_{a_s}$, $\underline{r}_{a_{s+1}}$	- Geometrical vectors which define the moment arm of $\underline{F}_s$, $\underline{F}_{s+1}$ with respect to the mass center.
$\underline{r}_{i_s}$, $\underline{r}_{i_{s+1}}$	- Geometrical vectors which define the moment arm of $\underline{P}_s$ and $\underline{P}_{s+1}$ with respect to the mass center.
ℓ_s, ℓ_{s+1}	- Geometrical vectors which define the position of the mass center with respect to an inertial system.
$\underline{\ell}_{a_s}$, $\ell_{a_{s+1}}$	- Geometrical vectors which define an application point of $\underline{F}_s$ and $\underline{F}_{s+1}$.
ℓ_{i_s}, $\ell_{i_{s+1}}$	- Geometrical vectors which define an application point of $\underline{P}_s$ and $\underline{P}_{s+1}$.
$\underline{q}_s$, $\underline{q}_{s+1}$	- The generalized coordinates of bodies s and s+1.

Table 1. The definitions of variables from Figure 1.

tions of system (1.5) are defined in Figure 1, in which the inertial forces for s and s+1 bodies are considered referred to a point other than the center of mass. However, we can write

$$\underline{f}_{a_p} - \underline{\ell}_{a_p} = \underline{0} \quad \underline{f}_{i_p} - \underline{\ell}_{i_p} = \underline{0} \quad p = s, s+1 \qquad (1.11)$$

where

$$\underline{f}_{a_p} = \underline{\ell}_p + \underline{r}_{a_p} \quad \text{and} \quad \underline{f}_{i_p} = \underline{\ell}_p + \underline{r}_{i_p}$$

By attaching to Equation (1.10) the linearized form of (1.11)

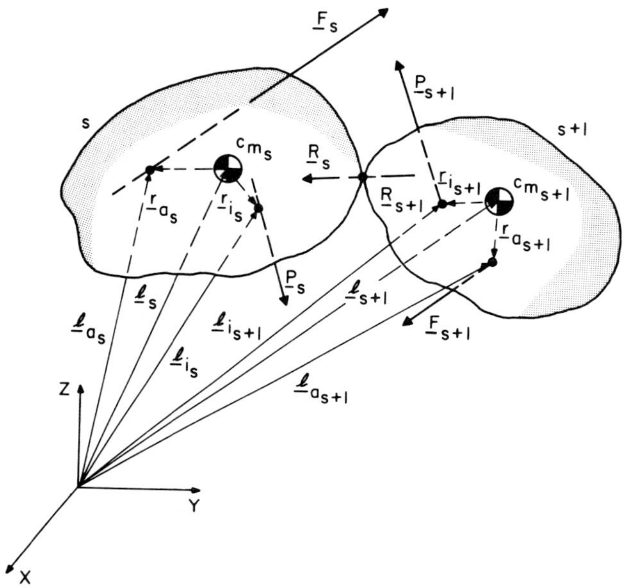

Figure 1. Definition of Geometrical Variables

where the Jacobian matrices are α_{a_p} and α_{i_p} contained in Equation (1.5), we can write

$$\begin{bmatrix} \beta_r & -I & 0 & 0 \\ \beta_a & 0 & -I & 0 \\ \beta_i & 0 & 0 & -I \end{bmatrix} \begin{bmatrix} \Delta q \\ \Delta \ell_r \\ \Delta \ell_a \\ \Delta \ell_i \end{bmatrix} = - \begin{bmatrix} f_c \\ f_a - \ell_a \\ f_i - \ell_i \end{bmatrix} \tag{1.12}$$

where

$$\beta_r = \begin{bmatrix} \alpha_{r_s} & -\alpha_{r_{s+1}} \end{bmatrix}$$

$$\beta_a = \begin{bmatrix} \alpha_{a_s} & 0 \\ 0 & \alpha_{a_{s+1}} \end{bmatrix} \qquad \beta_i = \begin{bmatrix} \alpha_{i_s} & 0 \\ 0 & \alpha_{i_{s+1}} \end{bmatrix}$$

$$\underline{f}_c = \underline{f}_{c_{s,s+1}}, \quad \underline{f}_a = [\underline{f}_{a_s}^T \ \underline{f}_{a_{s+1}}^T]^T, \quad \underline{f}_i = [\underline{f}_{i_s}^T \ \underline{f}_{i_s}^T \ \underline{f}_{i_{s+1}}^T]^T$$

$$\underline{\ell}_a = [\underline{\ell}_{a_s}^T \ \underline{\ell}_{a_{s+1}}^T]^T \qquad \underline{\ell}_i = [\underline{\ell}_{i_s}^T \ \underline{\ell}_{i_{s+1}}^T]^T$$

By analogy with equation formulation used in electrical circuits, Equation (1.12) represents the node voltage equations. By means of the equality (1.8) we can write for forces

$$\begin{bmatrix} \alpha_{r_s}^T & \alpha_{a_s}^T & 0 & \alpha_{i_s}^T & 0 \\ -\alpha_{r_{s+1}}^T & 0 & \alpha_{a_{s+1}}^T & 0 & \alpha_{i_{s+1}}^T \end{bmatrix} \begin{bmatrix} \underline{R}_s \\ \underline{F}_s \\ \underline{F}_{s+1} \\ \underline{P}_s \\ \underline{P}_{s+1} \end{bmatrix} = \underline{0}$$

or

$$[\beta_r^T \ \beta_a^T \ \beta_i^T] \begin{bmatrix} \underline{R}_s \\ \underline{F} \\ \underline{P} \end{bmatrix} = \underline{0} \tag{1.13}$$

where

$$\underline{F} = [\underline{F}_s^T \ \underline{F}_{s+1}^T]^T \quad \text{and} \quad \underline{P} = [\underline{P}_s^T \ \underline{P}_{s+1}^T]^T$$

Introducing the notation $[\beta_r^T \ \beta_a^T \ \beta_i^T] = \beta$, Equations (1.12) and (1.13) can be written

$$[\beta^T \ I] \begin{bmatrix} \Delta\underline{q} \\ \Delta\underline{c} \end{bmatrix} = -\begin{bmatrix} \underline{f}_c \\ \underline{f} - \underline{c} \end{bmatrix} \tag{1.14}$$

where

$$\underline{c} = [\underline{\ell}_a^T \ \underline{\ell}_i^T]^T \qquad \underline{f} = [\underline{f}_a^T \ \underline{f}_i^T]^T$$

and

$$[\beta] \begin{bmatrix} \underline{R}_s \\ \underline{F} \\ \underline{P} \end{bmatrix} = \underline{0} \tag{1.15}$$

β is called the incidence matrix [1], but is different from the incidence matrix used in electrical circuits analysis because its entries, in the most general case, are different from unity. Its simplicity depends upon the choice of the generalized coordinates. One example is the choice of the translational generalized coordinates to be the coordinates of the center of mass. In this case the part of $\underset{\sim}{\beta}$ given by these coordinates will contain only unit entries. $\underset{\sim}{\beta}$ becomes entirely a topological matrix, [2], when the mechanical system is formed from point masses connected with applied forces given by springs, the universal attraction law, etc.

In the case of constrained rigid body systems, rotational generalized coordinates introduce torques and constraints which are highly nonlinear functions of the rotational coordinates (mainly sin and cos). In this case the formulation of $\underset{\sim}{\beta}$ is difficult.

To automate the incidence matrix set up, we will define standard components of mechanical systems as bodies, joints, springs, dampers, etc., for which we can set standard submatrices of the incidence matrix. The submatrices are then arranged according to the topology of the physical system.

2. SPARSE MATRIX FORMULATION

The Gear integration algorithm [3] avoids numerical instability for systems with widely split eigenvalues. Given an equation of the form

$$\underline{G}(\underline{w},\dot{\underline{w}},t) = \underline{0}$$

the Gear corrector formula involves an equation of the form

$$\underset{\sim}{J}\Delta\underline{w} = \left(\frac{\partial \underline{G}}{\partial \underline{w}} + \frac{1}{K_o T}\frac{\partial \underline{G}}{\partial \dot{\underline{w}}}\right)\Delta\underline{w} = -\underline{G} \tag{2.1}$$

where K_o is a Gear coefficient and T is the time step. Introducing the notation $\dot{\underline{q}} = \underline{u}$, we define

$$\underline{w} = [\underline{u}^T \; \underline{q}^T \; \underline{\ell}_r^T \; \underline{\ell}_i^T \; \underline{\ell}_a^T \; \underline{R}^T \; \underline{P}^T \; \underline{F}^T]^T \tag{2.2}$$

so that

$$\Delta\underline{w} = [\Delta\underline{u}^T \; \Delta\underline{q}^T \; \Delta\underline{\ell}_r^T \; \Delta\underline{\ell}_i^T \; \Delta\underline{\ell}_a^T \; \Delta\underline{R}^T \; \Delta\underline{P}^T \; \Delta\underline{F}^T]^T \tag{2.3}$$

Now defining the inertial forces $\underline{P} = F_i(\dot{\underline{u}},\underline{u},\underline{q},t)$, the applied forces $\underline{F} = F_a(\underline{u},\underline{q},t)$, and $\underset{\sim}{\beta} = [\underset{\sim}{\beta}_r^T \; \underset{\sim}{\beta}_i^T \; \underset{\sim}{\beta}_a^T]$ we can write Equation (2.4), where $\underline{\gamma} = [\underline{R}^T \; \underline{P}^T \; \underline{F}^T]^T$ and where $\underline{f}_c$, $\underline{f} - \underline{c}$, and $\underline{A}$ are defined by Equations (1.2), (1.14), and (1.7). In Equation (2.4), a notational analogy is made with electrical circuit variables.

In the preceeding section, we termed $\underset{\sim}{\beta}$ to be the incidence matrix. In general the properties of the incidence matrices defined in [1] do not apply completely to $\underset{\sim}{\beta}$; in particular the part which is defined by the rotational coordinates and the torque equations has no incidence matrix properties.

For illustration, let us consider the seven degree of freedom linkage formed from four bodies and four spherical joints (Figure 2). The system is in equilibrium under the action of applied force $\vec{F}$ and applied torques $\vec{\zeta}_1$ and $\vec{\zeta}_3$. The graph of the system is represented in Figure 3.

The force equilibrium condition gives the following system of equations.

$$\begin{bmatrix} 1 & 1 & & & \\ & -1 & 1 & & 1 \\ & & -1 & 1 & \\ -1 & & & -1 & -1 \end{bmatrix} \begin{bmatrix} \vec{R}_1 \\ \vec{R}_2 \\ \vec{R}_3 \\ \vec{R}_4 \\ \vec{F} \end{bmatrix} = \underline{0}$$

or in condensed form

$$\underline{BQ} = \underline{0} \tag{2.5}$$

The matrix $\underset{\sim}{\beta}$ has all the usual properties of the incidence matrix defined in [1].

Column groups: Node voltages (columns 1–2), Branch voltages (columns 3–5), Branch currents (columns 6–8). Row groups: KVL (row 1), KCL (row 2), Constitutive Equations (rows 3–6).

$$
\begin{array}{l}
\text{KVL} \\ \text{KCL} \\ \\ \text{Constitutive} \\ \text{Equations} \\ \\
\end{array}
\begin{bmatrix}
\underset{\sim}{0} & \underset{\sim}{\beta}^{T} & \multicolumn{3}{c}{\longleftarrow -\underset{\sim}{I} \longrightarrow} & \underset{\sim}{0} & \underset{\sim}{0} & \underset{\sim}{0} \\
\underset{\sim}{0} & \dfrac{\partial(\underset{\sim}{\beta}\underline{\gamma})}{\partial \underline{q}} & \underset{\sim}{0} & \underset{\sim}{0} & \underset{\sim}{0} & \underset{\sim}{\beta}_{r}^{T} & \underset{\sim}{\beta}_{i}^{T} & \underset{\sim}{\beta}_{a}^{T} \\
\dfrac{\partial \underline{F}_{i}}{\partial \underline{u}} + \dfrac{1}{K_{o}T}\dfrac{\partial \underline{F}_{i}}{\partial \dot{\underline{u}}} & \dfrac{\partial \underline{F}_{i}}{\partial \underline{q}} & \underset{\sim}{0} & \underset{\sim}{0} & \underset{\sim}{0} & \underset{\sim}{0} & -\underset{\sim}{I} & \underset{\sim}{0} \\
-\underset{\sim}{I} & \dfrac{1}{K_{o}T}\underset{\sim}{I} & \underset{\sim}{0} & \underset{\sim}{0} & \underset{\sim}{0} & \underset{\sim}{0} & \underset{\sim}{0} & \underset{\sim}{0} \\
\dfrac{\partial \underline{F}_{a}}{\partial \underline{u}} & \dfrac{\partial \underline{F}_{a}}{\partial \underline{q}} & \underset{\sim}{0} & \underset{\sim}{0} & \underset{\sim}{0} & \underset{\sim}{0} & \underset{\sim}{0} & -\underset{\sim}{I} \\
\underset{\sim}{0} & \underset{\sim}{0} & \underset{\sim}{I} & \underset{\sim}{0} & \underset{\sim}{0} & \underset{\sim}{0} & \underset{\sim}{0} & \underset{\sim}{0}
\end{bmatrix}
\begin{bmatrix}
\Delta \underline{u} \\ \Delta \underline{q} \\ \Delta \underline{\ell}_{r} \\ \Delta \underline{\ell}_{i} \\ \Delta \underline{\ell}_{a} \\ \Delta \underline{R} \\ \Delta \underline{P} \\ \Delta \underline{F}
\end{bmatrix}
= -
\begin{bmatrix}
\left\{ \begin{array}{l} \underline{f}_{c} \\ \underline{f} - \underline{c} \end{array} \right. \\ \underline{A} \\ \underline{F}_{i} - \underline{P} \\ \dot{\underline{q}} - \underline{u} \\ \underline{F}_{a} - \underline{F} \\ \underline{0}
\end{bmatrix}
\tag{2.4}
$$

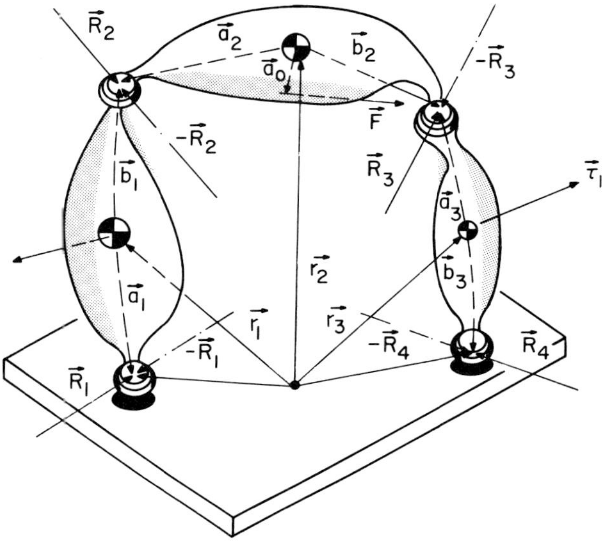

Figure 2. Four Body Example

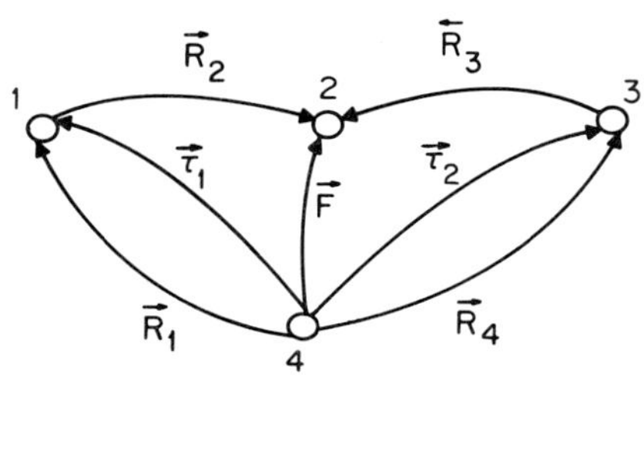

Figure 3. Graph of Four-Body Problem

The torque equilibrium condition gives

$$\underbrace{\begin{bmatrix} -(\times\vec{a}_1) & (\times\vec{b}_1) & 0 & 0 & \\ 0 & -(\times\vec{a}_2) & (\times\vec{b}_2) & 0 & (\times\vec{a}_o) \\ 0 & 0 & -(\times\vec{a}_3) & (\times\vec{b}_3) & \\ (\times\vec{b}_4) & 0 & 0 & -(\times\vec{a}_4) & \end{bmatrix}}_{\underset{\sim}{C}} \overset{\underset{\sim}{E}}{\begin{matrix} 1 & 0 \\ 0 & 0 \\ 0 & 1 \\ -1 & -1 \end{matrix}} \begin{bmatrix} \vec{R}_1 \\ \vec{R}_2 \\ \vec{R}_3 \\ \vec{R}_4 \\ \vec{F} \\ \vec{\zeta}_1 \\ \vec{\zeta}_3 \end{bmatrix} = 0$$

or in condensed form

$$\underset{\sim}{D} \begin{bmatrix} \underline{Q} \\ \underline{\zeta} \end{bmatrix} = \underline{0} \tag{2.6}$$

where $\underset{\sim}{D} = [\underset{\sim}{C}\ \underset{\sim}{E}]$. The matrix $\underset{\sim}{D}$ does not satisfy the incidence law because the sum of the terms of a column is non-zero. We can then define $\underset{\sim}{\beta}$ for the static equilibrium of the system as

$$\underset{\sim}{\beta} = \begin{bmatrix} \underset{\sim}{B} & \underset{\sim}{0} \\ \underset{\sim}{C} & \underset{\sim}{E} \end{bmatrix} \tag{2.7}$$

By writing the loop equations for velocities on the hypothesis that the coordinates for translations are the coordinates of the mass center, and by retaining all the kinematic variables for all bodies, we obtain the Jacobian of constraint functions as being $\underset{\sim}{\beta}^T$.

The simplicity of $\underset{\sim}{\beta}$ depends directly on the chosen set of the generalized coordinates. For one rigid body we will define

$$\begin{bmatrix} q_1 \\ q_2 \\ q_3 \end{bmatrix} = \begin{bmatrix} X_s \\ Y_s \\ Z_s \end{bmatrix} \quad \text{and} \quad \begin{bmatrix} q_4 \\ q_5 \\ q_6 \end{bmatrix} = \begin{bmatrix} \Psi_s \\ \phi_s \\ \theta_s \end{bmatrix} \tag{2.8}$$

where X_s, Y_s, Z_s represent the coordinates of the center of mass of s-th body and Ψ_s, ϕ_s, Θ_s are the Euler angles which determine the angular position of the principal system of inertia with respect to an arbitrary inertial system, as in Figure 4.

We will retain all the variables which define the system in non-reduced form. Therefore for N interconnected rigid bodies the total number of coordinates in the system will be 6N. Together with 6N coordinates we will retain the components of $\underline{R}$, $\underline{P}$, and $\underline{F}$ as explicit variables, obtaining a very sparse Jacobian $\underset{\sim}{J}$.

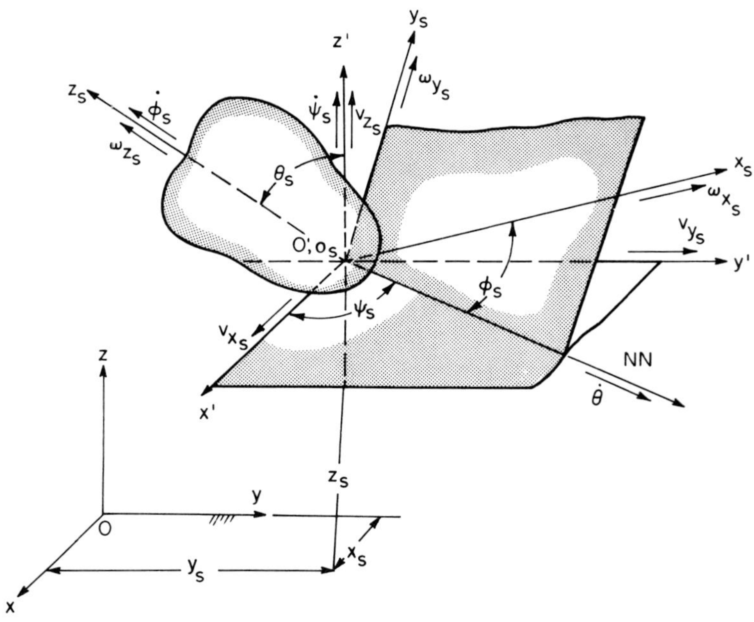

Figure 4. Definition of Generalized Coordinates

3. THE GENERAL EQUATIONS OF MOTION

The general form of the equations of motion can be given by the Lagrange equation

$$\frac{d}{dt}\left(\frac{\partial E}{\partial \dot{q}_j}\right) - \frac{\partial E}{\partial q_j} - Q_j = 0 \qquad (3.1)$$

where E is the kinetic energy of the system

q_j are the generalized coordinates,

$\dot{q}_j$ are the generalized velocities,

Q_j are the generalized forces (conservative and non-conservative)

$j = 1,\ldots,6$ where 6 represents the number of degrees of freedom. This equation is valid for constrained and unconstrained mechanical dynamic systems.

If we define the generalized momentum

$$p_j \triangleq \frac{\partial E}{\partial \dot{q}_j} \qquad (3.2)$$

Equation (3.1) takes the form

$$\dot{p}_j = \frac{\partial E}{\partial q_j} + Q_j \qquad j = 1,2,\ldots,6. \qquad (3.3)$$

It can be shown that (3.3) is a second-order equation in the variable q_j; we place this in a form containing only first derivatives by defining

$$u_j = \dot{q}_j \qquad (3.4)$$

In addition to the Lagrange equation above, mechanical systems must satisfy the connection equations.

$$f_{c_k} = 0 \qquad k = 1,\ldots,n \qquad (3.5)$$

which can be nonlinear functions of the generalized coordinates $\underline{q}$. Chace and Smith [4] have proposed that the combined set of differential-algebraic equations be re-formulated by introduction of Lagrange multipliers. Equations (3.2)-(3.5) can then be replaced by;

for each body

$$A_j = \dot{p}_j - \frac{\partial E}{\partial q_j} - Q_j + \sum_{k=1}^{n} \frac{\partial f_{c_k}}{\partial q_j} \lambda_k = 0; \quad j = 1,2,\ldots,6 \tag{3.6}$$

$$L_j = p_j - \frac{\partial E}{\partial \dot{q}_j} = 0 \tag{3.7}$$

$$u_j - \dot{q}_j = 0 \tag{3.8}$$

for each of n constraints:

$$f_{c_k} = 0 \qquad k = 1,\ldots,n$$

Although the efficiency of the numerical solution we will propose is not particularly sensitive to the number of equations, it appears worthwhile to note that the translational components of the momentum are trivally related to translational velocities; therefore we do not regard these translational components of momentum as explicit solution variables, leaving our set of free body equations in the final form for a body s,

translation (j = 1,2,3)

$$A_j = \frac{d}{dt}\left(\frac{\partial E}{\partial \dot{q}_j}\right) - \frac{\partial E}{\partial q_j} - \dot{Q}_j + \sum_{k=1}^{n} \frac{\partial f_{c_k}}{\partial q_j} \lambda_k = 0 \tag{3.9}$$

rotational

$$L_{j+3} = p_j - \frac{\partial E}{\partial \dot{q}_{j+3}} = 0 \tag{3.10}$$

$$A_{j+3} = \dot{p}_j - \frac{\partial E}{\partial q_{j+3}} - Q_{j+3} + \sum_{k=1}^{n} \frac{\partial f_{c_k}}{\partial q_{j+3}} = 0 \tag{3.11}$$

combined (j = 1,2,...,6)

$$u_j - \dot{q}_j = 0 \tag{3.12}$$

By dropping the subscripts j and k, Equations (3.9) and (3.11) can be rewritten

$$\underline{P} + \underset{\sim}{\beta}_a^T \underline{F} + \underset{\sim}{\beta}_r^T \underline{R} = \underline{0} \tag{3.13}$$

Letting $\underline{P} = [\underline{P}_t^T \ \underline{P}_r^T]^T$, we have for translational

$$\underline{P}_t = \frac{d}{dt}\left(\frac{\partial E}{\partial \dot{\underline{q}}}\right)^T - \left(\frac{\partial E}{\partial \underline{q}}\right)^T \qquad \text{where } \underline{q} = [X_s \ Y_s \ Z_s]^T \tag{3.14}$$

and for rotational

$$\underline{P}_r = \dot{\underline{p}} - \left(\frac{\partial E}{\partial \underline{q}}\right)^T \qquad \text{where } \underline{q} = [\Psi_s \ \phi_s \ \Theta_s]^T \tag{3.15}$$

$$\underline{L} = \underline{p} - \left(\frac{\partial E}{\partial \dot{\underline{q}}}\right)^T = \underline{0} \tag{3.16}$$

and

$$\underset{\sim}{\beta}_a^T \underline{F} = -\underline{Q} \qquad \underset{\sim}{\beta}_r^T \underline{R} = \underset{\sim}{\beta}_r^T \underline{\lambda} \tag{3.17}$$

We will assume that $\underset{\sim}{\beta}_r^T$, representing the linearized constraint equations, is invertible. Then we have

$$\underline{R} = \underline{\lambda}. \tag{3.18}$$

Equation (3.13) shows that the matrix $\underset{\sim}{\beta}_i^T = I$, which means that the inertial forces are reduced to the mass center of the body. Equation (3.18) gives the physical meaning of λ as being the reaction forces which act over the body and create the equilibrium of inertial and applied forces in the free body motion.

On the basis of these results the Jacobian matrix $\underset{\sim}{J}$ from Equation (2.4) can be written alternatively

$$
\begin{array}{l}
\\ \\ \\ \\ \\ \text{Translation} \\ \\ \\ \text{Rotation} \\ \\ \\ \\
\end{array}
\begin{array}{c}
\begin{array}{ccccccc}
\Delta\underline{u}^T & \Delta\underline{p}^T & \Delta\underline{q}^T & \Delta\underline{\ell}_a^T & \Delta\underline{R}^T & \Delta\underline{P}^T & \Delta\underline{F}^T
\end{array} \\
\left[\begin{array}{ccccccc}
\underset{\sim}{0} & \underset{\sim}{0} & \underset{\sim}{\beta}_r & \underset{\sim}{0} & \underset{\sim}{0} & \underset{\sim}{0} & \underset{\sim}{0} \\
\underset{\sim}{0} & \underset{\sim}{0} & \underset{\sim}{\beta}_a & -\underset{\sim}{I} & \underset{\sim}{0} & \underset{\sim}{0} & \underset{\sim}{0} \\
\underset{\sim}{0} & \underset{\sim}{0} & \dfrac{\partial(\underset{\sim}{\beta}\dot{\underline{\gamma}})}{\partial\underline{q}} & \underset{\sim}{0} & \underset{\sim}{\beta}_r^T & \underset{\sim}{I} & \underset{\sim}{\beta}_a^T \\
\underset{\sim}{I}\,\dfrac{m_s}{K_o T} & \underset{\sim}{0} & \underset{\sim}{0} & \underset{\sim}{0} & \underset{\sim}{0} & \underset{\sim}{M} & \underset{\sim}{0} \\
\dfrac{\partial\underline{P}}{\partial\underline{u}} & \underset{\sim}{I}\,\dfrac{1}{K_o T} & \dfrac{\partial\underline{P}}{\partial\underline{q}} & \underset{\sim}{0} & \underset{\sim}{0} & \underset{\sim}{N} & \underset{\sim}{0} \\
\dfrac{\partial\underline{L}}{\partial\underline{u}} & -\underset{\sim}{I} & \dfrac{\partial\underline{L}}{\partial\underline{q}} & \underset{\sim}{0} & \underset{\sim}{0} & \underset{\sim}{0} & \underset{\sim}{0} \\
-\underset{\sim}{I} & \underset{\sim}{0} & \underset{\sim}{I}\,\dfrac{1}{K_o T} & \underset{\sim}{0} & \underset{\sim}{0} & \underset{\sim}{0} & \underset{\sim}{0} \\
\dfrac{\partial\underline{F}_a}{\partial\underline{u}} & \underset{\sim}{0} & \dfrac{\partial\underline{F}_a}{\partial\underline{q}} & \underset{\sim}{0} & \underset{\sim}{0} & \underset{\sim}{0} & -\underset{\sim}{I}
\end{array}\right]
\end{array}
\qquad (3.19)
$$

where $\begin{bmatrix} \underset{\sim}{M} \\ \underset{\sim}{N} \end{bmatrix} = -\underset{\sim}{I}$ and m_s is the mass of body s.

4. THE EQUATIONS OF CONSTRAINTS

The numerical techniques used for simulation of mechanical dynamic systems are a consideration in selection of the form of the constraint functions. It has been customary in the formulations involving Lagrange multipliers to use nonholonomic functions of constraints of the form $\underline{f}_c = \underline{f}(\ddot{\underline{q}},\dot{\underline{q}},\underline{q})$. The purpose of using this type of constraint is to find a way of interpreting $\underline{f}_c$ as second order differential equations, thus avoiding the algebraic form. Any method of numerical integration could then be applied.

In the sparse matrix formulation, the Gear integration algorithm has the capability of solving a simultaneous set of differential and algebraic equations. In an effort to improve computational performance while preserving the sparsity of the Jacobian, the holonomic functions of constraints $\underline{f}_c = \underline{f}(\underline{q})$ are preferred. As a consequence, all the terms of the Jacobian derived from the partial derivatives of the functions $\underline{f}_c$ with respect to the generalized velocities are zero. Another reason for this choice is the capability of solving the static equilibrium using the same matrix $\underset{\sim}{J}$.

Let us consider two neighboring bodies s and s+1 represented in Figure 5. $o_s\ n_s\ y_s\ z_s$ and $o_{s+1}\ n_{s+1}\ y_{s+1}\ z_{s+1}$ represents the principal systems of inertia of the two bodies and the systems $o^*_s\ n^*_s\ y^*_s\ z^*_s$ and $o^*_{s+1}\ n^*_{s+1}\ y^*_{s+1}\ z^*_{s+1}$ are rigidly connected to the bodies s and s+1. The last two systems are defined with respect to the inertial system OXYZ by $\vec{\ell}_s$, Ψ^*_s, ϕ^*_s, Θ^*_s, and $\vec{\ell}_{s+1}$, Ψ^*_{s+1}, ϕ^*_{s+1}, Θ^*_{s+1}.

The direction of the axis of the relative rotation and the direction of the relative translation are given by $\hat{k}_s \equiv \hat{k}_{s+1}$. For the relative translational motion o^*_s and o^*_{s+1} are not required to be coincident points.

Using the notations from Figure 5, Table 2 gives the library of joints and their constraint equations which can be incorporated in sparse matrix formulation. The joint type number represents the number which is attached to each joint in the later computer implementation of the sparse formulation.

5. NUMERICAL SOLUTION

Recalling Equation (2.2) and the definition of solution vari-

No.	Name of joint	Graphical symbol	Equations of constraints	Joint type number
1.	Spherical		$\vec{\ell}_s - \vec{\ell}_{s+1} = 0$	1
2.	Universal		$\vec{\ell}_s - \vec{\ell}_{s+1} = 0$ $\hat{k}\cdot\hat{k}_s - \hat{k}\cdot\hat{k}_{s+1} = 0$	2
3.	Revolute		$\vec{\ell}_s - \vec{\ell}_{s+1} = 0$ $\hat{k}_s \times \hat{k}_{s+1} = 0$	3
4.	Cylindrical		$(\vec{\ell}_s - \vec{\ell}_{s+1}) \times \hat{k}_s = 0$ $\hat{k}_s \times \hat{k}_{s+1} = 0$	5
5.	Translational	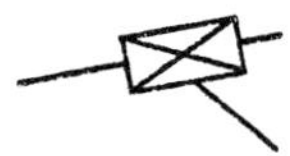	$(\vec{\ell}_s - \vec{\ell}_{s+1}) \times \hat{k}_s = 0$ $\hat{k}_s \times \hat{k}_{s+1} = 0$ $\hat{\ell}_s \cdot \hat{\ell}_{s+1} - \cos\alpha = 0$	4
6.	Screw		$(\vec{\ell}_s - \vec{\ell}_{s+1}) \times \hat{k}_s = 0$ $\hat{k}_s \times \hat{k}_{s+1} = 0$ $P\cdot(\hat{k}\cdot\hat{k}_s) \cdot \cos^{-1}(i_s \cdot i_{s+1} - \alpha) -$ $\hat{k}\cdot(\ell_s - \ell_{s+1}) = 0$ P is the parameter of helical motion	6

Table 2. Library of Joints

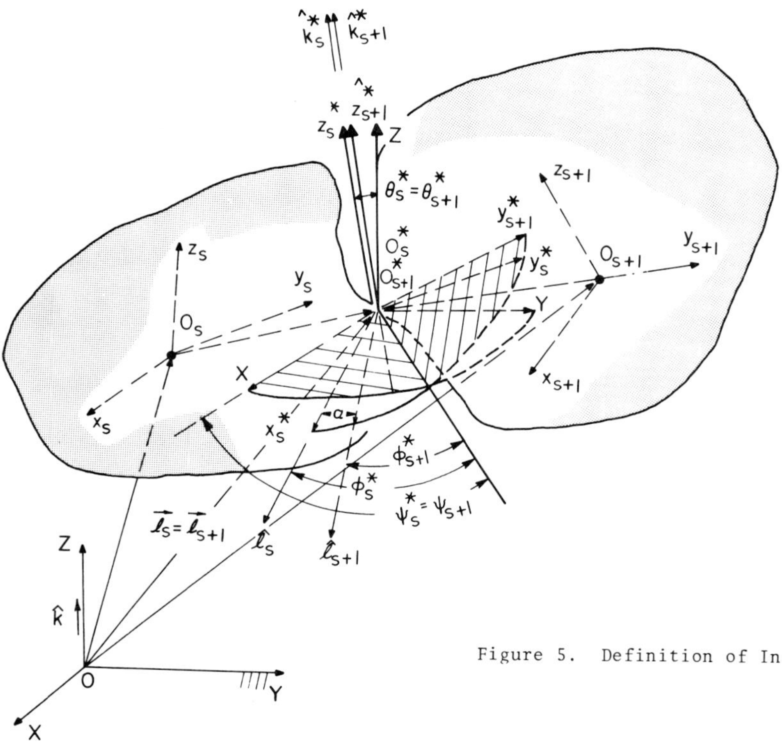

Figure 5. Definition of Inertial Coordinates

ables $\underline{w}$ and Equation (3.18) we can write

$$\underline{G}(\dot{\underline{u}},\underline{u},\dot{\underline{q}},\underline{q},\dot{\underline{p}},\underline{p},\underline{\ell}_a,\underline{\ell},\underline{F}_a,t) = \underline{0} \tag{5.1}$$

where $\underline{G}$ is a vector which includes the inertial force function $\underline{F}_i$, the geometrical constraints $\underline{f}_c$, the definition function for $\underline{\vec{\ell}}_a$, and the applied force functions $\underline{F}_a$.

These vector equations represent a mixture of differential and algebraic equations. Such a mixed system is conveniently solved by discretization of t and iterative (implicit) solution of the resulting equations using Newton's method at each time step. This procedure is appropriate for solutions of both the static and transient cases.

The implicit integration of (5.1) proceeds from corrector formula [1] [3].

$$\left(\left(\frac{1}{K_o T}\frac{\partial \underline{G}}{\partial \underline{\dot{u}}} + \frac{\partial \underline{G}}{\partial \underline{u}}\right)\Delta\underline{u} + \left(\frac{1}{K_o T}\frac{\partial \underline{G}}{\partial \underline{\dot{q}}} + \frac{\partial \underline{G}}{\partial \underline{q}}\right)\Delta\underline{q} + \left(\frac{1}{K_o T}\frac{\partial \underline{G}}{\partial \underline{\dot{p}}} + \frac{\partial \underline{G}}{\partial \underline{p}}\right)\Delta\underline{p} + \right.$$

$$+ \frac{\partial \underline{G}}{\partial \underline{\ell}_a}\Delta\underline{\ell}_a + \frac{\partial \underline{G}}{\partial \underline{\lambda}}\Delta\underline{\lambda} + \frac{\partial \underline{G}}{\partial \underline{F}_a}\Delta\underline{F}_a = -\underline{G}(\underline{w},t) \qquad (5.2)$$

The equation for static case results from setting $\frac{1}{T} = 0$ and $\underline{u} = \underline{0}$.

For a node (body) Equation (2.4) can be written as in Equation (5.3), or more compactly

$$\underset{\sim}{J}\left(\underline{w},\frac{1}{K_o T}\right)\Delta\underline{w} = -\underline{G} \qquad (5.4)$$

For a multi-body problem the Jacobian can be restructured in a form which aids its automatic generation, viz.

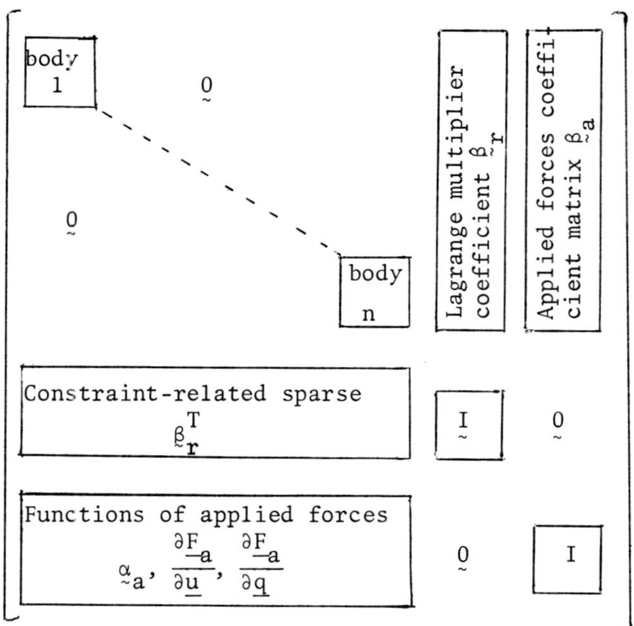

(5.5)

Motion equation for a typical body

Constraint equations

Applied forces

$$
\begin{bmatrix}
\frac{\partial \underline{F}_i}{\partial \dot{\underline{u}}} \frac{1}{K_o T} + \frac{\partial \underline{F}_i}{\partial \underline{u}} & \tilde{I} \frac{1}{K_o T} & \frac{\partial \underline{F}_i}{\partial \underline{q}} + \frac{\partial (\tilde{\alpha}^T \underline{\gamma})}{\partial \underline{q}} & \tilde{0} & \tilde{\alpha}_r^T & \tilde{\alpha}_a^T \\
\frac{\partial \underline{L}}{\partial \underline{u}} & -\tilde{I} & \frac{\partial \underline{L}}{\partial \underline{q}} & \tilde{0} & \tilde{0} & \tilde{0} \\
-\tilde{I} & \tilde{0} & \tilde{I} \frac{1}{K_o T} & \tilde{0} & \tilde{0} & \tilde{0} \\
\tilde{0} & \tilde{0} & \tilde{\alpha}_r & \tilde{0} & \tilde{0} & \tilde{0} \\
\tilde{0} & \tilde{0} & \tilde{\alpha}_a & -\tilde{I} & \tilde{0} & \tilde{0} \\
\frac{\partial \underline{F}_a}{\partial \underline{u}} & \tilde{0} & \frac{\partial \underline{F}_a}{\partial \underline{q}} & \tilde{0} & \tilde{0} & -\tilde{I}
\end{bmatrix}
\begin{Vmatrix}
\Delta \underline{u} \\ \Delta \underline{p} \\ \Delta \underline{q} \\ \Delta \underline{\ell}_a \\ \Delta \underline{\lambda} \\ \Delta \underline{F}
\end{Vmatrix}
= -G(\underline{w}, t) \tag{5.3}
$$

6. COMPUTER IMPLEMENTATION

The above formulational and numerical principles were implemented in the form of a general computer program for static, transient, modal, and vibrational types of analysis for nonlinear mechanical systems. The program is named ADAMS (*A*utomatic *D*ynamic *A*nalysis of *M*echanical *S*ystems).

From our experience, we can cite the advantages and disadvantages of the sparse matrix formulation.

ADVANTAGES

(1) No topclogical preprocessing is necessary to establish a set of independent variables; equations can be developed directly from the connection data, component-by-component.

(2) All angular and displacement variables are retained as solution variables; none are eliminated in the interest of producing a reduced set of equations with fewer variables. This aids the modeling process, since new interbody constraints can be accomodated merely by adding algebraic equations relating appropriate variables.

(3) All joint reaction forces are explicit solution variables and therefore the formulation is compatible with current methods of continuous mechanics for internal stress analysis. Also, friction effects are routinely related to reaction forces.

(4) Since angular momentums are explicit solution variables, impact is easily modeled.

DISADVANTAGES

(1) Time is wasted in solving for variables of no interest to the designer.

(2) State variables are not identified, increasing the numeri-

cal difficulty in solving for linearized modal or vibrational responses.

These procedures were implemented in the ADAMS program with the following features.

(1) Mechanical Component Descriptions

Descriptions of (1) linkages by masses, inertial moments, guess of initial generalized coordinates; (2) joints by type, and linkage adjacencies; (3) springs and dashpots by their force coefficients and their attachment points relative to links; and (4) force and displacement inputs.

(2) Design Capability (three dimensional)

- (a) Static analysis
- (b) Large displacement (nonlinear) transient analysis
- (c) Small displacement (linearized) analysis around a static solution or at any solution point in time. This includes
  1. vibrational analysis
  2. modal analysis, modal sensitivity, and modal optimization.

(3) Algorithms

- (a) Sparse matrix formulation
- (b) Sparse matrix compiler for static and transient analysis; sparse matrix interpreter for vibrational and modal analysis
- (c) Gear implicit integration for transient solution
- (d) Muller's method for modal analysis

(4) Typical Applications

- (a) Suspension system design of automative, aircraft and other transportation vehicles
- (b) Three-dimension dynamic analysis of vehicular motion
- (c) Analysis of biomechanical systems (e.g., collision analysis)

(d) High-speed mechanical device design (e.g., fast printers, textile machines)

(5) The types of joints which can be handled by ADAMS program are represented in Table 1.

7. EXAMPLE

Since the most distinctive characteristic of this formulation is its sparsity orientation, an example has been chosen that illustrates the growth of computation time as a function of problem size. A full matrix-based procedure should grow as $0(n^3)$; a sparsity-oriented method typically grows as $0(n^{1.5})$.

The problem chosen is a three-dimensional model of a Boeing 747 landing system (Fig. 6). The full five-strut, 18-wheel model was described by 41 bodies (links), several times larger than state-of-the-art landing system models. The wing and rear struts each have one translational, 5 revolute, 2 spherical, and 2 universal joints. In addition, a single wing strut was modeled in three dimensions by 11 bodies. Computation times for equation formation and numerical solution are given in Table 3 for each case. The computation time for a single corrector iteration- both formulation and solution - is shown to grow as $0(n^{1.4})$, an effort quite consistent with results from other classes of simulation problems.

No. of bodies	No. of equations	Matrix entries per row	Times (secs.) per equation formulation and solution	Total simulation costs per run
$n_1 = 11$ (single strut)	239	5.77	.060	\$45
$n_2 = 41$ (5 struts)	927	6.19	.380	\$250-\$400 (estimated)
n	$\approx 0(n)$	$\approx$ constant	$0(n^{1.4})$	

Table 3. Summary of Computation Times and Complexity Estimate (Times are for Amdahl 470 V/6)

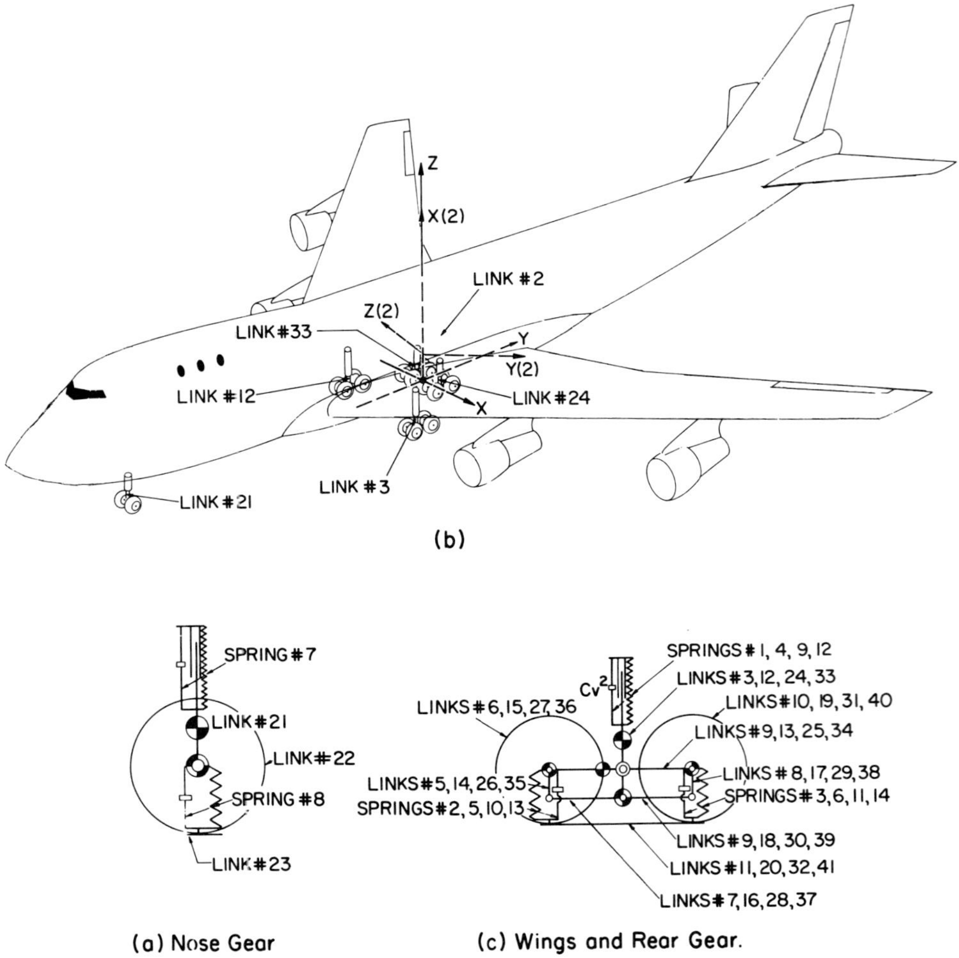

Figure 6. 41-Body Aircraft Landing System Example

REFERENCES

1. Calahan, D.A., *Computer Aided Network Design*, McGraw-Hill, 1972.
2. Andrews, G.C., and H.R. Kesavan, "The Vector-Network Model: A New Approach to Vector Dynamics", *Mechanism and Machine Theory*, Vol. 10, pp. 57-75, 1975.
3. Gear, C.W., "The Control of Parameters in the Automatic Integration of Ordinary Differential Equations", *Internal Report*, Department of Computer Science, University of Illinois, April, 1970.
4. Chace, M.A., and D.A. Smith, "DAMN - A Digital Computer Program for the Dynamic Analysis of Generalized Mechanical Systems", *SAE Paper 710244*, January, 1971.

ELEMENTARY CATASTROPHE THEORY

M. J. SEWELL
Department of Mathematics
University of Reading
England

SUMMARY

An introductory account is given of some of the basic ideas in elementary catastrophe theory, with particular reference to applications. The list of section headings summarizes the list of ideas highlighted in the article. Some new remarks are included with the review material.

1. INTRODUCTION

One of the explicit objectives of this Conference was to

'cross disciplinary boundaries' and 'to consider the possible use of radically different forms of mathematics than are currently being used' in various fields of applied science, with particular reference to problems amenable to solution by discretization.

It seemed appropriate and consistent with this objective to offer an introductory explanation of some ideas of catastrophe theory. A portion of the material described in the lecture may be found in an article on some mechanical examples of catastrophe theory by Sewell [1], and detailed specific examples with diagrams are given there which will not be repeated here.

Lighthill begins his book 'Fourier Analysis and Generalised Functions' [2] with a dedication to Paul Dirac 'who saw that it must be true', to Laurent Schwartz 'who proved it', and to George Temple 'who showed how simple it could be made'. Perhaps it is appropriate to say, with regard to catastrophe theory, that René Thom first saw that it must be true (circa 1963?), that B. Malgrange and J. Mather contributed substantially to proving it (by 1968) and that Christopher Zeeman should have the credit for showing how simple it could be made (1969-74).

The book by Thom [3], subtitled An Outline of a General Theory of Models, is remarkable in its sweep of ideas and it has been compared in reviews with Newton's Principia. It does, however, make substantial demands on the reader, and a beginner may wish to complement it with some of Zeeman's original articles such as the Vancouver address [4], or one of the references contained therein , until he has learnt enough to construct examples of his own. A third viewpoint, less oriented towards applications but containing statements of underlying mathematical results, has recently been put by Arnol'd [5] in a survey of the 'Normal forms of functions' in neighbourhoods of what will here be called critical equilibrium points.

2. FAST AND SLOW TIME SCALES

Catastrophe theory gives, among other things, a qualitative description of how sudden effects can arise from smooth causes. This facet of the theory may therefore illuminate situations where the presence of two time scales is intrinsic. The sudden effect, or change of 'behaviour' takes place on a fast time scale and is called a 'catastrophe'. The smooth cause is a sequence of continuous adjustments, on a slow time scale, of certain 'control' parameters. The model shows how a catastrophe might have to occur during such a sequence and, on the other hand, suggests geometrically those sequences which would avoid a catastrophe.

3. TERMINOLOGY

Within the context of a single application, the terminology is not always very suitable. A 'catastrophe' is not necessarily disastrous in the everyday sense - the word is merely a shorthand for any sudden change. For example, there may be 'good' and 'bad' catastrophes changing the level of one's mathematical enjoyment [6] and a good catastrophe would be a contradiction in terms if one were restricted to everyday language.

Again, the values of every 'control' parameter need not always be susceptible to easy manipulation from outside - 'controls' are intended to represent merely independent variables, upon which the behaviour of the system will depend. For example, the geometry variable describing a shallow shell susceptible to snap buckling is fixed for a *given* 'system' but can be imagined varying to generate a *family* of systems. Nevertheless it enters the potential energy, as also does the assignable load on a given shell, in exactly the mathematical way required for interpretation as an 'essential control' variable in the cusp catastrophe model of buckling, as is shown in [1].

On the other hand, of course, there are contexts where the terminologies are suitable: mechanical snap-buckling *can* be truly a catastrophe as can the crash of a stock exchange [7] or the onset of a riot as a sudden change in the level of disorder in any institution, such as a university or a prison [8].

'Catastrophe' and 'control', notwithstanding their ambiguity, have therefore become well-established technical terms. They are certainly graphic.

4. EQUILIBRIUM SURFACE

In the context of a system containing a function $V(x_i, \lambda_\alpha)$ of n 'behaviour' variables x_i and of k 'control' parameters λ_α, the n gradient equations

$$\frac{\partial V}{\partial x_i} = 0 \tag{4.1}$$

define a k-dimensional surface in the (n + k)-dimensional $x_i - \lambda_\alpha$ space.

In the mechanics of engineering structures this surface has been called the 'equilibrium surface' by Sewell [9], because the vanishing gradient represents zero resultant force in that context. Equilibrium surface points which have the additional property that the determinant

$$\left| \frac{\partial^2 V}{\partial x_i \partial x_j} \right| = 0 \tag{4.2}$$

are sometimes called 'critical'. This is because, if the system has suitable dynamical properties, its equilibrium might become unstable there.

Of particular interest would be any critical equilibrium point which terminates a track on the equilibrium surface consisting entirely of stable points at which all the eigenvalues of the matrix $\left[\partial^2 V/\partial x_i \partial x_j\right]$ are positive, *and* whose instability is confirmed by the necessary extra considerations - such as explicit verification of the absence of an isolated local minimum of V with respect to the x_i at the given λ_α (e.g. by Taylor expansion of V to third or higher powers).

Then the instability will be manifested (if the fast-time dynamics allows it) by a sudden jump in the x_i-values (at constant λ_α-values) to another and distant equilibrium 'configuration' if it exists, and otherwise to the boundary of the behaviour space.

The standard introductory illustration combining a balance of simplicity and complexity is the so-called 'cusp catastrophe' deriving from the quartic

$$V = \tfrac{1}{4}x^4 + \tfrac{1}{2}ux^2 + vx \tag{4.3}$$

having one behaviour variable x and two control parameters u and v.

The equilibrium surface $\partial V/\partial x = 0$ is the smooth folded cubic surface

$$x^3 + ux + v = 0 \tag{4.4}$$

shown in the upper half of Fig. 1, and the critical equilibrium points on it have the additional property

$$3x^2 + u = 0 \tag{4.5}$$

according to (4.2). These critical points correspond to the line of the fold in the surface, where its smoothly turning tangent plane happens to be parallel to the x-axis.

The fold-line itself is a smooth space curve in three dimensions (Fig. 4 of [9]) whose parametric equations are

$$u = -3x^2, \qquad v = 2x^3. \tag{4.6}$$

Any particular value x_o therefore specifies a point on the fold-line about which the x-expansion of V is

$$V - \frac{3}{4}x_o^4 = x_o(x - x_o)^3 + \frac{1}{4}(x - x_o)^4, \qquad (4.7)$$

so that all the equilibrium configurations on the fold-line are in fact unstable except the origin (which is stable). The fold-line divides the equilibrium surface into two parts, of stable points outside the fold-line and of unstable points 'within' it (see Fig. 1). The set of unstable equilibrium surface points is thus 'closed almost everywhere' by the fold-line. Any system represented by (4.3), and by suitable fast-time dynamics, will therefore experience a catastrophe when any stable equilibrium path on the surface arrives at the fold. There will be a sudden jump or fall in the x-value, at constant u and v values, onto the more distant stable part of the equilibrium surface.

5. BIFURCATION SET

The projection of the fold-line onto the control space (the u, v plane in Fig. 1) is obtainable by eliminating x from (4.6), giving

$$4u^3 + 27v^2 = 0 \qquad (5.1)$$

which is the cusp shown in the lower half of Fig. 1. This cusp curve manifests the catastrophe points in the control parameter space. It may be thought of as a 'failure locus', for this specific type of failure. Points outside the cusp are certainly 'safe', being the projection of unique and stable equilibrium configurations; but points inside it may be unsafe, being the simultaneous projection of two stable and one unstable equilibrium points. Note in passing that the safe region is neither concave nor convex.

The cusped curve is an example of what is called the 'bifurcation set' in catastrophe theory. Equations (4.1) and (4.2) are n + 1 equations relating n + k variables, and if the n x_i are

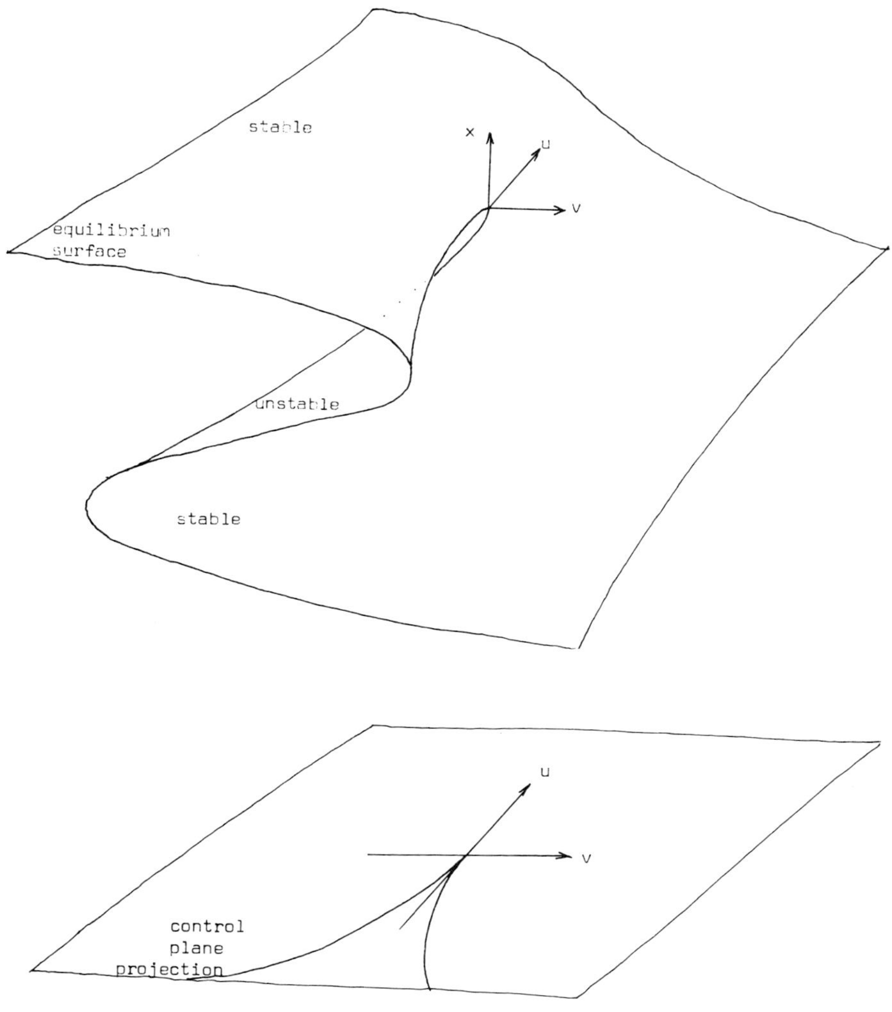

Fig. 1 - Equilibrium Surface and Projected Bifurcation Set for the Cusp Catastrophe

imagined to be eliminated from these n + 1 equations, there will remain a (k - 1)-dimensional hypersurface in the *control space*. This hyper-surface can also be thought of as obtained by a 'projection mapping' onto the control space. The hypersurface is called the *bifurcation set.*

6. THE BASIC CLASSIFICATION THEOREM

A primary result in catastrophe theory is a theorem (due essentially to Thom, but with contributions by Malgrange and Mather) that the possible local shapes of such bifurcation sets can be specified in an explicit way *independently* of the dimension n of the behaviour space. The specification depends on the dimension k of the control space, in the following way.

When n is large, direct determination of the bifurcation set by eliminating all of the n x_i will obviously be a forbiddingly intricate procedure. The basic theorem says that if the original potential $V(x_i, \lambda_\alpha)$ is smooth enough, the *local* shape of the bifurcation set is qualitatively equivalent to that obtainable by applying the same procedure to one of a small number K (depending on k) of simple polynomial potentials involving only *one* (x) *or two* (x and y) behaviour variables.

This number K (of so-called 'elementary catastrophes') is related to the dimension of the control space k by Table 1.

k	1	2	3	4	5	6
K	1	2	5	7	11	∞

Table 1. Number of elementary catastrophes.

The list of the first seven of these polynomials (which are linear in the control parameters) is given in Table 2 for convenience.

Name	Universal unfolding
Fold	$V = \frac{1}{3}x^3 + ux$
Cusp	$V = \frac{1}{4}x^4 + \frac{1}{2}ux^2 + vx$
Swallow tail	$V = \frac{1}{5}x^5 + \frac{1}{3}ux^3 + \frac{1}{2}vx^2 + wx$
Butterfly	$V = \frac{1}{6}x^6 + \frac{1}{4}ux^4 + \frac{1}{3}vx^3 + \frac{1}{2}wx^2 + tx$
Hyperbolic umbilic	$V = x^3 + y^3 + wxy - ux - vy$
Elliptic umbilic	$V = x^3 - 3xy^2 + w(x^2 + y^2) - ux - vy$
Parabolic umbilic	$V = xy^2 + x^4 + wx^2 + ty^2 - ux - vy$

Table 2. Elementary catastrophes for ≤ 4 controls.

A connected account of the proof of the basic theorem has been given by Zeeman and Trotman [10]. The shapes of the bifurcation sets for at least the first seven polynomials or elementary catastrophes have been obtained geometrically once and for all (e.g. see the diagrams in Woodcock and Poston [11] or Thom [3]).

The theorem is a result about the persistence (or 'structural stability') of singularities of projection mappings, under certain kinds of perturbation of potential functions.

7. APPLICATION OF THE THEOREM

The existing strict proof of the theorem applies for any finite n, which may therefore be very large, but not infinite. Nevertheless governing equations of type (4.1) also frequently emerge whenever discrete approximations (such as the finite-element method) can be reliably applied to infinite-dimensional systems characterizable by a variational principle. We could therefore anticipate, for example, applications of catastrophe theory in the field of optimization.

In any event, the first decision to be made in approaching an application is whether it is characterized by a gradient description or not. If not, we may be in the realm of 'generalized' catastrophes to which the larger part (Chapter 6 onwards) of Thom's book is devoted [3]. For gradient systems we may hope to apply one of the K 'elementary' catastrophes.

The next decision, for a supposed gradient system, is how many 'essential' control parameters influence the behaviour of the system. This is a key issue, and not always a straightforward one because an objective criterion for the meaning of 'essential' may not be available. There may be feedback in the mind with the simultaneous issue of what is going to be the interpretation of the single (or pair of) 'macroscopic' or gross behaviour variables x (and y). Some care is needed to avoid giving the impression that one is merely setting up the problem to fit the theory in a facile way.

On the other hand, a successful interpretation of k control variables means, according to Table 1, that the local bifurcation set will be found among those obtained from the corresponding number K of elementary catastrophes; and in these latter there exist only either one or two gross behaviour variables whose use may convert the problem into one which is *accessible* at the experimental level. This is in comparison with the original n-dimensional space of 'microscopic' behaviour variables, such as the n amplitudes of a truncated eigenfunction expansion, associated for example with the eigenvalue problem

$$\left[\frac{\partial^2 V}{\partial x_i \partial x_j} - \mu \delta_{ij} \right] \xi_j = 0. \qquad (7.1)$$

These n eigenfunctions may not be easily measurable *individually*.

At a point on the equilibrium surface (4.1) where (7.1) has m coincident *zero* eigenvalues $\mu = 0$ $(1 \leq m \leq n)$, the basic classification theorem appears to offer (for the purpose of describing the local bifurcation set) a condensation of the n behaviour variables into one or two 'overall' variables x and y; and hence a *transition* from the microscopic inaccessible level to a macroscopic accessible level. For example, we might point the issue by imagining that $n - m$ of the x_i, corresponding to the non-zero eigenvalues of (7.1), are discarded altogether (closer examination shows this to be an unnecessary over-simplification), and the remaining m analytically combined in some way which distributes them between x and y. But the precise nature of this condensation mechanism (the issue is raised by Sewell in [1]) is not understood. If $m = 1$ one can expect that only one of x and y will be required, but if $m > 1$ it is not even clear whether the condensation *necessarily* requires both of x and y, or if only one of them will suffice.

Some insight might be gained by studying Hansen's analysis [12] of the buckling of an axially compressed circular cylindrical shell. This analysis is based on an explicit potential energy functional, and the first eigenvalue has multiplicity m of order 100. There appears to be condensation not only of the behaviour variables, but also of the effective control parameters.

8. TWO ESSENTIAL CONTROLS

For a definite example, suppose it is decided that the problem is a gradient problem in which there are just two essential control variables. Then according to Tables 1 and 2 the local bifurcation set is to be found from among two elementary catastrophes (normally $K > k$ except when $k \leq 2$). These two polynomials are the cusp catastrophe (4.3) whose bifurcation set is illustrated in Fig. 1, and the fold catastrophe

$$V = \frac{1}{3}x^3 + ux \qquad (8.1)$$

where u is a single control parameter. The equilibrium surface for this is the parabola or 'fold' $x^2 + u = 0$ in the x, u plane (Fig. 2) whose critical point satisfies $2x = 0$, i.e. is the vertex of the parabola. The bifurcation set in the control space (the u-axis) is thus just the origin $u = 0$, which is also the projection of the vertex. We therefore are assured that we have *at most* these two bifurcation sets to choose from if our original decision of $k = 2$ essential control variables was right. In fact the cusp catastrophe organizes or subsumes two folds (approaching the cusp from either side, as Fig. 1 shows), since any vertical plane transverse to the fold in Fig. 1 will intersect the equilibrium surface locally in a curve like that of Fig. 2 (except at the cusp point itself). For this reason it will be adequate to concentrate upon the cusp catastrophe whenever $k = 2$. (The theorem says that it is adequate to adopt (8.1) alone whenever $k = 1$ is assured a priori).

Fig. 2 - Fold Equilibrium Surface

9. BUCKLING OF STRUCTURES WITH $k = 2$

The buckling of structures is a field offering many examples of elementary catastrophes. This can be inferred from the fact that the equilibrium of elastic structures capable of large geometrical changes under conservative loading can be characterized by variational principles containing non-quadratic functionals of the displacement. Among the possible control parameters are load (e.g. if dead loading), imperfection (e.g. in a distribution of load or of material), position of unassigned load (e.g. if rigid loading), initial structure geometry (e.g. depth of a shell), moduli. These are listed by Sewell [1, 9]. If we can decide in advance that only at most two of these are 'essential', then we can apply the cusp catastrophe.

For a straight strut under compression by an assigned axial load, it is known (with the aid of much hindsight) that load and imperfection are essential control parameters (normalized by a suitable combination of elastic moduli and strut dimensions). Real continuous engineering structures are often represented, in the introductory engineering literature, by simple 'thought models' made out of springs and rigid bars. This type of modelling seems really to be aimed at elucidating precisely that 'essential qualitative behaviour' of the real thing (in that particular field) which is also the more explicit object of catastrophe theory (in much wider fields). It is therefore perhaps not surprising to find that where such spring models are successful and informative, they have local potential energies corresponding rather closely to those polynomials in the list of elementary catastrophes. For example, in models of the strut (see [1,9]) the potential energy emerges precisely in the form of the cusp catastrophe (4.3) in which

x = buckling deflection

$-u$ = load in excess of buckling load

-v = imperfection due (e.g.) to eccentricity of loading.

Other spring models containing two essential control variables can lead in the first instance to a cubic energy

$$V(\theta) = \theta V'(0) + \frac{1}{2}\theta^2 V''(0) + \frac{1}{6}\theta^3, \tag{9.1}$$

and Budiansky [13] gives an example in which the normalized variables here have the interpretations

θ = buckling deflection

$V''(0)$ = load

$V'(0)$ = imperfection.

The equilibrium surface is an anticlastic fold in the space of these three variables. This fold can be viewed as locally equivalent to part of the fold in Fig. 1 away from the cusp point, so that any transverse section reproduces Fig. 2. Analytically the mapping $x = \theta + V''(0)$ onto a canonical behaviour variable x converts (9.1) into (8.1) in which the single *effective* control parameter is

$$u = 2V'(0) - [V''(0)]^2. \tag{9.2}$$

As remarked by Sewell [14], this single control has a complicated physical interpretation of imperfection minus $(\text{load})^2$.

This example illustrates how *either* of the $K = 2$ elementary catastrophes predicted by catastrophe theory can emerge from a decision that there exist $k = 2$ essential controls having similar physical interpretations.

What else might catastrophe theory have predicted ahead of time from this particular interpretation for $k = 2$, i.e. before the formulation and solution of partial differential equations which have in fact been studied for many years in the field of elastic buckling? Given an adequate justification of the discretization, the basic classification theorem imposes the cusp-shaped bifurcation set onto the *many*-dimensional problem, with the buckling-load axis tangential to the curve (*at* the cusp in the quartic case or *away* from it in the cubic) and imperfection axis perpendicular to it, as illustrated in Fig. 3. A small change in imperfection can therefore imply a large change in buckling load. If the latter change is a decrease (increase) the system is imperfection sensitive (insensitive). The direction of increasing buckling load varies from problem to problem. This is the criterion of imperfection-sensitivity which in fact was propounded by Koiter ([15], reviewed in [14]) on the basis of explicit solutions of the differential equations.

When *is* it reliable to apply the n-dimensional basic classification theorem to infinite-dimensional problems? Chillingworth [16] has studied this problem in the context of an infinite-dimensional formulation of the buckling of a strut. He deploys a lemma of Gromoll and Meyer [17]. The upshot appears to support the notion that infinite-dimensional eigenvalue problems having *discrete* (as opposed to continuous) *spectra* could be amenable to existing catastrophe theory for finite-dimensional behaviour spaces - without having to wait for a full classification theorem for the infinite-dimensional case.

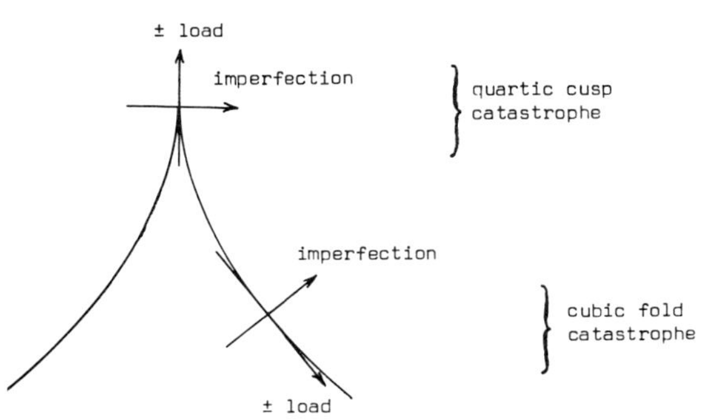

Fig. 3 - Sensitivity of Buckling Load to Imperfections in Structures

10. EQUILIBRIUM PATHS

These are tracks on the equilibrium surface. The plane $v = 0$ intersects the smooth folded surface of Fig. 1 in the branched equilibrium paths shown in Fig. 4. As u decreases from the right, a stable equilibrium path is followed until the branching point is reached (this is the only stable point on the whole fold-line of the equilibrium surface), where two stable (full line) and one unstable (dashed line) equilibrium paths branch off as u continues to decrease. In the mechanical examples this is quasi-static (the limit of slow-time) bifurcation buckling. For example, as load is slowly increased in a single test, the equilibrium of a straight strut or flat plate will begin to follow the stable deflected configuration after the load exceeds the point where the equilibrium deflection is no longer unique. Theories determining the limit of such uniqueness have been constructed for the relevant partial differential equations in elasticity and plasticity by Hill [18], and calculations of the ensuing finitely branched solutions have been carried out by Hutchinson [19]. Some of this work is reviewed in [14] and [20]. Parallel planes $v = \text{constant} \neq 0$ will intersect Fig. 1 in equilibrium paths near to those of Fig. 4 but without the branching point, thus showing the effect of (for example) imperfections.

Planes $u = \text{constant}$ will intersect Fig. 1 in quite different equilibrium paths. For $u < 0$ the intercept is the S-shaped curve shown in Fig. 5, where again the full line is stable and the dashed line is unstable. Suppose v is slowly altered in the direction of the single arrow, for example the change of transverse load on a shallow arch or shell like a biscuit-tin lid [1], or the change of chemical control in a heart muscle fibre [21]. The system eventually reaches a limit point of locally maximum $\pm v$ corresponding to the unstable fold-line of Fig. 1, and the sudden jump or catastrophe in fast-time takes place, marked

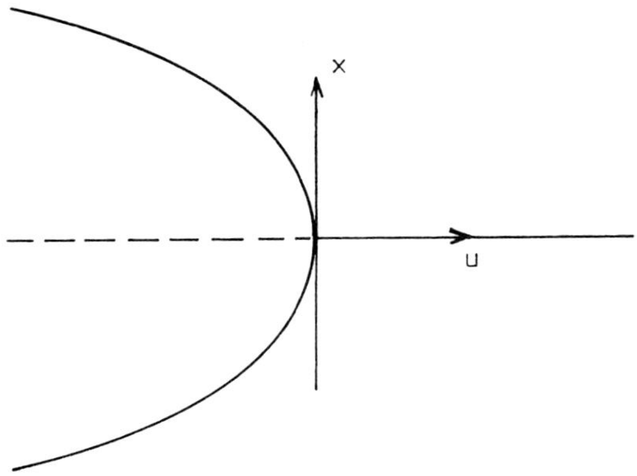

Fig. 4 - Trident Branching of an Equilibrium Path

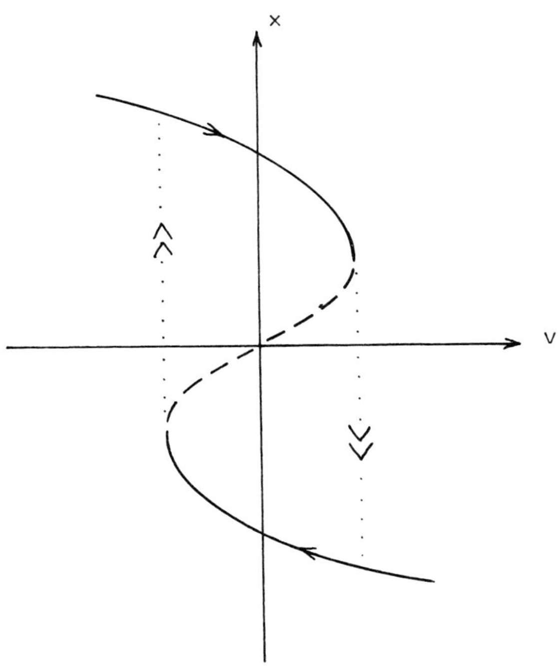

Fig. 5 - Hysteresis Cycle in Slow and Fast Times

by the double arrows. This represents snap-buckling of the configuration of the biscuit-tin lid, and contraction or expansion of the heart muscle fibre in the heart beat. Different levels of $-u$ (initial shallowness of the tin lid, or tension in the muscle fibre) will alter the amplitude of the hysteresis cycle shown in Fig. 5 - for example when the heart is bypassed in an operation the tension is reduced so much that the heart only heaves and does not beat (section of Fig. 1 with $u > 0$). Zeeman [21] remarks that as u is decreased through zero, the consequent appearance of the cycles shown in Fig. 5 represents the formation of a 'clock', of which the emergence of a Hopf bifurcation is one example. He calls the appearance of a new biological clock a 'change in phenotype'.

Planes tangential to the fold-line in Fig. 1, whose trace on the control plane is tangential to the cusp curve *away* from the cusp-point itself (cf. Fig. 3), will intersect the equilibrium surface in two equilibrium paths exhibiting the Poincaré 'exchange of stabilities' (Fig. 6) as they cross the fold line. This type of unstable branching point occurs, for example, in certain types of frame structures [22], but it cannot be precisely realized experimentally because of the imperfection insensitivity implied for the fold catastrophe by Fig. 3.

In other non-mechanical applications of the cusp catastrophe equilibrium surface, different kinds of equilibrium path are appropriate which are *not* necessarily obtained by plane sections of the surface as in the three cases just described. The basic point of interest is only how the equilibrium path arrives at (or avoids) the fold-line on the equilibrium surface, and so initiates (or avoids) a catastrophe. These other types of path might be proposed ab initio, or as the solutions of a system of differential equations (the 'slow equations') proposed on the surface. Both types will be illustrated in what follows.

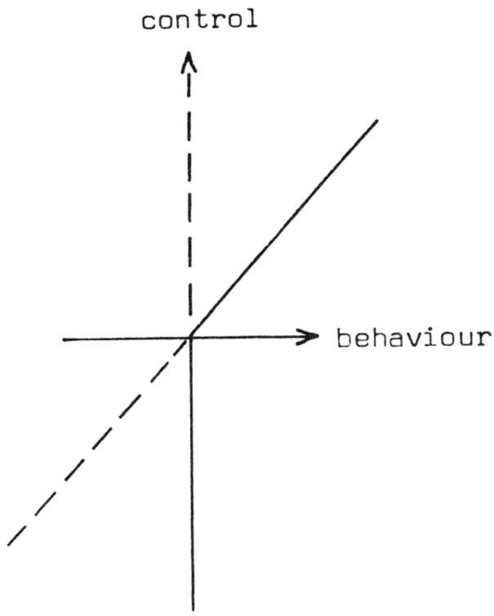

Fig. 6 - Poincaré Exchange of Stabilities

11. COMPLETENESS IN QUALITATIVE MODELLING

The *completeness* of the list of 11 elementary catastrophes in Tables 1 and 2 is the justification for confidence in offering geometrical models of qualitative situations in fields where no mathematical models were previously thought possible, because of the difficulty of quantification. Such fields include embryology, evolution, physiology, neurology, psychology, sociology, economics. For example, if we can be sure that at most three essential control factors are present, then Tables 1 and 2 show that we can confine attention to at most five polynomials V. And that these require at most two overall behaviour variables to describe the overall observable effect, even if the behaviour of the system at a substrate level really requires very many individual behaviour variables.

We have not the space here to do more than make a list of examples of these applications, mostly of the cusp catastrophe, in an attempt to convey the flavour and variety which has so far been achieved in non-physical contexts (i.e. where there is *not* also much detailed knowledge of governing partial differential equations etc.).

In the following Table 3 we incorporate the distinction which Zeeman [23] draws between pairs of control parameters which are either (a) 'normal' and 'splitting' or (b) 'conflicting'. In Fig. 1 v is a 'normal' factor because the behaviour x depends monotonically on v (actually increases with -v) on the stable part of the surface; whereas -u is a 'splitting' factor because the u-axis lies *within* the cusp projection, above which the equilibrium value of x is not unique, but 'split' between two stable and one unstable values. On the other hand, in some contexts it is really v - u and -v - u which are the natural controls to employ, and they are called conflicting factors because the whole cusp lies between them, and therefore so does the control area of 'conflict', where the behaviour has to choose between two different stable equilibrium positions. In case (a) we append n. or s. in Table 3, and in case (b) we append neither. The relevant references are entered in the Table, and study of these reveals a variety of different types of track on the folded equilibrium surface, including equilibrium paths which avoid the catastrophic fold by semi-spiralling round the cusp points as in Fig. 11.

The combustion catastrophe is from the area of chemical kinetics where more is known about the underlying differential equations than in the other examples; but there may be several controls and the choice of those two cited has recently been considered by Mehra and Blum [27]. One may also ask whether 'explosion boundaries' in pressure-temperature planes, of the kind pictured for example by Dainton ([28], Fig. II 4), can usefully be regarded as a bifurcation set in a catastrophe theory model.

Model name and reference		Control parameters	Description of sudden behaviour change
Aggression [24]		Rage Fear	Attack Flight
Manic depressive [23]	n. s.	Environment Chemical imbalance	Mania Depression
Committee behaviour [23]	n. s.	Support Opposition	Opinion switch
Military action [25]	n. s.	Threat Cost	Escalation Disengagement
Censorship [25]	n. s.	Erotic content Aesthetic value	Opinion jump ~Pendulum of fashion
Proverbs [4] 'More haste less speed'		Skill Haste	Gear change, or performance jump
Prison or campus riot [8]	n. s.	Tension Alienation	Riot Sudden calm
Scientific enjoyment [6]	n. s.	Creativity Technical skill	Take-off Relapse
Kitchen catastrophe [26]		Hunger Unpalatability	Eat Desist
Proposal Fig. 8	n. s.	Content Form	Acceptance Rejection
Combustion [27]	n. s.	Heat flux Oxygen partial pressure	Ignition Quenching

Table 3. Examples of the cusp catastrophe

12. CONFLICT

The notion of a 'conflict' between alternative stable equilibrium values of behaviour has proved to be a very graphic aid to the imagination in model building with elementary catastrophes. When the control point in the cusp catastrophe lies within the cusp, there are two such stable alternatives.

If a lecturer will construct a Zeeman catastrophe machine [29] he will readily be able to convey this sense of conflict to his audience in the following way. As the control point is brought inside the cusp and moved towards the opposite boundary, one can halt the control temporarily; one can then point out to one's audience that they are holding their breath. Why are they doing so? - in apprehension of the fact, which they have sensed, that more than one solution has become possible (thus illustrating the discomfort induced in a mathematician by the knowledge that uniqueness of his solution is no longer assured); and in apprehension also of the impending catastrophe. If the movement of the control is resumed and the snap buckling is allowed to occur, the audience then breathes out again and relaxes in the knowledge that the worst has occurred, and only one stable solution is again assured.

For an explicit example of how to construct a simple model embodying conflict, let us consider an issue which arises in relation to the danger's inherent in a supposed optimal design. Although it has lately become fashionable again, this is an old subject in terms of intuitively held knowledge. In using catastrophe theory to articulate this or other old knowledge in a new (geometrical) way, one may sometimes gain advantage as a consequence of *highlighting the options*.

Explicit proposals in optimal design date back at least as far as 1858, when Oliver Wendell Holmes [30] proposed* in 'The Deacon's Masterpiece', which he called 'a logical story', that a 'one-hoss shay' should be built such

"Thut the weakes' place mus' stan' the strain;
'n' the way t' fix it, uz I maintain,
T' make that place uz strong uz the rest."

In other words, to optimize the use of material one should design so that every part can only fail simultaneously - 'nothing local, as one might say', but

"All at once and nothing first, -

---

* I owe this reference to Professor Niels Lind.

Just as bubbles do when they burst."

Evidently Holmes realized that such a global collapse would be more catastrophic than any partial failure of an under-optimized design, and would certainly be harder to mend. Perhaps it was fortunate that the 'Masterpiece' lasted 100 years, which is longer than some recent box-girder bridges.

Suppose we could agree that long-term *viability* of a product (e.g. motor car, nylon stockings) depended essentially on just the two control factors of optimality of functional design *and* accessibility to repair. Then the cusp catastrophe applies (*if* the system is a gradient system), with conflicting factors as shown in the control plane of Fig. 7 where the viability levels are also indicated. That is, a good (bad) design which is easy (hard) to mend will be assigned a high (low) long-term viability; but whereas a poor design which is easy to mend might be tolerated for a while, a good design which is hard to mend could understandably be assigned either high or low viability by different individuals. The smooth folded surface whose height represents viability will be like that of Fig. 1, but with the control axes differently oriented so that its projection onto the control plane is Fig. 7. Track (1) represents an individual car of reasonable design which, as time goes on, becomes harder to mend until its owner suddenly decides (at the blob) that it is no longer viable and sells it. Track (2) represents a range of other cars available on the market, whose manufacturer is seen to be improving both the basic design *and* the after-sale spare parts service (so that the repair facility is improving). At the blob the potential customer suddenly decides that the model is viable and buys. I was told by a former Yugoslav partisan that they had access to both German and Russian weapons. The German guns were superbly engineered to fine tolerances, but in the field were hard to mend when they did go wrong. By contrast the Russian weapons were more crudely made, but easier to mend. Hindsight shows that, while both the Deacon's shay and the box-girder bridge

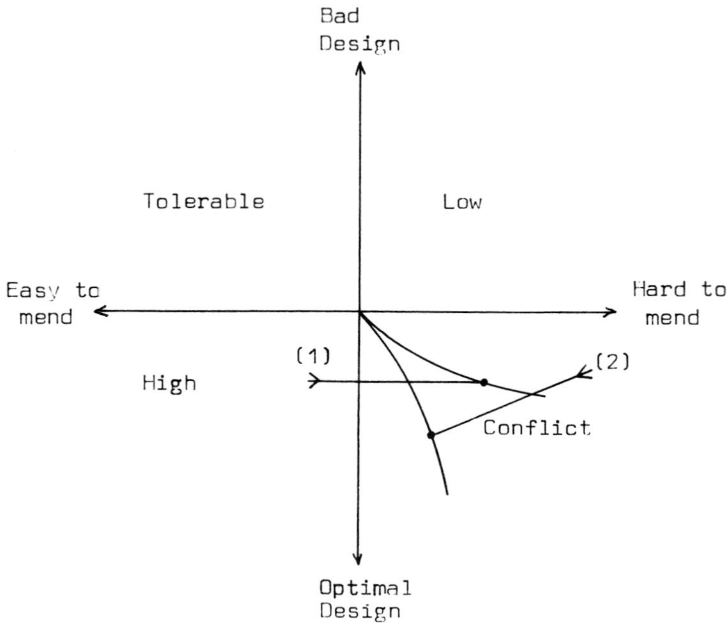

Fig. 7 - Dependence of Long-Term Viability on Two Controls

lie *within* the cusp of Fig. 7, the shay should have been assigned to the upper part of the surface, and the bridge to the lower part.

A policy proposal, or a scientific article, or an artistic object, is often judged according to both its form and content. In Table 3 content was labelled as a normal factor influencing acceptability, but form was labelled as a splitting factor. Fig. 8 sums up the smooth folded surface of acceptability by its projection onto the control plane. This model contains the notion that, if content is only moderately impressive or 'arguable', a conflict can be sown in the assessor's mind by improving the form or manner or presentation or type of advocacy. It also illustrates the moderate success sometimes achieved by one said to be 'more interested in form than content'. The downward catastrophe takes

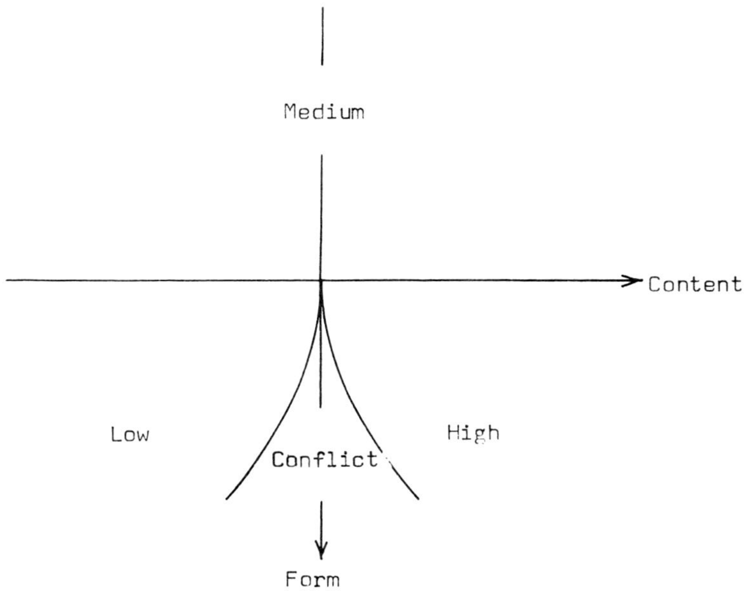

Fig. 8 - Dependence of Acceptability on Form and Content

place when a plagiarism or pretence is 'rumbled' - as happened in Hans Andersen's tale of 'The Emperor's New Clothes'. The well-known ambivalent effect of form in splitting opinion about abstract artistic objects of little discernible content is also included.

The innumerable conflicts of everyday life can be used as a vehicle for discussing catastrophes and the highlighting of options even with children (the 'brake and axselerator catastrofee' was propounded to me on the New York Thruway), but the wisdom of doing so will depend on maintaining a sense of the conflict [31, Fig. 1] between the trivial and the profound.

13. INTERPRETATIONS OF THE GENERATING POTENTIAL

The most obvious interpretation to expect for the potential function $V(x_i, \lambda_\alpha)$ whose gradient generates the equilibrium surface (4.1) is that of potential energy of the system. This is indeed an appropriate interpretation in, for example, the buckling of elastic structures, possibly with due allowance for rotating reference frames (e.g. see [9]). Lyapunov functions or cost functions are other possibilities, or indeed any functional generating a variational principle.

In many of the non-physical examples of catastrophe theory constructed from the conflict viewpoint, however, the actual potential generating the gradient field plays a minor role and is often not mentioned at all. This trend in the literature is unfortunate, because it induces justified sceptism in the reader who is looking for illustrations of the various features of the main classification theorem. There are remarks in references [8] and [25] suggesting that in many cases the potential will be related to the probability $P(x)$ that the behaviour will be at the level x, with the 'equilibrium surface' generated by $\partial P/\partial x = 0$. The maxima (minima) of the probability distribution for a given control then would correspond to the stable (unstable) equilibrium states. For example, the probability distribution along the vertical trace $v = 0$ of the S-shaped equilibrium curve of Fig. 5 (rotated in Fig. 9(a)) would correspond to the distribution with two maxima shown in Fig. 9(b).

This problem of identifying the existence and interpretation of the generating potential arises also in another more physical context, namely the theory of phase transitions. Boiling or condensation of a gas might be viewed as catastrophes in the sense of sudden changes of density $1/\nu$ at suitable temperature t and pressure p, and Fowler [32] has shown that the variables in van der Waals' and Berthelot's equations of state can be simply

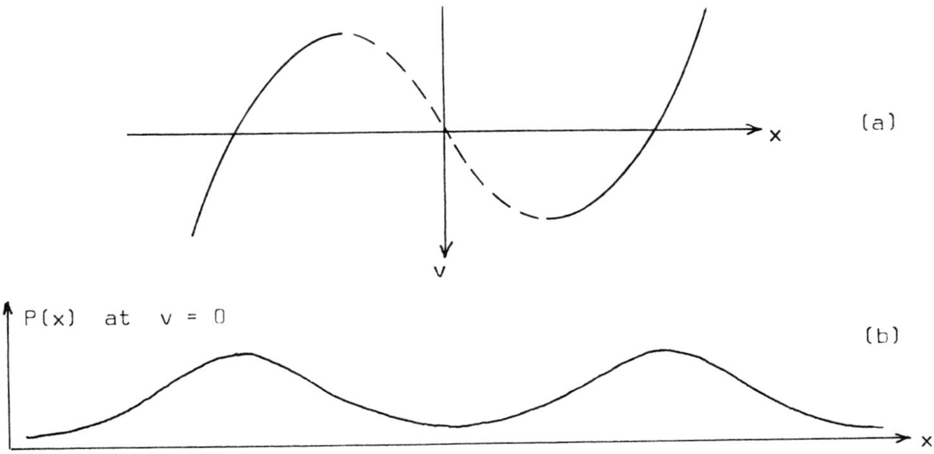

Fig. 9 - Probability Distribution with Two Conflicting Maxima

transformed so as to express these equations as examples of the cubic equation of (in this context thermal) equilibrium (4.4), by taking origin at the normalized liquid/gas critical point $(\nu, t, p) = (1, 1, 1)$. He remarks, however, that 'there appears to be no physically relevant potential'. In fact the potential exists and, if thermal equilibrium is viewed in the right way, it is interpretable as proportional (with ν-dependent proportionality factor) to the well-known 'thermomechanical potential' function, e.g. see Sewell [33, equation (23)]. The latter has the values of $\nu p + \Psi(t, \nu)$ + arbitrary constant, where $\Psi(t, \nu)$ is the free energy. Van der Waals' equation emerges by inserting into $-p = \partial\Psi/\partial\nu$ the particular free energy

$$\Psi(t, \nu) = \nu - \frac{3}{\nu} - \frac{8}{3}(t + 1)\log(3\nu - 1) \tag{13.1}$$

in the domain $\nu > \frac{1}{3}$ (to avoid the logarithmic singularity).

The transformation $\frac{1}{\nu} = x + 1$ can then be used to construct a function

$$V(x, t, p) = \frac{1}{3}(1 + x)^2(2 - x)[\Psi(t, \nu(x)) + \frac{p}{x + 1} + \text{constant}] \tag{13.2}$$

whose gradient is Fowler's canonical cubic (with controls $\frac{1}{3}(8t + p)$ and $\frac{1}{3}(8t - 2p)$), provided the datum constant is always chosen to make the thermomechanical potential zero on the equilibrium surface. The x-dependent proportionality factor is never zero because the domain of the free energy is $-1 < x < 2$, which includes the liquid/gas critical point $x = 0$.

This example shows that in examples of the general type

$$V(x_i, \lambda_\alpha) = \Phi(x_i)\Psi(x_i, \lambda_\alpha) \tag{13.3}$$

with bounded non-zero $\Phi(x_i)$, a clear interpretation may attach to Ψ rather than V.

14. COMPROMISE

The 'butterfly' catastrophe

$$V = \frac{1}{6}x^6 + \frac{1}{4}ux^4 + \frac{1}{3}vx^3 + \frac{1}{2}wx^2 + tx \tag{14.1}$$

has four controls, and so in general a four-dimensional equilibrium surface in 5-space. For each *fixed* u and v we can represent the surface as a folded two-dimensional surface in (x, w, t)-space, and if $u < 0$ this surface is not unlike that shown in Fig. 1, except that it has two extra folds smoothly joining an intermediate 'stable' platform to those already present in Fig. 1.

As explained and illustrated by Zeeman and Isnard ([23], [25]), there are applications in which this intermediate level can be interpreted as a 'pocket of compromise'. Instead of Fig. 9, there

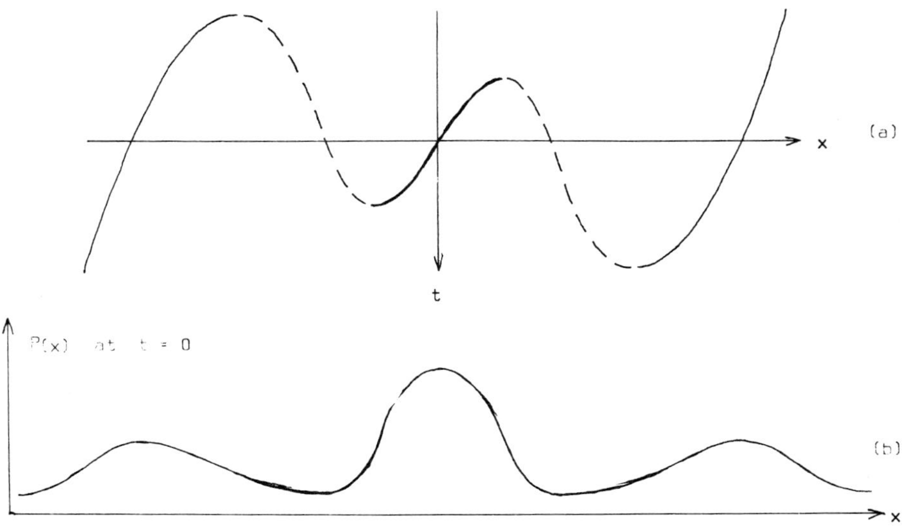

Fig. 10 - Probability Distribution with Compromise Maximum

exist cross-sections of the surface (e.g. with $u < 0$, $v = 0$, and some $w > 0$) such that a convoluted S-shape contains the intermediate compromise level. A probability distribution along the horizontal trace $t = 0$ in Fig. 10(a) would then show three (stable) maxima and two (unstable) minima, and the intermediate platform may correspond to the greatest probability maximum - hence the notion of compromise.

Examples necessarily require at least three essential control factors, and some examples of the cusp catastrophe can be refined to illustrate this evolution of a compromise. The 'butterfly factor' $-u$ may correspond to real time ('time heals'), for example in the problem of military action or national defence or wage bargaining. External factors (of which time is only one illustration) often act as 'the checks and balances' to induce the emergence of a pocket of compromise. For example, consider the student experiencing a conflict caused by obscure presentation to him of intrinsically fascinating material. He will often have the conflict resolved for him by the existence of an examination later on. This induces an intermediate level as the most probable level of attention to the lecture course.

15. SLOW EQUATIONS

It was remarked in §10 that tracks on an equilibrium surface might be specified as the solutions of differential equations assigned on the surface, that is as orbits on the surface, instead of being given ab initio, e.g. by certain cross-sections as illustrated earlier. *Such* differential equations may be called the 'slow equations' describing a slow-time development *on* the surface, in contrast to the more familiar fast-time equations describing fast-time return or homeostasis of the system to the equilibrium surface.

For example, Zeeman's model of the heartbeat [21] in its simplest form is built on the pair of equations

$$\varepsilon \frac{dx}{dt} = - \frac{\partial V}{\partial x}, \qquad \frac{db}{dt} = \frac{\partial V}{\partial b} \tag{15.1}$$

with $V = \frac{1}{4}x^4 + \frac{1}{2}ax^2 + b(x - x_a)$.

Here the behaviour x is length of muscle fibre, and b is an electro-chemical control which is altered in situ by the pacemaker wave, so triggering a heart beat cycle which is a refinement of that in Fig. 5. Control $-a$ is tension in the muscle fibre (provided in situ by blood pressure), x_a is a constant characterized by a single equilibrium point (diastole) on the cycle, ε is a small constant and t is real time.

The first equation of (15.1) is the fast equation, and from it we can deduce the equilibrium surface (4.4) of the cusp catastrophe by letting the 'slow-time parameter' $\varepsilon \to 0$. The second equation is the slow equation, whose function is to describe a slow feedback along the equilibrium surface after the jump contractions and relaxations. To the extent that there is only one genuine equilibrium point (at diastole) in this example the term 'equilibrium surface' is strictly speaking a misnomer, and may be replaced if desired by 'slow manifold'.

Hamiltonian mechanics, e.g. in its very simplest form

$$m \frac{dq}{dt} = p, \qquad \frac{dp}{dt} = -\frac{\partial V}{\partial q}, \tag{15.2}$$

has two coupled fast equations and no slow equation. A scale factor m in the mass distribution can be used like ε in (15.1) to obtain the equilibrium surface $\partial V/\partial q = 0$ (or at least an organizing centre for an unfolding into such a surface) by letting $m \to 0$.

Slow equations have to be hypothesized separately to suit each desired context. Equations with spiral-type flow solutions on the cusp catastrophe surface can be used to model situations where a cycle involves only *one* jump followed by *smooth* return, instead of a jump return (and hence two jumps in the cycle) as in Fig. 5. This smooth return occurs in the propogation of nerve impulses, and Zeeman's starting point [21] for a modelling of experimental data on this problem is the fast equation

$$\varepsilon\dot{x} = -(x^3 + ax + b) \tag{15.2}$$

and the pair of slow equations

$$\begin{aligned} \dot{a} &= -2a - 2x \\ \dot{b} &= -a - 1. \end{aligned} \tag{15.3}$$

There is a sharp rise in 'sodium conductance' $-x$ at the beginning of a nerve impulse triggered by a change in 'membrane potential' b, followed by a slower rise and fall in 'potassium conductance' a during the smooth return (round the cusp point) of the sodium conductance to its former value. Such a track is sug-

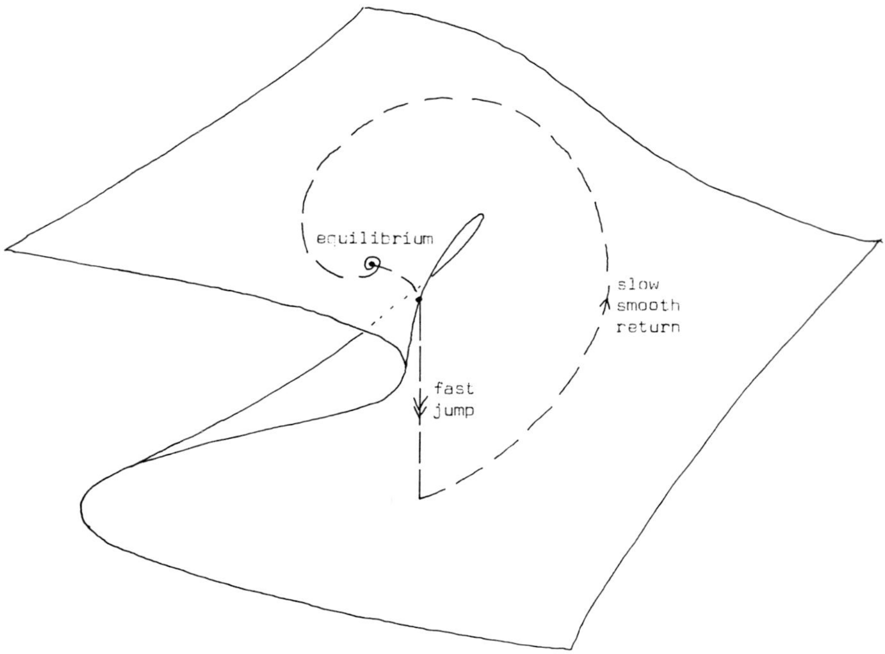

Fig. 11 - A Single Fast Jump Followed by Slow Smooth Return

gested in Fig. 11.

The simplest model of an earthquake would presumably include the quality of sudden release of shear stress preceded by a slow build-up. It may be asked whether behaviour on, for example, the San Andreas fault can usefully be viewed in this particular type of periodic way.

16. OTHER ASPECTS OF CATASTROPHE THEORY

There are other major areas of catastrophe theory which we have not space to describe in this review. They include the role played by the three umbilics, and the diversity of ideas and applications (especially to embryology and morphogenesis) to be found in

Thom's book [3]. We conclude with the briefest of remarks.

In many of Thom's applications the control variables are actually the three space variables and time. This is an insistent theme in the book, and for that reason alone provides a contrast with the type of examples described here. Also, Thom spreads himself far beyond the confines of the local results predicted by a strict interpretation of the main classification theorem, pointing the way in various directions and taking the view that the degree of interest in an idea rarely corresponds to its formal rigour. As remarked in §7, he also devotes much space to generalized catastrophes in non-gradient systems. The book is now widely available and we refer the reader to it. The English edition contains a new discussion of the parabolic umbilic.

Outside of Thom's book, the literature does not convey a very clear sense of perspective in looking at the particular question of the role of the umbilics, certainly by comparison with that of the cusp which I have tried to indicate in this article. In optics there are some examples. Partial differential equations in (for example) elasticity and fluid mechanics are often solved approximately by variational methods, and these solutions can be examined to see if they reveal umbilics. In a few cases they have appeared to do so, whereas in other cases such an attempted identification does not give a clear answer.

Suppose that a discretization of a continuum variational problem yields an energy approximation which is *not* locally that of one of the universal unfoldings listed in Table 2. Suppose, for example, that it differs from one of them only in having a 'structurally unstable' organizing centre (as pointed out in [1], Hutchinson's [34] approximate solution for spherical shell buckling under external pressure has organizing centre xy^2). Would this fact have any useful implication for the adequacy of the original approximation?

On the other hand, if an approximate solution does lead to an energy immediately identifiable with a 'structurally stable' organizing centre, is that approximation thereby endowed with special significance? Or do we have an easily identifiable catastrophe just by a lucky fluke? By adopting slightly different approximating procedures could the investigator have arrived at a significantly different organizing centre? Presumably a structurally stable organizing centre should not be sensitive to 'imperfections' in this sense. Is the role of the idea of structural stability in approximation theory well understood?

REFERENCES

1. Sewell, M.J., "Some Mechanical Examples of Catastrophe Theory", to appear in *Bulletin of the Institute of Mathematics and its Applications*, June/July 1976.
2. Lighthill, M.J., *Introduction to Fourier Analysis and Generalised Functions*, Cambridge University Press, 1964.
3. Thom, R., *Structural Stability and Morphogenesis*, W.A. Benjamin, Inc., Reading, Mass., 1975.
4. Zeeman, E.C., "Levels of Structure in Catastrophe Theory", International Congress of Mathematicians, Vancouver, 1974.
5. Arnol'd, V.I., "Normal Forms of Functions in Neighbourhoods of Degenerate Critical Points", *Russian Mathematical Surveys* 29, 10-50, 1974.
6. Zeeman, E.C., "Research, Ancient and Modern", *Bulletin of the Institute of Mathematics and Its Applications*, 10, 272-281, 1974.
7. Zeeman, E.C., "On the Unstable Behaviour of Stock Exchanges", *J. Math. Economics* 1, 39-49, 1974.
8. Zeeman, E.C., Hall, C.S., Harrison, P.J., Marriage, G.H. and Shapland, P.H., "A Model for Prison Disturbances", Warwick University Preprint, 1975. To appear in *J. Math. and Stat. Psychology*.

9. Sewell, M.J., "On the Connexion Between Stability and the Shape of the Equilibrium Surface", *J. Mech. Phys. Solids* 14, 203-230, 1966.

10. Zeeman, E.C. and Trotman, D.J.A., "The Classification of Elementary Catastrophes of Codimension ≤ 5", Warwick University Preprint, 1974.

11. Woodcock, A.E.R. and Poston, T., *A Geometrical study of the Elementary Catastrophes*, Springer-Verlag, Berlin, 1974.

12. Hansen, J., "Influence of General Imperfections in Axially Loaded Cylindrical Shells", Report 76, Danish Center for Applied Mathematics and Mechanics, November 1974.

13. Budiansky, B., "Theory of Buckling and Post-Buckling of Elastic Structures", *Advances in Applied Mechanics* 14, 1-65, 1974. Academic Press.

14. Sewell, M.J., "Elastic and Plastic Bifurcation Theory", Reading University Mathematics Department, Preprint of Lectures to the Seminar on the Plasticity Theory and its Application in Techniques, University of Niš, Yugoslavia, July 1975.

15. Koiter, W.T., "On the Stability of Elastic Equilibrium", NASA Tech. Trans. F10-833, 1967 (translation of a 1945 Thesis in Dutch).

16. Chillingworth, D.R.J., "The Catastrophe of a Buckling Beam", Proceedings of the Symposium on Applications of Topology and Dynamical Systems, ed. A.K. Manning, University of Warwick, 1973/4.

17. Gromoll, D. and Meyer, W., "On Differentiable Functions with Isolated Critical Points", *Topology* 8, 361-369, 1969.

18. Hill, R., "Bifurcation and Uniqueness in Nonlinear Mechanics of Continua" (Muskhelishvili volume), *Soc. Ind. App. Math.*, Philadelphia, 155-164, 1961.

19. Hutchinson, J.W., "Plastic Buckling", *Advances in Applied Mechanics* 14, 67-144, 1974. Academic Press.

20. Sewell, M.J., "A Survey of Plastic Buckling", pp. 85-197 of *Stability* (ed. H. Leipholz), University of Waterloo, 1972.

21. Zeeman, E.C., "Differential Equations for the Heartbeat and Nerve Impulse", *Towards a Theoretical Biology* 4 (ed. C.H. Waddington), 8-67, 1972.
22. Roorda, J., "Stability of Structures with Small Imperfections", *J. Engng. Mech. Div., Am. Soc. Civ. Engrs.* 91, 87, 1965.
23. Zeeman, E.C., "Applications of Catastrophe Theory", Warwick University Preprint, Tokyo International Conference on Manifolds, 1973.
24. Zeeman, E.C., "The Geometry of Catastrophe", *Times Literary Supplement*, 10th December 1971.
25. Isnard, C.A. and Zeeman, E.C., "Some Models from Catastrophe Theory in the Social Sciences", *Use of Models in the Social Sciences* (ed. L. Collins) Tavistock, London, 1974.
26. Sewell, M.J., "Kitchen Catastrophe", *Mathematical Gazette*, 59, 246-249, 1975.
27. Mehra, R.K. and Blum, E.H., "On the Dynamics of the Ignition of Paper and Catastrophe Theory", Report RM-75-42, International Institute for Applied Systems Analysis, Schloss Laxenburg, Austria.
28. Dainton, F.S., *Chemical Reactions*, 2nd edition, Methuen, 1966.
29. Zeeman, E.C., "A Catastrophe Machine", *Towards a Theoretical Biology* 4 (ed. C.H. Waddington), 276-282, 1972.
30. *The Poetical Works of Oliver Wendell Holmes*, Routledge, London, 1896.
31. Zeeman, E.C. and Thom, R., "Catastrophe Theory: Its Present State and Future Perspectives", Warwick University Preprint, 1974.
32. Fowler, D.H., "The Riemann-Hugoniot Catastrophe and van der Waals' Equation", *Towards a Theoretical Biology* 4 (ed. C.H. Waddington), 1-7, 1972.
33. Sewell, M.J., "On Applications of Saddle-Shaped and Convex Generating Functionals", *Physical Structure in Systems Theory*, (ed. J.J. van Dixhoorn and F.J. Evans), 219-245, Academic Press, 1974.
34. Hutchinson, J.W., "Imperfection Sensitivity of Externally Pressurized Spherical Shells", *J. Appl. Mech.* 34, 49-55, 1967.

ON THE EXTENSION OF COMPLEMENTARY VARIABLES MODELING THEORY TO BIOLOGY AND ECOLOGY

G. H. SMERAGE
University of Florida
Institute of Food and Agricultural Sciences
Gainesville, Florida

SUMMARY

Although they exhibit other, strictly biological properties, biosystems and ecosystems are substantially physicochemical. After a comparative review of the main properties of the complementary variables representation of physical systems and the compartment representation of biological and ecological systems, a compatible complementary variables representation of chemical systems is presented. Application of this representation to the physicochemical aspects of biological systems is discussed with illustrative examples. The main conclusions are that a general complementary variables theory of physicochemical systems is needed and that this theory could contribute significantly to the advancement of biological and ecological sciences.

1. INTRODUCTION

A major consideration in system science is representation or, simply, modeling. The complementary (conjugate in thermodynamics) variables theory of representation is used for physical systems. The foundations of complementary variables (CV) modeling are physical science and mathematics, and the strength and success of this theory stem from the fact that it has captured the essence of physical phenomena and systems. Mathematical statements of the concepts permit unified, operational treatment of electrical, mechanical, fluid, thermal, and compound systems. CV theory is analytical; it consists of concepts by which (1) a total system can be decomposed into a set of basic components and structure, (2) the components and structure can be described, and (3) the total system can be described in terms of its parts. Conversely, the theory may be used to synthesize new systems with prescribed functions and characteristics.

During the past decade interest has become strong in a systems approach to biological and ecological systems [1-9]. The International Biological Program focused attention on the representation and behavior analysis of such systems[10,11,12]. The traditional approach to representing biological and ecological systems is compartment modeling, which is a method of dynamic bookkeeping for the extensive quantities of a system. It lacks a solid foundation of biological or physicochemical fundamentals, and compartment models generally fail to reflect the biological or physicochemical essence of a system. Consequently, a compartment model is little more than a postulated system of equations that is designed to reproduce the time series of extensive quantities observed in real biological or ecological systems. Interpretation of model-predicted system behavior in terms of biological and physicochemical concepts is generally limited.

A plausible view of biological and ecological systems is that they involve two broad classes of phenomena--physicochemical and cybernetic--and that physiology, population dynamics, competition, preditor-prey interactions, and other specific topics can be analyzed and represented by these sciences. This view is most plausible for physiological systems, which are predominantly physicochemical but also exhibit cybernetic properties.

A few scientists are proceeding along this path [13-21]. Attention has been focused on representing the physicochemical aspects of biological and ecological systems directly in terms of the relevant physicochemical science. The purpose of this paper is to discuss CV representation of the physicochemical aspects of biological and ecological systems.

2. CV MODELS OF PHYSICAL SYSTEMS

The CV theory of representation is a unified approach to electrical, mechanical, fluid, thermal, and compound systems. Its strength and success undoubtedly stem from its attention to and direct expression of the fundamentals of physical science. In this section several important properties of physical systems and the CV representation of physical systems are summarized. The reader is referred to the extensive literature for details [22-32].

1. The most important and powerful property of CV theory is the decomposition with sufficient resolution of a total system into its elementary components and structure followed by the derivation of a total system mathematical model from the component and structural descriptions.
2. The dynamic behavior of a physical system is the result of imbalances in the energy intensities within the system and the properties of the components and structure

of the system. A small set of basic component processes is the basis for all complex systems, and complex processes and total systems are networks of these basic components.

3. Physical processes comprise three classes: transport, storage, and transformation. Transport, storage, and transformation are fundamental concepts that span the four energy domains: electrical, mechanical, fluid, and thermal. Thus, basic components such as resistors, inertias, capacitors, transformers, and transducers are found in all energy domains. The invariance of these concepts over the four energy domains is manifest in a comparison of the constitutive relations for the basic components of the four energy domains.
4. The variables in physical systems comprise two principal classes: the extensive (through) variables and the intensive (across) variables. The principal extensive and intensive variables are denoted generically by γ and ξ, respectively, and they exist in pairs (ξ,γ). This is the reason for the term complementary (conjugate). The principal extensive variables are flow quantities: mass or volume flow, entropy flow, charge flow, and force. The principal intensive variables are measures of energy intensity, i.e., potentials: pressure, temperature, voltage, and velocity. Integrals of these variables yield two classes of secondary variables, the integrated extensive variables and the integrated intensive variables [27]. Mass, charge, volume, and entropy are integrated extensive variables. Momentum and magnetic flux are integrated intensive variables.
5. The essence of a component process is expressed by relations between its relevant intensive and extensive vari-

ables. These relations may be linear or nonlinear. The component variables also enter into the description of system structure.

6. Often it is possible to express resistance, capacitance, compliance, inductance, inertance, and other parameters of basic components in terms of geometric and material properties [25]. For example, the resistance and inertance in the pressure-volume flow relationship for a capillary of radius r and length ℓ are given, respectively, by

$$R = \frac{8\eta\ell}{\pi r^4} \quad \text{and} \quad I = \frac{\rho\ell}{\pi r^2} \tag{2.1}$$

where ρ and η are the mass density and viscosity of the flow material. The resistance of an electrical resistor of length ℓ, uniform cross area A, and resistivity ρ is

$$R = \rho \frac{\ell}{A} \tag{2.2}$$

The capacitance of a parallel plate electrical capacitor with area A, plate separation ℓ, and permittivity ε of the intervening material is

$$C = \varepsilon \frac{A}{\ell} \tag{2.3}$$

7. A system is a network of basic components; as such, it exhibits structure. A description of system structure consists of two sets of equations: the interface set in the extensive variables and the circuit set in the intensive variables. The interface set is found by applying the *interface law* for extensive variables

$$\sum_{\forall k} \gamma_{jk} = 0 \tag{2.4}$$

to the interfaces between the components of the system [26,27,20], where j is the interface index, k is the component index, and γ_{jk} is the extensive variable from the j-th interface to the k-th component. The circuit set is found by applying the *circuit law* for intensive variable changes to circuits (closed sequences of interfaces) of the same energy type in the system [ibid.]

$$\sum_{\text{circuit}} \xi_{ij} = 0 , \tag{2.5}$$

where $\xi_{ij} \triangleq \xi_i - \xi_j$ is the difference between the intensive variables at the i-th and j-th interfaces. From (2.4) and (2.5), system structure is linear.

8. A mathematical model for a total system is derived by any of several methods [26,27] that combine the component and structural descriptions.
9. Since system structure is linear, the linearity or nonlinearity of a total, fixed-structure system depends directly on the linearity or nonlinearity of the components in the system.
10. At the j-th interface, the energy flow into the k-th component is given by

$$p_{jk} = \xi_j \gamma_{jk} , \tag{2.6}$$

where ξ_j is the intensive variable associated with the j-th interface. The energy flows at an interface are constrained by

$$\sum_{\forall k} p_{jk} = 0 \tag{2.7}$$

11. The state of a system is the instantaneous distribution of energy in the system. The intensive variables of storage components and the through variables of inertial components are appropriate variables by which to measure the instantaneous state of a physical system.

12. Mathematical CV models of physical systems are dynamical. Therefore, they are amenable to a wide range of system behavior analyses, including modes of behavior, responses to specific inputs and initial states, and stability [25,33,34]. The components and structure in which a system has been conceptualized provide a basis for interpreting and understanding system behavior.

3. COMPARTMENT MODELS

Compartment models are the traditional approach in biology and ecology to represent matter and energy flows, population dynamics, and other system considerations [35,36,37]. A compartment is defined to be a quantity of some entity, i.e., an extensity. A conceptual model of the compartment type for a particular system consists of a network of compartments for all relevant entities in the system. Paths linking compartments represent flows of contents between compartments. In addition, there may be independent flows from the environment to certain compartments in the system. A hypothetical compartment model is illustrated in Fig. 1. The x_i and f_{ij} in the figure indicate, respectively, the contents of the i-th compartment and the flow from the i-th compartment to the j-th compartment. The contents of and flows incident to a compartment are homogeneous.

The mathematical representation of a compartment model is developed by applying an appropriate conservation law, e.g., of mass, energy, or population, to each compartment and the flows incident to it. A set of first-order differential equations is obtained in the form

$$\frac{dx_i}{dt} = \sum_{\forall j} f_{ij} \tag{3.1}$$

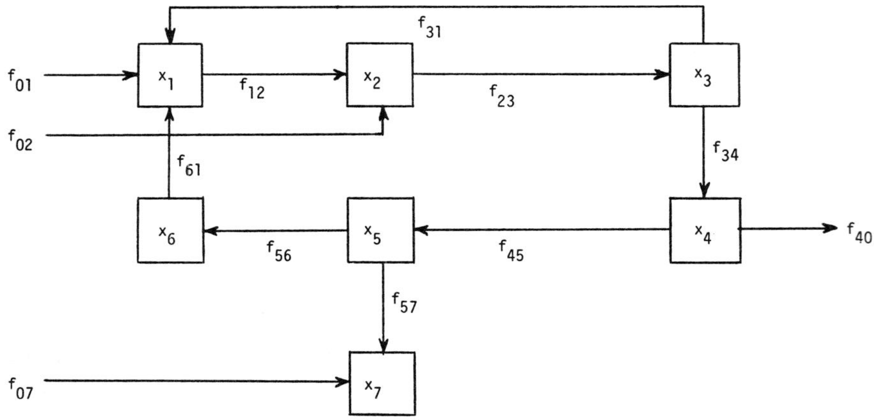

Fig. 1. A hypothetical compartment model.

The $\{x_i\}$ are the state variables of the system. Equation (3.1) indicates that compartment models have linear structure.

The mechanisms of the $\{f_{ij}\}$ are postulated as relationships between the contents of compartments i and j. Occasionally the choice of relationships is guided by biological or ecological rationale; it may be more common for the choice to be empirical. Common mechanisms are

$$\begin{aligned} f_{ij} &= \text{independent function of time} \\ f_{ij} &= k_{ij}\, x_i \qquad \text{donor control} \\ f_{ij} &= k_{ij}\, x_j \qquad \text{acceptor control} \\ f_{ij} &= k_{ij}\, x_i\, x_j \qquad \text{joint control} \end{aligned} \tag{3.2}$$

where k_{ij} is a constant. Substitution of the flow descriptions in (3.1) yields a set of coupled first-order differential equations which is the mathematical model of the total system. Clearly, the model is dynamical and its linearity depends on the flow mechanisms postulated.

The physical systems analyst will recognize a similarity between compartment models and CV models of transport-storage networks, e.g., R-C electrical, fluid, or thermal networks. However, the two modeling theories differ markedly in their approaches to system conceptualization and in their scientific foundations. CV theory has a strong, broad foundation; it is fundamental in its treatment of physical systems. It is comprehensive and apparently applicable to all physical systems. In contrast, the foundation of compartment theory is limited and weak; it contains few biological or physicochemical fundamentals. In particular the following points are stressed about compartment modeling.

1. As used in biology and ecology, compartment models represent systems consisting of transport, storage, and transformation processes.
2. A system is not conceptualized as a network of basic biological or physicochemical components that represent the transport, storage, and transformation processes in the system. Indeed, a set of basic components has not been defined for compartment modeling. It follows that compartment models are not process oriented and that biology and ecology have not benefited from the sciences of transport, storage, and transformation phenomena.
3. There is only one class of variables--the contents of and flows between compartments, which are extensive quantities.
4. There are no intensive variables. This is a major defect of compartment modeling, for it corresponds to neglect of the central role of energy as the effort by which processes occur. Dynamic imbalance of the spatial intensities of energy is of prime importance to system behavior; energy flow is a secondary consideration and concomitant with flows of extensive quantities.
5. Without intensive variables, compartment models can neither adequately nor properly represent biological and ecological systems and the basic phenomena they comprise. It

follows that discovery of global properties of these systems [38] using compartment models will be difficult if not impossible. Furthermore, understanding and solving environmental problems, which pertain to biophysicochemical systems, will be less than satisfactory using compartment models.

6. From the preceding points we conclude that compartment modeling is merely a form of dynamic bookkeeping on the extensive quantities of biological and ecological systems. It does not adequately capture the essence of these systems and their underlying biological and physicochemical processes. It can also be characterized as dynamic curve fitting or empirical modeling i.e., postulating the form and coefficients for a system of first order differential equations that closely reproduces observed behaviors of the variables in a real system.

4. CV REPRESENTATION OF CHEMICAL SYSTEMS

Examples of physicochemical subsystems in biology and ecology are diffusion, bulk flow, muscle contraction and gross movement, photosynthesis, respiration, and nerves. Some of these systems involve chemical and thermal types of energy only. Others include electrical and/or mechanical energies as well. In this paper consideration is limited to chemical systems, and a common artifice in the modeling of electrical, mechanical, and fluid systems is employed for thermal energy--thermal effects are accounted for only by temperature dependence of the parameters of the chemical components. It is assumed that the heat associated with chemical energy dissipation is readily absorbed by the system's environment and that the temperatures of chemical components are relatively constant and known.

Chemical systems pertain to transport, storage, and transformations of matter. The relevant variables for a CV theory of chemical systems are [39,40,41,42]

extensive: flow of a species γ (moles/sec.)

integrated extensive: quantity of a species n (moles)

intensive: chemical potential of a species μ (joules/mole)

Alternate choices of units may be used. The concepts, basic components, and structure of chemical systems are to be ascertained and described in terms of these variables.

The development parallels, as it must, the existing CV theory of electrical and fluid systems. More accurately, the development is a generalization of the existing theory of fluid systems, which pertains to bulk flows of matter and is described with pressure and volume flow as the intensive and extensive variables. The chemical approach, using chemical potential μ as the intensive variable, is the general approach due to the dependence between changes in μ, pressure p, and temperature T. For a single species

$$d\mu = -s\,dT + v\,dp \qquad (4.1)$$

where s and v are the specific entropy and specific volume of the species, respectively [ibid.]. Thus, bulk flow is contained in chemical theory, and it is manifest in those special cases when gradients of μ in a system depend on pressure gradients only.

The reader may be less familiar with chemical potential than voltage and pressure, but it is the chemical analog of them. Thus μ of a species is the instantaneous energy of one unit of that species at a location in space and in a chemical state. Since changes in energy rather than absolute energy are measurable, an alternate definition is: μ of a species at a point in space and time and chemical state is the change in energy of one unit of that species when it is instantaneously transferred from a reference point of space and/or chemical state to the point and/or state in question. We indicate the arbitrary reference by ρ. Note that, like other potentials, μ is a point function of space and time.

4.1 Chemical components

Transport, storage, and transformation are special cases defined on the general chemical object illustrated in Fig. 2a. The object is a bounded region of space that contains e joules of energy and n moles of matter consisting of M component species in amounts $n_1,\ldots,n_M$. The component species flow across the boundary at N distinct ports at rates $\gamma_1,\ldots,\gamma_N$ and chemical potentials $\mu_1,\ldots,\mu_N$. The set of complementary variables is $\{(\mu_i,\gamma_i)\}$. Potentials are measured with respect to the reference denoted by ρ. Depending on the nature of a specific chemical object, the flows may be species distinct or several flows of the same species may occur at different potentials.

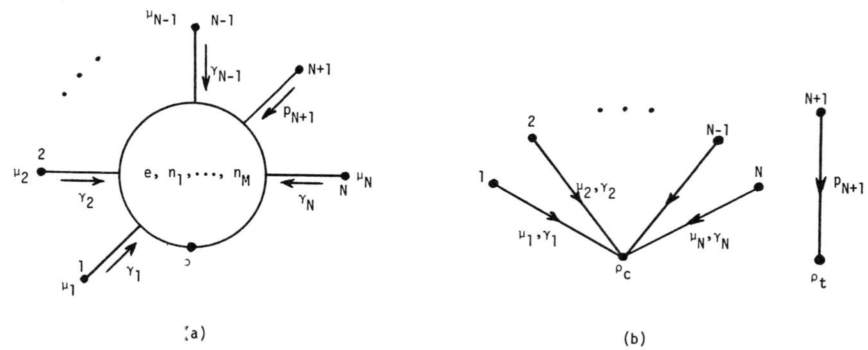

Fig. 2. (a) A general (N+1)-port chemical object and (b) its linear graph.

Flow γ_i at the i-th port is accompanied by energy flow p_i, where

$$p_i = \mu_i \gamma_i \tag{4.2}$$

The (N+1)-th port indicated in Fig. 2 is the thermal port with energy flow p_{N+1} (heat). Applying the law of energy conservation to the object,

$$\frac{d}{dt} e = \mu_1 \gamma_1 + \ldots + \mu_N \gamma_N + p_{N+1} \tag{4.3}$$

The law of conservation of matter constrains the extensive variables

$$\frac{d}{dt} n = \frac{d}{dt} (n_1 + \ldots + n_M) = \gamma_1 + \ldots + \gamma_N \tag{4.4}$$

Equations (4.3) and (4.4) only partially describe the object. The constitutive relations between the $\{\gamma_i\}$ and the $\{\mu_i\}$ provide the dynamic description of the chemical object. We define now a few basic chemical components. In each case, the discussion is with respect to the figures and equations summarized in Table 1.

4.1.a Transport

Passive transport of a single chemical species is the three-port object illustrated in Table 1; ports 1 and 2 are chemical and the third port is thermal. Flow of the species between ports 1 and 2 is driven by $\mu \triangleq \mu_1 - \mu_2$. A pure transport object exhibits no dynamic storage or transformations. In this case $dn/dt = 0$ and from (4.4) $\gamma_1 = -\gamma_2 \triangleq \gamma$. Dissipation R and inertia I are the component processes, the properties of which are expressed by the constitutive relation

$$\mu = \mu(\gamma, \frac{d\gamma}{dt}) = \mu_R(\gamma) + \mu_I \left(\frac{d\gamma}{dt}\right) \tag{4.5}$$

In the linear case

$$\mu = R \gamma + I \frac{d\gamma}{dt} \tag{4.6}$$

process	process diagram	component diagram	graph	description
passive transport	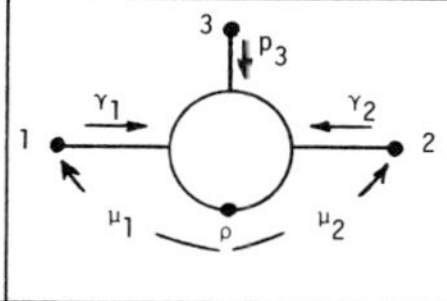	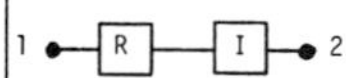	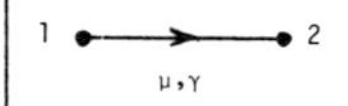	$\mu = \mu_1 - \mu_2$ $\gamma = \gamma_1 = -\gamma_2$ $\mu = \mu_R(\gamma) + \mu_I(\frac{d\gamma}{dt})$ $\mu = R\gamma + I\frac{d\gamma}{dt}$ linear case
1-port storage	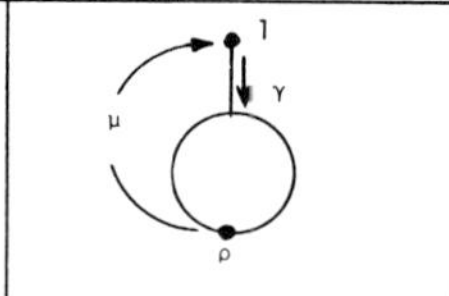	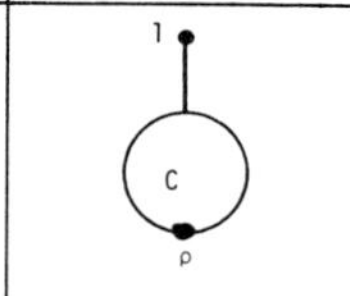	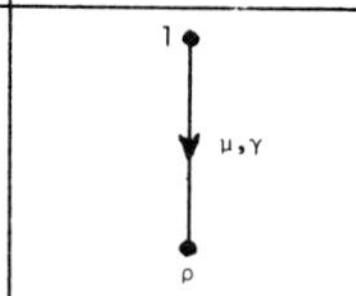	$\mu = \mu(n)$ $\gamma = \frac{\partial n}{\partial \mu}\frac{d\mu}{dt}$ $\gamma = C\frac{d\mu}{dt}$ linear case
M-port storage	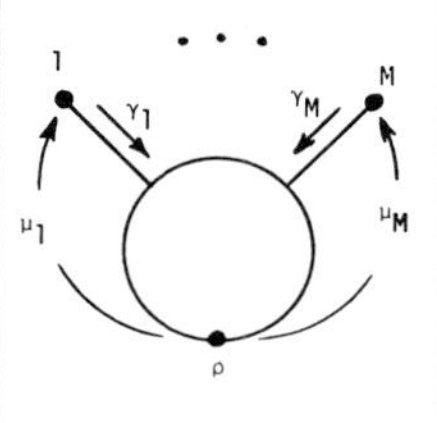	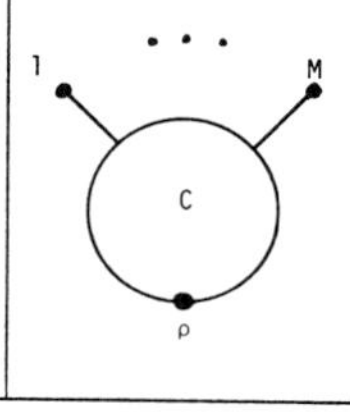	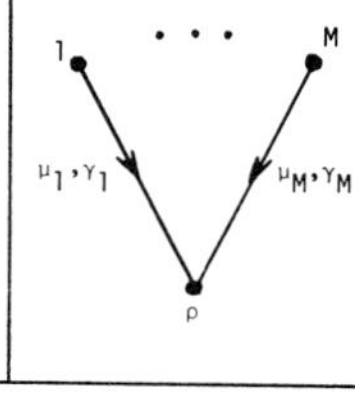	$n = [n_1, \cdots, n_M]^T$ $\mu = [\mu_1, \cdots, \mu_M]^T = \mu(n)$ $\gamma = [\gamma_1, \cdots, \gamma_M]^T$ $\gamma = C^{-1}\frac{d\mu}{dt}$ linear case $C^{-1} = [\frac{1}{C_{ij}}]$
ideal pure N-port trans-formation	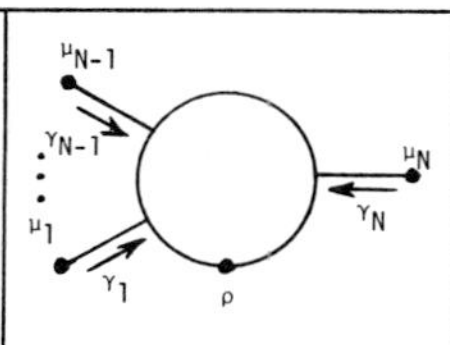	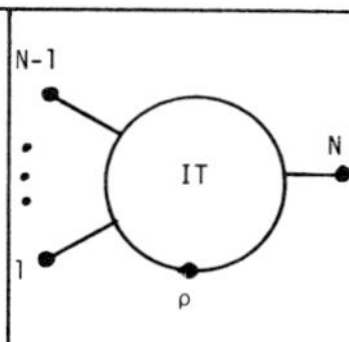	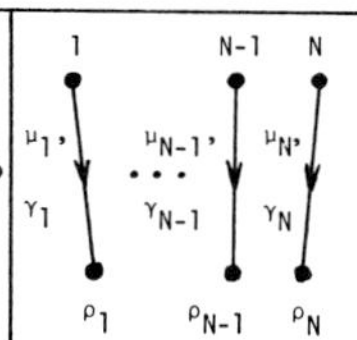	$\gamma_i = -k_i\gamma_N \quad i = 1, \cdots, N-1$ $\mu_N = k_1\mu_1 + \cdots + k_{N-1}\mu_{N-1}$
pure (N+1)-port trans-formation	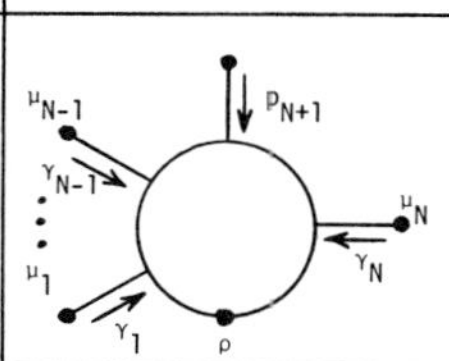	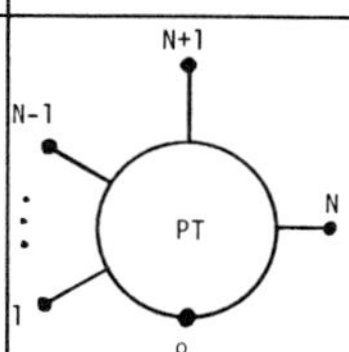	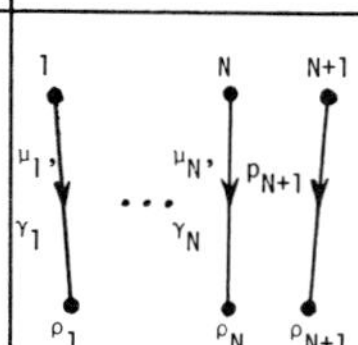	$\gamma_i = -k_i\gamma_N \quad i = 1, \cdots, N-1$ $p_{N+1} = -\mu_{N+1}\gamma_N$ $\mu_{N+1} = \mu_{N+1}(\gamma_N)$ $\mu_N = k_1\mu_1 + \cdots + k_{N-1}\mu_{N-1} + \mu_{N+1}$ $\mu_{N+1} = R_o\gamma_N$ linear case
intensive source	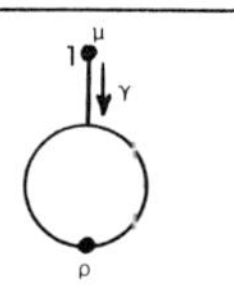	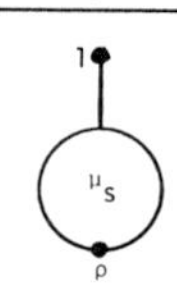	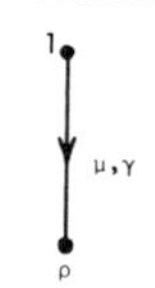	μ_s = arbitrary time function independent of γ
extensive source	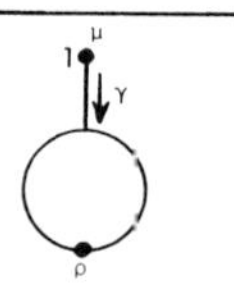	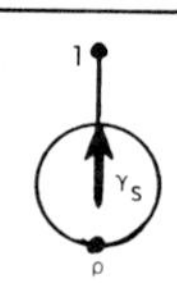	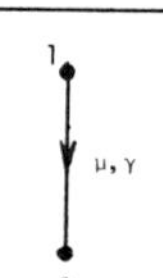	γ_s = arbitrary time function independent of μ

Table 1. Chemical processes and components.

where R and I are the resistance and inertance of the dissipative and inertial components. For most biological and ecological transport processes the dissipative component is expected to dominate.

Multicomponent chemical transport objects exist, and active transport is common in biology [43,44,45]. Active transport is defined to be the movement of a species against its concentration gradient, a process requiring energy input. CV representation of these transport phenomena is incomplete and will not be considered. The main problem with CV representation of active transport is that the process itself is not well understood.

4.1.b Storage

Pure storage of a single chemical species is the 1-port chemical capacitor C illustrated in Table 1. Chemical potential energy is stored in the configuration of accumulated matter. The constitutive relation involves the moles stored n and the chemical potential μ,

$$\mu = \mu(n) \tag{4.7}$$

From (4.4) the flow at the port is

$$\gamma = \frac{dn}{dt} = \frac{\partial n}{\partial \mu}\frac{d\mu}{dt} \tag{4.8}$$

In the linear case $\partial n/\partial \mu = C$, the capacitance of the 1-port store, and

$$\gamma = C\frac{d\mu}{dt} \tag{4.9}$$

M-port storage of several species is common; it is illustrated in Table 1. When the species do not interact, the off-diagonal terms in the C^{-1} matrix are zero and the M-port capacitor is equivalent to M independent 1-port capacitors. The off-diagonal terms in the C^{-1} matrix are non-zero when the stored species interact.

4.1.c Transformation

Chemical transformations include phase changes, reactions, and transductions of chemical energy into thermal, electrical, or mechanical energy. Research on the dynamic CV representation of chemical transformations is recent and limited; therefore, the discussion will be limited to the hypothesis of ideal and practical transformations.

The ideal, pure chemical transformation illustrated in Table 1 is an N-port object that contains no mass or energy, involves no thermal energy, and independently constrains its intensive and extensive variables. Let the transformation in Table 1 consist of N-1 reactants and one product. The material flows at the ports are

$$\gamma_i = -k_i \gamma_N \qquad i = 1,\ldots,N-1 \tag{4.10}$$

where $\{k_i\}$ are the stochiometric ccefficients [40,46]. From (4.4) with dn/dt = 0

$$\sum_{i=1}^{N-1} k_i = 1 \tag{4.11}$$

No heat and no energy storage imply that $p_{N+1} = 0$, de/dt = 0, and from (4.3)

$$\mu_1 \gamma_1 + \ldots + \mu_N \gamma_N = 0 \tag{4.12}$$

Substituting (4.10) in (4.12), we obtain the constraint between the potentials

$$\mu_N = k_1 \mu_1 + \ldots + k_{N-1} \mu_{N-1} \tag{4.13}$$

A more realistic representation includes the energy (heat) of reaction. This is defined to be the pure (N+1)-port chemical transformation in Table 1 for which (4.10) and (4.11) apply but $p_{N+1} \neq 0$. A description of p_{N+1} is required. We hypothesize that

$$p_{N+1} = p_{N+1}(\gamma_N) = -\mu_{N+1}\gamma_N \tag{4.14}$$

where μ_{N+1} is the thermal energy input to the transformation per unit of product species created (output). The μ_{N+1} is negative when the forward transformation is exothermic and positive when endothermic. The signs reverse for the reverse reaction. From (4.3) with de/dt = 0, (4.10), and (4.14)

$$\mu_N = k_1 \mu_1 + \ldots + k_{N-1}\mu_{N-1} + \mu_{N+1} \tag{4.15}$$

Considering the fact that (4.13) is the equilibrium ($\gamma_N = 0$) condition between the potentials and the hypothesis in (4.14), we hypothesize further that

$$\mu_{N+1} = \mu_{N+1}(\gamma_N) \tag{4.16}$$

Taken together, (4.15) and (4.16) imply a family of curves like Fig. 3, which is for a hypothetical exothermic transformation of two reactants into one product. Each curve corresponds to a constant value for the parameter $\mu^i \triangleq k_1\mu_1 + k_2\mu_2$. The μ_3 axis intercept of each curve is $\mu_3 = \mu^i$. Consistent with the exothermic assumption, the curves have positive slope. The slope would be negative for an endothermic transformation.

With respect to chemical energy, a pure transformation is equivalent to an ideal transformation plus a resistor R_o attached to port N. The resistor is positive for exothermic transformations and negative for endothermic cases. The degree of nonlinearity of this resistor is currently unknown. If it is approximately linear with value R_o, then

$$\mu_{N+1} = R_o \gamma_N \tag{4.17}$$

The physical systems engineer will notice the similarity between the ideal chemical transformation and ideal electrical and mechanical transformers [25]. The pure chemical transformation is similar to models of transistors and vacuum tubes used in

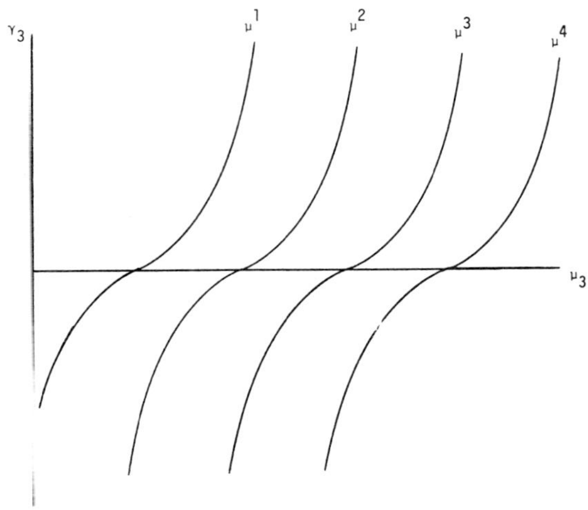

Fig. 3. Characteristic curves for the output of a hypothetical 4-port pure chemical transformation. Product is generated at rate γ_3 and potential μ_3 in an exothermic transformation: species 1 + species 2 → species 3.

electronics [47]. The usefulness of the ideal and pure transformations in representations of real chemical reactions is unknown. Much more research, experimental as well as theoretical, is necessary to fully assess these models and to study and represent other important factors such as activation energy, catalysis, entropy production, and exothermy/endothermy [44,48-52]. Also, multiple reactant-multiple product transformations must be represented. Finally, other forms of energy transfer in chemical reactions must be considered. Chemical-mechanical, chemical-electrical, and chemical-chemical transfers of reaction energy, as well as the chemical-thermal case considered above, are important. Indeed, these other modes of energy transfer are more import than heat in biology. Metabolic processes depend on the slow, enzyme-controlled transfer of energy in sequences of endothermic and exothermic chemical reactions.

4.1.d Sources

A 1-port chemical store of large capacitance C can deliver low level material flow with little change in its potential. An idealization of such a chemical store and other chemical subsystems exhibiting this property is the ideal 1-port chemical potential source μ_s in Table 1. This component develops arbitrary potential μ_s independent of its flow γ. The dual source is the ideal 1-port chemical flow source γ_s in Table 1 that develops arbitrary flow γ_s independent of its potential μ.

Ideal sources are frequently used to represent system environments. The relationship between the intensive and extensive variables of an ideal source is determined by the properties of the system driven by that source. Extensive coverage of equivalent representations, including the Thevanin and Norton theorems, may be found in the literature [25,30].

4.2 The structure of chemical systems

A chemical system is conceptualized as a network of storage, transport, and transformation processes. As with all networks, chemical networks contain interfaces between components and circuits (closed sequences of interfaces). Note that individual interfaces and circuits pertain to single species.

Two laws are the basis for structural descriptions. The *interface law* for material flows is a statement of matter conservation that establishes the dependencies between the extensive variables (species flows) in chemical systems. At the i-th interface

$$\sum_{\forall j} \gamma_{ij} \tag{4.18}$$

where γ_{ij} is the flow from the i-th interface to the j-th component. This is illustrated in Fig. 4a.

The *circuit law* for changes in potential of a single chemical species is a statement of energy conservation that establishes the dependencies between intensive variables (chemical potentials) in a chemical system. Let μ_i and μ_j be the potentials of a spec-

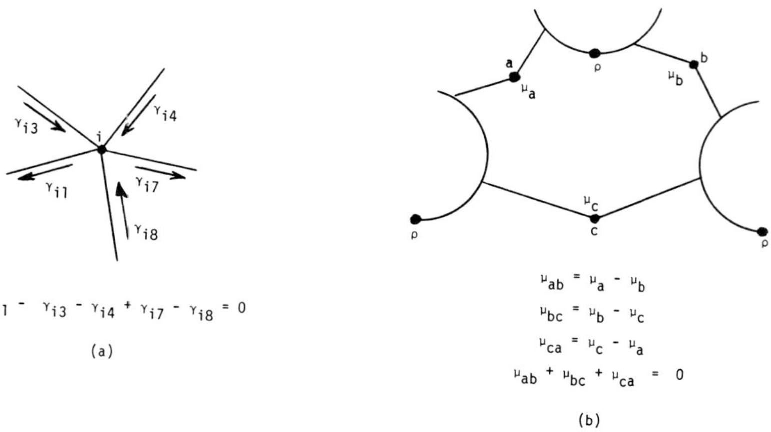

Fig. 4. Illustrations of (a) the interface law at the i-th interface and (b) the circuit law for the circuit a,b,c,a.

ies at interfaces i and j in a system and define

$$\mu_{ij} \overset{\Delta}{=} \mu_i - \mu_j \tag{4.19}$$

About a closed sequence of interfaces for the species

$$\sum_{\text{circuit}} \mu_{ij} = 0 \tag{4.20}$$

This is illustrated in Fig. 4b. Circuits are readily identified from the system graph.

Although redundant with (4.18), (2.7) also applies to the interfaces in chemical systems.

4.3 Component graphs

Analysis of system structure and formulation of a system mathematical model are facilitated by linear graph theory and the system graph [26,27,30,53,54]. The graph of a chemical system is readily determined once the graphs of the chemical components have been identified. The graph for the general chemical object is given in Fig. 2b, and the graphs for the components discussed in Section 4.1 are given in Table 1. As one might expect from the generality of graph theory, it applies to chemical and physicochemical systems.

4.4 Formulation of chemical system models

Conceptual and mathematical CV representations of chemical systems are developed by application of the same concepts and procedures used for physical systems. Component descriptions are developed from appropriate measurements and/or theoretical considerations. System structure is described by application of (4.18) and (4.20) according to established procedures. A total system mathematical model is derived from the component and structural descriptions by any of several methods. The reader is referred to the extensive literature for details [22,25,26,27].

5. CV MODELING IN BIOLOGY AND ECOLOGY

Anticipating successful development of a CV theory of representation for physicochemical systems, it is important to begin studying and modeling physicochemical systems in biology and ecology from the CV viewpoint. A few models are presented in this section. Due to a lack of authentic component descriptions and for ease of presentation, linear component models are used in the mathematical models. However, the examples illustrate the approach and the potential for CV representation of physicochemical systems in biology and ecology.

Diffusion, osmosis, and bulk flow of chemical species are very common in biological and ecological systems. Figure 5 presents a representation that applies in each case; it consists of two stores connected by a transport path. Chemical potentials μ_1 and μ_2 of storage components C_1 and C_2 act at the ends of the transport path. The species flow is driven by and in the direction of

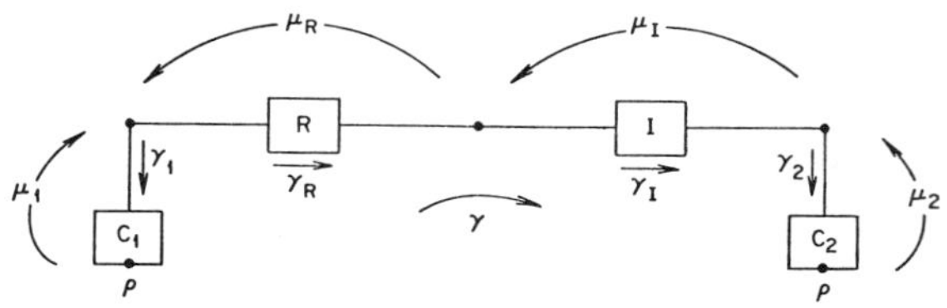

Components:

$$n_1 = C_1\mu_1 \Rightarrow \gamma_1 = \frac{d}{dt}n_1 = C_1\frac{d}{dt}\mu_1 \quad \text{or} \quad \mu_1 = \frac{1}{C_1}\int \gamma_1$$

$$n_2 = C_2\mu_2 \Rightarrow \gamma_2 = \frac{d}{dt}n_2 = C_2\frac{d}{dt}\mu_2 \quad \text{or} \quad \mu_2 = \frac{1}{C_2}\int \gamma_2$$

$$\mu_R = R\gamma_R$$

$$\mu_I = I\frac{d}{dt}\gamma_I$$

Structure:

$$-\gamma_1 = \gamma_R = \gamma_I = \gamma_2 \triangleq \gamma$$

$$\mu_1 - \mu_R - \mu_I - \mu_2 = 0$$

Formulation 1:

$$\frac{d^2}{dt^2}\gamma + \frac{R}{I}\frac{d}{dt}\gamma + \frac{C_1 + C_2}{IC_1C_2}\gamma = 0$$

Formulation 2: state variables $\gamma_I = \gamma$ and $\Delta\mu_C \triangleq \mu_1 - \mu_2$

$$\frac{d}{dt}\gamma = -\frac{R}{I}\gamma + \frac{1}{I}\Delta\mu_C$$

$$\frac{d}{dt}\Delta\mu_C = -\frac{C_1 + C_2}{C_1C_2}\gamma$$

Fig. 5. Transport and storage of a single chemical species.

the potential difference $\mu = \mu_1 - \mu_2$. Assuming linear components, a mathematical model is given in Fig. 5. Whether the system is one of diffusion, osmosis, or bulk flow depends on the nature of the transport path, the chemical species considered, and the relative importance of the concentration, temperature, and pressure components of μ_1 and μ_2.

Chemical reaction systems are very important in biology and ecology. A closed chemical reaction with two reactants and one product is represented in Fig. 6. It consists of a chemical store for each of the three species and a pure chemical transformation. Assuming linear components, the model for the system is the first order differential equation given in Fig. 6. A concep-

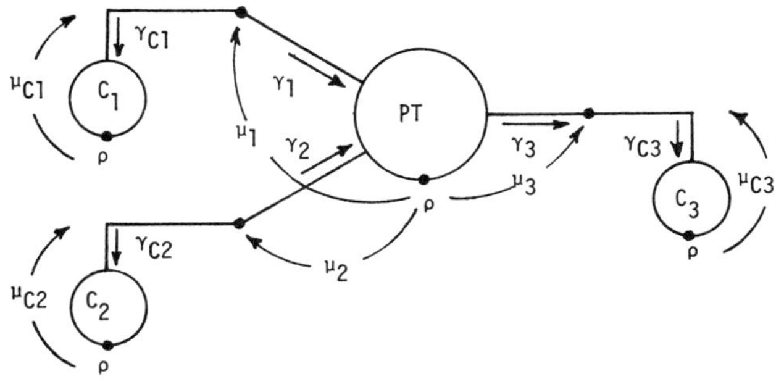

Components:

$$\gamma_{C1} = C_1 \frac{d}{dt}\mu_{C1} \qquad \gamma_1 = -k_1\gamma_3$$

$$\gamma_{C2} = C_2 \frac{d}{dt}\mu_{C2} \qquad \gamma_2 = -k_2\gamma_3$$

$$\gamma_{C3} = C_3 \frac{d}{dt}\mu_{C3} \qquad \mu_3 = k_1\mu_1 + k_2\mu_2 + R_o\gamma_3$$

Structure:

$$\mu_{C1} = \mu_1 \qquad \mu_{C2} = \mu_2 \qquad \mu_{C3} = \mu_3$$

$$\gamma_{C1} = -\gamma_1 \qquad \gamma_{C2} = -\gamma_2 \qquad \gamma_{C3} = -\gamma_3$$

Model:

$$\frac{d}{dt}\gamma_3 + \frac{1}{R_o}\left(\frac{k_1^2}{C_1} + \frac{k_2^2}{C_2} + \frac{1}{C_3}\right)\gamma_3 = 0$$

Fig. 6. Closed chemical reaction: species 1 + species 2→species 3.

tual representation of a hypothetical open chemical system with flow sources, storage, transport, and chemical reaction is presented in Fig. 7.

The physiology and physiological ecology of plants provides a very complex and interesting biological problem for CV representation. Although much research has been done and continues on the cellular and biochemical details of photosynthesis, respiration, nutrition, growth, translocation, and transpiration [55-61], attempts to synthesize dynamic, total plant physiology models are recent and few. However, total plant models would be useful to biology, agriculture, and environmental science. It appears that physicochemical plant models have been limited to water transport and the translocation of organic nutrients [45,62-66]; these

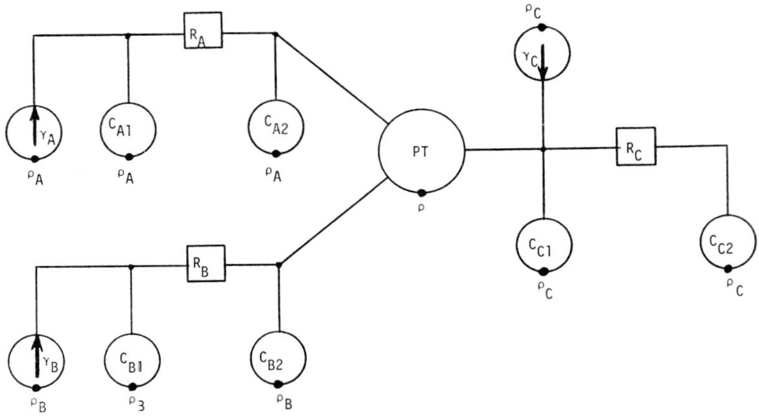

Fig. 7. Hypothetical open chemical system with storage, transport, and transformation of species A, B, and C.

models have been informally of the CV type.

A conceptual model for CV representation of the transport and storage of water in a plant is proposed in Fig. 8. The model is

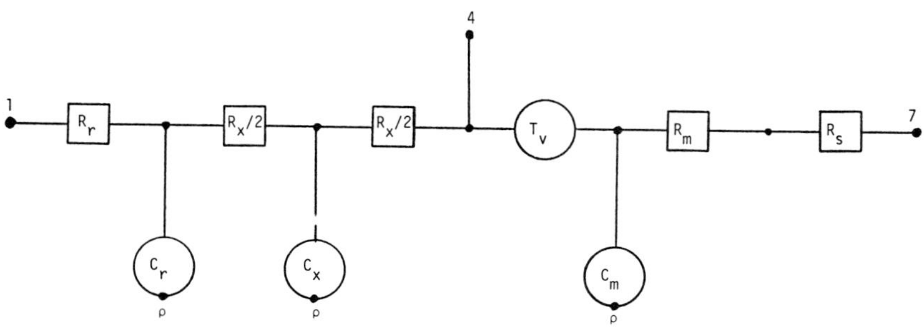

Fig. 8. Model of the flow and storage of water in a plant.

for the plant alone, so total system analysis would require the addition of a representation of the plant's environment, which consists of soil and atmospheric water subsystems. Water is stored in the living tissues of the roots, stem, and leaves. These stores are represented by C_r, C_x, and C_m, respectively. Water flows from root surface (point 1) to leaf surface (point 7) through several resistors: R_r of the porous root cortex, R_x of the xylem capillaries in the stem (Poiseuille resistance), R_m of the porous leaf mesophyll, and R_s of the leaf stomata. Transformation T_v represents the evaporation of water into the intercellular spaces (C_m) of the mesophyll.

The dynamic flow and storage of water in a plant arise from the drop in chemical potential of water between the root and leaf surfaces. This potential drop exhibits considerable diurnal variation. Derivation of the system mathematical model is straightforward, and with specific values for the parameters the model could be analyzed to predict system behavior. Except for R_s all resistors and capacitors are expected to be reasonably linear. The stomatal aperture varies diurnally, being closed at night, open during the day, and often closed during mid-day. Thus R_s is a controlled resistor that modulates water flow; this complicates the analysis somewhat. With respect to point 6 in Section 2, preliminary analysis by the author indicates that it may be possible to describe the resistors and capacitors of this model in terms of the material and geometric properties of a plant.

The photosynthetic subsystem of a plant transforms radiant energy into chemical energy that is stored in carbohydrates. This is performed in the chloroplasts of cells located primarily in the palisade layer of leaves. It consists of two main parts--the photochemical and carbon fixation systems. A cursory conceptual model of photosynthesis is proposed in Fig. 9. Again, it refers to the plant alone and representations of the environmental CO_2, O_2, and light sources must be added to give a total system model. Light absorbed by chlorophyll in the chloroplasts

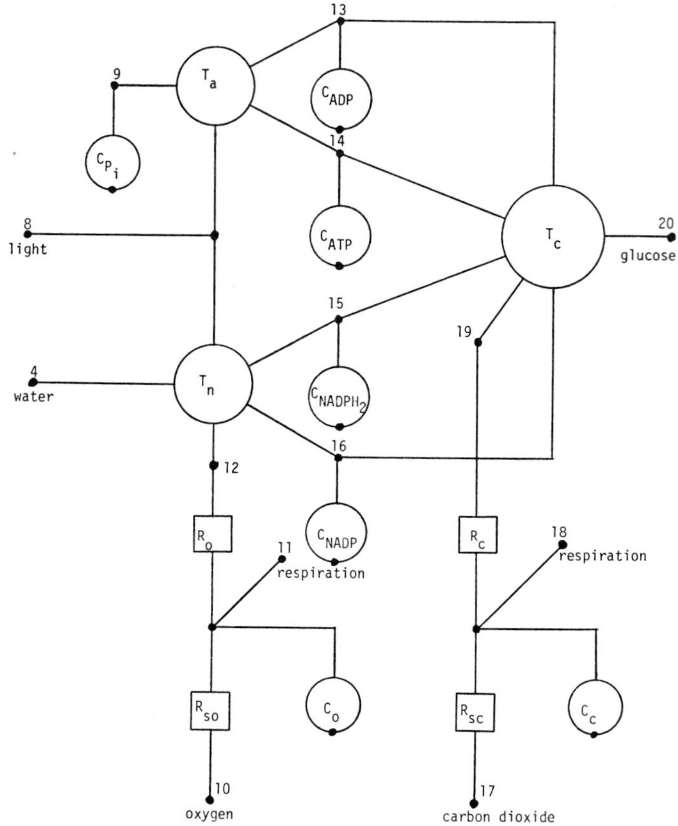

Fig. 9. A model of photosynthesis.

(point 8) drives a series of reactions that convert ADP (adenosine diphosphate) and NADP (nicotinamide adenine dinucleotide phosphate) into their high energy relatives ATP and $NADPH_2$. The ADP to ATP conversion T_a involves an inorganic phosphate denoted by P_i. In transformation T_n light drives the decomposition of water with the evolution of oxygen and subsequent reduction of NADP to $NADPH_2$ by the hydrogen atoms. The water is supplied from point 4 in Fig. 8.

Carbon fixation is non-photochemical; the energy required to reduce CO_2 to carbohydrate is provided by the energy captured during the photochemical formation of ATP and $NADPH_2$. The carbon fixing transformation T_c in Fig. 9 represents the overall result of a sequence of many reactions known as the Calvin cycle--the conversion of CO_2 into glucose, $C_6H_{12}O_6$, through energy-absorbing reactions that consume ATP and $NADPH_2$ and regenerate ADP and NADP. The glucose output appears at point 20.

Fig. 9 indicates stores of P_i, ADP, ATP, NADP, and $NADPH_2$. The CO_2 fixed by T_c comes from the store C_c in the intercellular spaces of the leaf mesophyll. This store is supplied by respiration (point 18) and atmospheric CO_2 that diffuses through the stomatal resistance R_{sc} from the atmosphere (point 17). Resistance to CO_2 diffusion into the chloroplasts is indicated by R_c. Oxygen released by the T_n transformation diffuses from the chloroplasts through R_o into storage C_o in the intercellular spaces of the mesophyll. From there it diffuses through stomatal resistance R_{so} to the atmosphere (point 10). In addition there is an oxygen link to respiration (point 11).

The glucose produced by photosynthesis is used by the plant in respiration. Since respiration occurs in the leaves, stem, and roots, glucose must be moved to these sites. A conceptual model for the system that accomplishes this is proposed in Fig. 10. Glucose enters point 20 from the carbon fixing transformation T_c in Fig. 9. From there it goes into leaf respiration and storage and into the translocation system. Carbohydrate is translocated as sucrose, and its long-term storage form is starch. The model includes the necessary carbohydrate transformations: T_{gs} for glucose to starch, T_{gsu} for glucose to sucrose, and T_{sus} for sucrose to starch. Stores of glucose and starch in the leaves are indicated by C_g and C_{sl}. Translocation of sucrose occurs in the phloem, which is represented by the resistor R_p. In the stem and roots sucrose is converted to starch for storage in C_{ss} and C_{sr}, respectively. Respiratory utilization of starch is indicated by points 24 and 26 in Fig. 10.

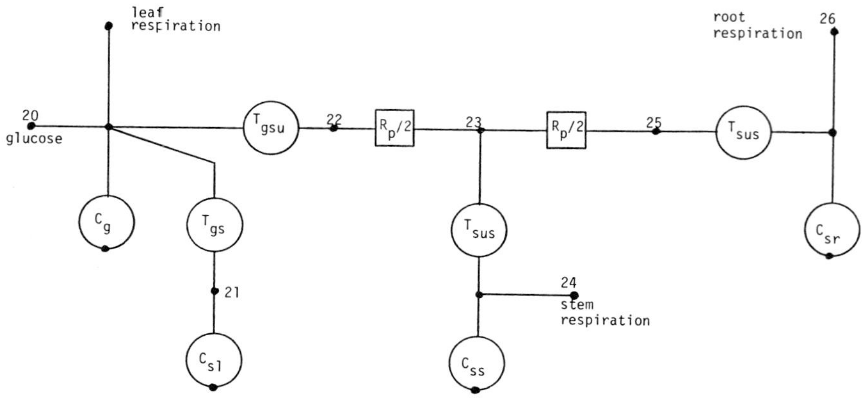

Fig. 10 A model of the storage and translocation of carbohydrate in a plant.

Conceptual models have been presented for three major subsystems in the physiology of plants. By the addition of representations for respiration, inorganic nutrition, and growth mechanisms a total plant physiology model could be obtained. The next step is to obtain CV descriptions of the components in the conceptual models. This most crucial step will be very difficult, and it will require a major commitment to experimental and theoretical research. The conceptual basis for research already exists in CV theory and models like those in Figs. 8, 9, and 10. The greatest difficulty is expected to lie in the availability of suitable instrumentation and experimental methods to implement the research. This in general is expected to be the key to successful application of CV theory to physicochemical systems in biology and ecology.

Ecosystems consist of plants, animals, micro-organisms and abiotic physicochemical environments [67,68,69]. CV representation and analysis of the physicochemical aspects of ecosystems will require development of CV models for each of the ecosystem components. This development would proceed in a manner similar to that used above for plant physiology.

6. GENERAL CV REPRESENTATION OF PHYSICOCHEMICAL SYSTEMS

The research upon which the preceding sections were based has convinced the author that a general theory for CV representation of physicochemical systems is needed. Although many physical systems exhibit a dominant type of energy--electrical or mechanical or fluid--transformations between the dominant type and thermal energy are always present. In addition there are numerous cases of conversion between the electrical, mechanical, and fluid energy domains, and occasionally thermal energy may be as important in determining system properties as the other types of energy. In these cases direct CV representation of processes in multiple energy domains is required, and the artifice used in Section 4 for thermal energy is inappropriate.

The CV theory of physical systems adequately treats electrical, mechanical, and fluid transformations. However, the treatment of transformations involving thermal energy is limited. The complicity of thermal, mechanical, and electrical energies in chemical systems points to the need to consider multiple energy domains during the development of CV representations of the transport, storage, and transformation processes and the structure of chemical systems.

While developing a general CV theory of physicochemical systems, it seems appropriate to attempt to correct a long-standing inconsistency in the CV theory of physical systems--the use of heat as the extensive variable of thermal systems. While temperature is the proper intensive thermal variable, the product $T\gamma$ does not equal energy flow under the current designation for γ.

A better choice for γ is entropy flow; then $T\gamma$ does equal energy flow [39,40,70,71]. It remains to build a complete, consistent and meaningful thermal theory using temperature and entropy flow as the complementary variables and to incorporate it in a general CV theory of physicochemical systems.

Current research by the author is directed toward the development of a general CV theory for physicochemical systems. Descriptions are sought for transport, storage, and transformation processes and system structure in the complementary variables for four energy domains--μ and n for chemical energy, T and S (temperature and entropy) for thermal energy, p and V (pressure and volume) for mechanical energy, and v and q (voltage and charge) for electrical energy. The development is based on classical thermodynamics, irreversible, non-equilibrium thermodynamics [39, 40,72], chemical kinetics, and the existing CV theory of physical systems. At this time the results are too limited to warrant discussion in this paper.

7. CONCLUSIONS

This research has led to several conclusions.

1. There is a need for a general CV theory of dynamic physicochemical systems. At a minimum the theory must be a synthesis of classical thermodynamics, irreversible thermodynamics, chemical kinetics, and the existing CV theory of physical systems.
2. Physicochemical systems in biology and ecology should be represented and analyzed directly in terms of physicochemical science.
3. A CV theory of physicochemical systems would provide a unified, fundamental approach to the representation and analysis of physicochemical systems in biology and ecology.
4. Development and application of a CV physicochemical theory for

biology and ecology is a sound objective. It could contribute to advances in biological and ecological sciences and to applied problem solving in medicine, agriculture and environmental affairs.

5. The practicability of CV representation of physicochemical systems in biology and ecology will depend mainly on successful development of suitable instrumentation and experimental methods to implement the theory.
6. A CV physicochemical approach would make available to biology and ecology the extensive concepts and methodologies of physical systems engineering, including modeling, behavior analysis, design, optimization, and control.
7. Development and application in biology and ecology of a CV theory of physicochemical systems will require extensive research.

REFERENCES

1. Patten, B. C., ed. (1971a). *Systems Analysis and Simulation in Ecology*, Vol. I. New York: Academic Press.
2. Patten, B. C., ed. (1972). *Systems Analysis and Simulation in Ecology*, Vol. II. New York: Academic Press.
3. Odum, H.T. (1972). *Environment, Power, and Society*. New York: Wiley.
4. Watt, K. E. F. (1966). *Systems Analysis in Ecology*. New York: Academic Press
5. Van Dyne, G. M. (1969). *The Ecosystem Concept in Natural Resource Management*. New York: Academic Press.
6. Schwan, H. P. (1969). *Biological Engineering*. New York: McGraw-Hill.
7. Clynes, M. & J. H. Milsum (1970). *Biomedical Engineering Systems*. New York: McGraw-Hill.

8. Levin, S. A. (1975). *Ecosystem Analysis and Prediction.* Philadelphia: Society for Industrial and Applied Mathematics.

9. (1976) *Proc. 1st Intn'l. Congress on Ecology.* The Hague, Netherlands.

10. Worthington, E. B. (1975). *The Evolution of IBP.* Cambridge: Cambridge Univ. Press.

11. Patten, B. C., ed. (1975). *Systems Analysis and Simulation in Ecology*, Vol. III. New York, Academic Press.

12. Reichle, D. E. (1975). Advances in ecosystem analysis. *Bio Science* *25*(4), 257.

13. Auslander, D. M., G. F. Oster, A. Perelson & G. Clifford (1972). On systems with coupled chemical reaction with diffusion. *J. Dyn. Syst. Meas. and Control.* *94(G)*, 239.

14. Oster, G. F., A. S. Perelson & A. Katchalsky (1973). Network thermodynamics: dynamic modeling of biophysical systems. *Q. Rev. Biophys.* *6(1)*,1.

15. Caswell, H., H. E. Koenig, J. A. Resh & Q. E. Ross (1972). Introduction to systems science for ecologists. In *Systems Analysis and Simulation in Ecology*, Vol. II (B. C. Patten, ed.). New York: Academic Press.

16. Koenig, H. E., A. M. Hilmersen & L. Yuan (1969). Modern system theory in the agricultural industry. *Trans. Am. Soc. Agric. Engrs.* *12*,190.

17. Koenig, H. E. & R. L. Tummala (1972). Principles of ecosystem design and management. *IEEE Trans. Syst. Man. & Cybern.* *2*,449.

18. DeMichele, D. W. & P. J. H. Sharpe (1973). An analysis of the mechanics of guard cell motion. *J. Theor. Biol.* *41*,77.

19. DeMichele, D. W. & P. J. H. Sharpe (1974). A parametric analysis of the anatomy and physiology of the stomata. *Agric. Meteorol.* *14*,229.

20. Smerage, G. H. (1976). Matter and energy flows in biological and ecological systems. *J. Theor. Biol.* 57, 203.
21. Smerage, G. H. & W. F. Campbell (1976). Toward a capability to predict seed yield. Submitted for publication.
22. Koenig, H. E. & W. A. Blackwell (1961). *Electromechanical System Theory*. New York: McGraw-Hill.
23. Paynter, H. M. (1961). *Analysis and Design of Engineering Systems*. Cambridge, Mass.: MIT Press.
24. MacFarlane, A. G. J. (1964). *Engineering Systems Analysis*. Reading, Mass.: Addison-Wesley.
25. Shearer, J. L., A. T. Murphy & H. H. Richardson (1967). *Introduction to System Dynamics*. Reading, Mass.: Addison-Wesley.
26. Koenig, H. E., Y. Tokad & H. K. Kesavan (1967). *Analysis of Discrete Physical Systems*. New York: McGraw-Hill.
27. Martens, H. R. & D. R. Allen (1969). *Introduction to Systems Theory*. Columbus, Ohio: C. E. Merrill.
28. Cannon, R. H. (1967). *Dynamics of Physical Systems*. New York: McGraw-Hill.
29. Karnopp, D. & R. C. Rosenberg (1968). *Analysis and Simulation of Multiport Systems*. Cambridge, Mass.: MIT Press.
30. Blackwell, W. A. (1968). *Mathematical Modeling of Physical Networks*. New York: Macmillan.
31. Karnopp, D. & R. C. Rosenberg (1975). *System Dynamics: A Unified Approach*. New York: Wiley.
32. Dixhoorn, J. J. van & F. J. Evans, eds. (1974). *Physical Structure in Systems Theory*. New York: Academic Press.
33. Cooper, G. R. & C. D. McGillem (1967). *Methods of Signal and System Analysis*. New York: Holt, Rinehart & Winston.
34. Desoer, C. A. (1970). *Notes on a Second Course on Linear Systems*. New York: Van Nostrand Reinhold.
35. Atkins, G. L. (1969). *Multicompartment Methods for Biological Systems*. New York: Barnes & Noble.

36. Patten, B. C. (1971b). A primer for ecological modeling and simulation with analog and digital computers. In *Systems Analysis and Simulation in Ecology*, Vol. I. (B. C. Patten, ed.). New York: Academic Press.

37. Ulanowicz, R. E. (1972). Mass and energy flow in closed ecosystems. *J. Theor. Biol.* *34*,239.

38. Reichle, D. E., R. V. O'Neill & W. F. Harris (1976). Principles of energy and material exchange in ecosystems. In *Proc. 1st Intn'l. Congress on Ecology*. The Hague, Netherlands.

39. Katchalsky, A. & P. F. Curran (1965). *Nonequilibrium Thermodynamics in Biophysics*. Cambridge, Mass.: Harvard Univ. Press.

40. Callen, H. B. (1960). *Thermodynamics*. New York: Wiley.

41. Spanner, D. C. (1964). *Introduction to Thermodynamics*. New York: Academic Press.

42. Linford, J. A. (1966). *An Introduction to Energetics*. London: Butterworths.

43. Christensen, H. N. (1975). *Biological Transport*, 2nd ed. New York: W. A. Benjamin.

44. Lehninger, A. L. (1971). *Bioenergetics*, 2nd ed. New York: W. A. Benjamin.

45. Peel, A. J. (1974). *Transport of Nutrients in Plants*. New York: Wiley.

46. Aris, R. (1965). *Introduction to the Analysis of Chemical Reactions*. Englewood Cliffs, N. J.: Prentice-Hall.

47. Millman, J. & C. Halkias (1967). *Electronic Devices and Circuits*. New York: McGraw-Hill.

48. Eyring, H. & E. M. Eyring (1963). *Modern Chemical Kinetics*. New York: Reinhold.

49. Laidler, K. J. (1965). *Chemical Kinetics*, 2nd ed. New York: McGraw-Hill.

50. Warn, J. R. W. (1969). *Concise Chemical Thermodynamics*. London: Van Nostrand Reinhold.

51. Thompson, J. J. (1967). *An Introduction to Chemical Kinetics*. Boston: Houghton Mifflin.

52. Williams, V. R. & H. B. Williams (1973). *Basic Physical Chemistry for the Life Sciences*, 2nd ed. San Francisco: W. H. Freeman.

53. Seshu, S. & M. B. Reed (1961). *Linear Graphs and Electrical Networks*. Reading, Mass.: Addison-Wesley.

54. Chan, S. P. (1969). *Introductory Topological Analysis of Electrical Networks*. New York: Holt, Rinehart & Winston.

55. Bassham, J. A. & M. Calvin (1957). *The Path of Carbon in Photosynthesis*. Englewood Cliffs, N. J.: Prentice-Hall.

56. Nobel, P. S. (1974). *Introduction to Biophysical Plant Physiology*. San Francisco: W. H. Freeman.

57. Levitt, J. (1969). *Introduction to Plant Physiology*. St. Louis, Mo.: C. V. Mosby.

58. Goss, J. A. (1973). *Physiology of Plants and Their Cells*. New York: Pergamon.

59. Gerking, S. D. (1974). *Biological Systems*, 2nd ed. Philadelphia: Saunders.

60. Leopold, A. C. & P. E. Kriedemann (1975). *Plant Growth and Development*, 2nd ed. New York: McGraw-Hill.

61. Zelitch, I. (1971). *Photosynthesis, Photorespiration, and Plant Productivity*. New York: Academic Press.

62. Kramer, P. J. (1969). *Plant and Soil Water Relationships: A Modern Synthesis*. New York: McGraw-Hill.

63. Crafts, A. S. (1961). *Translocation in Plants*. New York: Holt, Rinehart & Winston.

64. Cowan, I. R. (1965). Transport of water in the soil-plant-atmosphere system. *J. Appl. Ecol.* 2,221.

65. Tyree, M. T., C. Caldwell & J. Dainty (1975). The water relations in hemlock *(Tisuga canadensis)*. V. The localization of resistances to bulk water flow. *Can. J. Bot.*

53,1078.

66. Taylor, H. M. & B. Klepper (1975). Water uptake by cotton root systems: An examination of assumptions in the single root model. *Soil Science* *120(1)*,57.

67. Kormondy, E. J. (1969). *Concepts of Ecology*. Englewood Cliffs, N. J.: Prentice-Hall.

68. McNaughton, S. J. & L. L. Wolf (1973). *General Ecology*. New York: Holt, Rinehart & Winston.

69. Odum, E. P. (1971). *Fundamentals of Ecology*. 3rd ed. Philadelphia: W. B. Saunders.

70. Thoma, J. U. (1974). Models, bond graphs and entropy. In *Physical Structure in Systems Theory* (J. J. van Dixhoorn & F. J. Evans, eds.). New York: Academic Press.

71. Rietman, J. (1974). New system variables for the flow of thermal energy based on the concept of exergy. In *Physical Structure in Systems Theory* (J. J. van Dixhoorn & F. J. Evans, eds.). New York: Academic Press.

72. Priggogine, I. (1961). *Introduction to the Thermodynamics of Irreversible Processes*, 2nd ed. New York: Wiley.

THE REASON OF THE ANALOGIES IN PHYSICS

E. TONTI
Istituto di Matematica del Politecnico di Milano
Milan, Italy

SUMMARY

An investigation on the reason of the analogies between physical theories is carried out. The reason is found in the fact that in every physical theory there are basic physical quantities that are referred to geometrical and chronometrical elements. The analysis is pursued further using the rudiments of algebraic topology, i.e. the notions of "chain" and that of "coboundary". A rational classification scheme for the physical variables and equations of whatever physical theory is then constructed. Some of these schemes close the paper.

1. INTRODUCTION

It is common experience that there are different physical theories that exhibit formal similarities, the well-known analogies. Analogies exist between physical theories that may differ even profoundly in the physical content.

Essentially an analogy consists in a one-to-one mapping between the physical quantities and the equations of a theory and those of another theory. This is usually described giving a correspondence table like Table 1.

It may happen that the physical variables that correspond to one another are of the same mathematical nature, both scalars, both vectors, both tensors, etc. In this case the equations connecting two corresponding physical variables are not only similar but are the *same* equations.

This is the case of analogies among the stationary thermal conduction, the electric conduction, the electrostatic field and the hydrodynamics (stationary, irrotational flow), that are all described by the Laplace equations. This leads to the use of the harmonic functions and, for plane problems, to the use of the functions of one complex variable . These analogies are largely employed as a device to obtain the solution of a problem using the known solution of the corresponding problem of another theory.

But there are analogies between physical theories in which the one-to-one mapping is between two physical variables that are not of the same mathematical nature. So one may be a scalar the other a vector, one may be a vector and the other a tensor, etc. Of this kind is the analogy between the theory of dislocations in solid mechanics and the magnetism. It happens that these analogies cannot be used for solution purposes because the corresponding equations are not the same. Nevertheless they can be used, and in fact they are used, as a means to transfer mathematical methods and numerical techniques from one theory to another.

3

Table 1 - Comparison among some physical theories.

physical theory	equations of structure	constitutive eqs.	equations of structure
electrostatic field	$E_k = -\partial_k \varphi$	$D^h = \epsilon^{hk} E_k$	$\partial_h D^h = \rho$
magnetostatic field	$B^h = \varepsilon^{hkl} \partial_k A_l$	$B^h = \mu^{hk} H_k$	$\varepsilon^{mlk} \partial_l H_k = j^m$
electromagnetic field	$\begin{cases} E_k = -\partial_k \varphi - \partial_t A_k \\ B^h = \varepsilon^{hkl} \partial_k A_l \end{cases}$	$\begin{cases} D^h = \epsilon^{hk} E_k \\ B^h = \mu^{hk} H_k \end{cases}$	$\begin{cases} \partial_h D^h = \rho \\ -\partial_t D^m + \varepsilon^{mlk} \partial_l H_k = j^m \end{cases}$
perfect fluid motion	$\begin{cases} \alpha = -\partial_t \varphi \\ v_k = \partial_k \varphi \end{cases}$	$\begin{cases} \rho = f(\alpha) \\ p^h = \rho(\alpha) a^{hk} v_k \end{cases}$	$\partial_t \rho + \partial_h p^h = 0$
thermal field	$p_k = \partial_k (\frac{1}{T})$	$\begin{cases} q^h = K(T) a^{hk} p_k \\ \sigma = \rho c_v \partial_t T \end{cases}$	$\partial_h q^h = \sigma$
irreversible thermodynamics	$F_{Ak} = \nabla_k f_A$	$J^{Bh} = L^{BhAk} F_{Ak}$	$\nabla_h J^{Bh} = G^B$
analytical dynamics	$v^A = \frac{D}{Dt} q^A$	$\begin{cases} p_B = \gamma_{BA} v^A \\ Q_B = Q_B(q) \end{cases}$	$\frac{D}{Dt} p_B = Q_B$
longitudinal vibrations	$\begin{cases} v = \partial_t s \\ \varepsilon = \partial_x s \end{cases}$	$\begin{cases} p = \rho v \\ \sigma = E \varepsilon \end{cases}$	$\partial_t p - \partial_x \sigma = f$
elastostatics (small displacem.)	$\gamma^A{}_h = \nabla_h u^A$	$p_B{}^k = A_B{}^k{}_A{}^h \gamma^A{}_h$	$-\nabla_k p_B{}^k = f_B$
elastodynamics (small displacem.)	$\begin{cases} \gamma^A{}_h = \nabla_h u^A \\ v^A = \partial_t u^A \end{cases}$	$\begin{cases} p_B{}^k = A_B{}^k{}_A{}^h \gamma^A{}_h \\ p_B = \rho a_{BA} v^A \end{cases}$	$\partial_t p_B - \nabla_h p_B{}^h = f_B$
Schrödinger field	$u_k = \partial_k \psi$	$v^h = \frac{\hbar^2}{2m} a^{hk} u_k$	$\partial_h v^h = \sigma$
meson field	$u_\alpha = \partial_\alpha \psi$	$\begin{cases} q^\beta = \hbar^2 g^{\beta\alpha} u_\alpha \\ \sigma = -(m_0 c)^2 \psi \end{cases}$	$\partial_\beta q^\beta = \sigma$

(*) Simbols and names are explained in the corresponding tables at the end of the paper. $h,k,l,i,\ldots = 1,2,3$; $\alpha, \beta, \gamma, \ldots = 0,1,2,3$; $x^0 = ct$; ∇_k = covariant derivative; D = absolute differential ; Dt = dt ; a_{hk} = metric tensor of the space ; $g_{\alpha\beta}$ = metric tensor of space-time with signature $(+---)$; $\partial_t = \frac{\partial}{\partial t}$, $\partial_k = \frac{\partial}{\partial x^k}$; ε^{hkl} is the Ricci tensor $(+\sqrt{a}, 0, -\sqrt{a})$.

In the study of a physical theory we are more or less consciously inspired by the experience made in other physical theories. This is typical of all sciences but is especially valid for physics.

It is apparent that if we had a more depth knowledge of the origin of the analogies between different physical theories we could make a systematic use of them. This would permit us to transfer information, mathematical techniques, numerical experiences, and so on from one theory to another. Moreover one may construct a unique mathematical model for many physical theories in which the common features are exhibited once and for all, the main properties are investigated with a great economy of mental labour, of time and as far as concerns the computer, of money. This apart from intellectual pleasure of the synthesis.

All these facts lead us to raise the question: *what is the origin of the analogies?*

In this paper we give an answer to this question. Moreover we show how one may construct a classification scheme for the physical quantities and equations of every physical theory. This classification scheme has lead in a natural way to the construction of a mathematical model valid for a large class of physical theories [6][7].

2. THE ORIGIN OF THE ANALOGIES

At the foundation of every physical theory we introduce some measurable physical quantities from which, in subsequent stages, other physical quantities are derived. It is at this initial stage that one makes the important observation contained in the following statement:

A: *in every physical theory there are basic physical quantities that are naturally referred to the most simple geometrical and chronometrical elements like points, lines, surface, volumes, time instants and time intervals and combinations of them.*

Let us give some examples: we speak of mass or electric charge contained in a *volume* at a given *time instant*, of the probability of finding a particle in a *volume* at a given *time instant*. We speak of the electric potential at a *point* at a given *time instant*, of the displacement of a *point* of a material continuum during a *time interval*. We speak of the position vector of a particle at a given *time instant*, of the impulse given to the particle during a *time interval*. We speak of the electric flux through a *surface*, of the energy and momentum flux through a *surface* during a *time interval*, of the internal energy and entropy production in a *volume* during a *time interval*.

We emphasize the fact that the physical quantities of different physical theories that are associated with same geometrical entity may have different mathematical nature. So with a surface one may associate respectively

magnetostatics:	→	magnetic flux	→	scalar
continuum mechanics:	→	contact force	→	vector
mechanics of polar continua	→	{ contact force and contact couple	→	{ motor or complex vector

As a natural consequence of the statement A there is the following statement:

B: *in every physical theory there are basic physical laws that state a physical quantity referred to a p-dimensional manifold* ω *like lines, surfaces, volumes, time interval, etc. is equal to a physical quantity referred to its boundary* $\partial\omega$.

Typical laws of this kind are those expressed by *balance equations*, in particular *continuity equations*, *equilibrium equations*, *equations of motion*, *circuital equations*, in particular *compatibility equations*; the equations that give the *general solution* of one of the preceding equations; the *equations defining the gradients*; and so on.

Examples of balance laws are: in magnetostatics the statement that sum of the magnetic fluxes through the *boundary of a volume* vanishes; in continuum mechanics the law of equilibrium that states that the sum of the forces acting on the *boundary of a volume* and of the external forces acting on that *volume* vanishes. The principle of conservation of energy states that the outgoing energy flux through the *boundary of a volume during a time interval* plus the energy stored in the *volume* in the same *time interval* is equal to the energy produced in the *volume* in the *time interval* considered. This law can be restated, with reference to space-time, by saying that the outgoing energy flux through the three-dimensional *boundary of a four-dimensional region* is equal to the energy produced inside it.

An example of circuital law we mention the Ampère circuital law: it states that the magnetomotive force along the *boundary of a surface* is equal to the current flowing through the *surface*. Often circuital equations arise as *compatibility conditions* of gradient-like equations. So in fluid dynamics the condition that the circulation of the velocity vector along a *closed line* vanishes implies that the velocity vector is the gradient of a scalar function (velocity potential). In the thermodynamic configuration space the statement of the vanishing of the circuital of the vector field, whose components are the "intensive" variables, along a *closed line* (Maxwell reciprocity relations) amounts to the statement of the existence of the entropy.

Perhaps this connection between physics and geometry is well known for a long time. But *it does not seem that it has been realized how far one can go in the understanding of the formal structure of physical theories by exploring the consequence of this remarkable connection.*

One of the remarkable consequences of this connection is the possibility of doing the rational investigation of the analogies between two physical theories according to the criterion that follows: to every physical quantity of one theory there corresponds that physical quantity of the other theory that is referred to *the same geometrical entity*. So comparing the perfect fluid motion with the magnetic field one may see that the analog of the vorticity vector $\vec{w}$ is the magnetic induction $\vec{B}$ because both are associated with a surface. To the velocity vector $\vec{v}$ in fluid dynamics corresponds the magnetic vector potential $\vec{A}$ of the magnetism because both are referred to lines. And so on. Such analogies are easily detected comparing the classification schemes of the various physical theories exibited in the tables that close this paper.

3. A PRELIMINARY CLASSIFICATION OF PHYSICAL VARIABLES

3.1 Configuration-kind variables

Since we propose to study the link between physical variables and basic geometrical elements we need a preliminary examination of the main physical variables and equations of a physical theory. This is a difficult task because of the great difference in terminology of different physical theories: whatever name we will choose to denote a typical entity of a theory we are sure to be exposed to criticism. Since from some part we must start, we decide to use, as far as possible, the terminology of mechanics, both analytical mechanics and continuum mechanics. The reason is that many physical theories are modelled on mechanics, among them field theories.

One of the basic notions of mechanics is that of "configuration" of a mechanical system. In discrete mechanical systems the configuration is described by a set of n generalized coordinates, universally denoted by q^k. In continuum statics the configuration is described giving the displacement vector of every point of the continuum: this is almost universally denoted by the letter $\vec{u}$. The coordinates q^k and the vector $\vec{u}$ will be called "configuration variables" of the respective mechanical systems. In field theories the analog of the mechanical configuration variables are the field functions (also called field variables). The variables that describe the configuration of a system or of a field will be called *configuration variables.*

So the configuration variable of an electromagnetic field is the vector potential A_μ; the configuration variable of the gravitational field in the relativistic description is the metric tensor g_μ. The configuration variables of a thermodynamical system may be the intensive variables. In quantum mechanics the configuration variable of a quantum-mechanical system is the state vector Ψ that gives the probability amplitude.

In a physical theory, once we have chosen the configuration variables and the independent variables, we may consider those variables that are linked to the configuration variables by means of operations of sum and difference, of total or partial derivatives and total or partial integrals with respect to the independent variables. All this without the intervention of physical constants, material parameters, coupling constants, phenomenological coefficients and other parameters linked to the geometry or the physics of the system. Exception is made for the light speed in vacuo. These variables, including the configuration variables themselves, will be called *configuration-kind variables.*

In continuum mechanics typical configuration-kind variables are the geometrical and kinematical variables like the displacement, strain, velocity, deformation gradient, angular velocity , vorticity, rate of deformation, etc.

3.2 Source-kind variables

Another basic entity of mechanics is the force concept. In continuum statics the body force is considered as a source of the change of the configuration of the system. In particle mechanics the force is the source of motion of the particle. The notion of "force" is replaced in other physical theories by that of "source" of a field. So the electric charges are the sources of the electrostatic field, the electric currents are the source of the magnetic field. The variables that describe the source of a field or of a phenomenon will be called *source variables*.

Let us consider those variables that are linked to the source variables by means of operations of sum and difference, of total or partial derivative and total or partial integrals with respect to the independent variables. All this without the intervention of physical constants, material parameters, coupling coefficients, phenomenological constants or any other parameter linked to the geometry or the physics of the system, with the exception of the light speed in vacuo. All these variables, including the source variable themselves, will be called *source-kind variables*.

In continuum mechanics typical source-kind variables are statical and dynamical variables like body force, body couple, momentum, angular momentum, stress, stress functions, stream functions, etc.

The choice of a set of configuration variables and a set of source variables in a physical theory is not unique: reasons of convenience may lead to prefer one choice to another.

Typical pairs of configuration and source variables are the lagrangian coordinates q^k and the generalized forces Q_k of analytical mechanics; the electrostatic potential and the electric charge density in electrostatics; the displacement vector and the body force vector in continuum mechanics, the metric tensor $g_{\mu\nu}$ and the stress energy-momentum tensor $T^{\mu\nu}$ in the relativistic gravitational theory; the state vector Ψ and the interaction term σ in quantum mechanics.

But other choices are possible: so in analytical mechanics one may choose the momenta p_k and the generalized velocities ν^k as configuration and source variables respectively. In continuum mechanics one may choose the stress potentials as configuration variables and the dislocation tensor as source variable. In magnetostatics one may choose the pair vector potential A_μ - current density J^μ or the pair scalar potential X - monopole charge density $\rho_{(m)}$ as configuration and source variables respectively.

In the *general field theory*, a theory which essentially rests upon formal properties common to many field theories, the configuration variables are the field functions, usually denoted by Ψ_α. Usually no explicit mention is made of the source variables that, in the case of interacting fields are implicitly included in the interaction lagrangian.

About the *mathematical nature* of the configuration and source variables we find the largest variety of types: they may be real or complex numbers, vectors, tensors, quaternions, motors, multivectors, Clifford numbers, matrices, spinors, operators, and so on.

3.3 Independent variables

In continuous systems and fields the configuration and source variables depend on some *independent variables* like the time and space coordinates. When the physical system under study exhibits some symmetry (like plane symmetry, spherical symmetry, etc.) one may use a lower number of variables. In other cases one may treat one or more variables as parameters typically the time. One may consider those independent variables as coordinates of a point of a space.

When we say that in every theory there are some physical quantities that are naturally referred to the geometrical elements of a space we refer to the *space of independent variables*. Since the choice of independent variables is not unique it follows that *the same physical quantity may be referred to different geometrical elements in different spaces*. So the electric potential Ψ will be referred to points of the three-dimensional physical space or to time lines of the four-dimensional space-time.

3.4 Other physical variables

In every physical theory one encounters other physical variables that are defined as functions of the configuration-kind and source-kind variables. So in particle mechanics one defines the potential energy V and the kinetic energy T respectively as

$$V \triangleq -\int_{0}^{\vec{r}} \vec{f}(\vec{r})\cdot d\vec{r} \qquad T \triangleq \int_{0}^{\vec{V}} \vec{V}\cdot d\vec{p}(\vec{V}) \tag{3.4.1}$$

They depend from the source-kind variables $\vec{f}$, $\vec{p}$ and from the configuration-kind variables $\vec{r}$, $\vec{V}$. Once the constitutive equations are given one may evaluate these physical variables. So when

$$\vec{p} = m\vec{V} \qquad \text{or} \qquad \vec{p} = m_o\left(1 - \frac{v^2}{c^2}\right)^{-1/2}\vec{V} \tag{3.4.2}$$

we obtain

$$T = \frac{1}{2}\,mv^2 \qquad \text{and} \qquad T = m_o c^2\left[\left(1 - \frac{v^2}{c^2}\right)^{-1/2} - 1\right] \tag{3.4.3}$$

that are the typical expressions of the kinetic energy of classical and relativistic mechanics respectively. In electrostatics and in magnetostatics one introduces the variables

$$W_{(e)} \triangleq \int_{0}^{\vec{D}} \vec{E}(\vec{D})\cdot d\vec{D} \qquad W_{(m)} \triangleq \int_{0}^{\vec{B}} \vec{H}(\vec{B})\cdot d\vec{B} \tag{3.4.4}$$

that are defined as functions of the source and configuration-kind variables of the respective theories. Once we specify the constitutive equations we obtain these new variables as functions of the configuration-kind variables only. Another example of physical variables of the electrostatic field is the Maxwell tensor giving the maxwellian stresses

$$t^h_k = D^h E_k - \frac{1}{2}\,(D^s E_s)\,\delta^h_k \tag{3.4.5}$$

This physical variable is a function of the two variables $\vec{E}$ and $\vec{D}$. Once we specify the phenomenological equation $\vec{D} = \vec{D}(\vec{E})$ the stress tensor becomes a function of the configuration-kind variable only.

3.5 Equations of structure

The equations that link the various configuration-kind variables of a single physical theory do not contain physical constants, neither material parameters, as we have said. These equations contain algebraic sums, time or space derivative, time or space integrals, and their *linear* combinations. Typical are the equations defining the gradients, those arising from circuital laws and compatibility equations. Of this kind are the equations

$$\vec{B} = \text{curl } \vec{A}, \quad \vec{V} \triangleq \frac{d\vec{r}}{dt}, \quad \varepsilon_k^h \triangleq \nabla_k u^h \qquad (3.5.1)$$

Also the equations that link the various source-kind variables of a theory are of this kind. Typical are the balance equations and the conservation laws, e.g. the equations

$$-\nabla_k t_h^k = f_h, \quad \frac{d\vec{p}}{dt} = \vec{f}, \quad \iint_{\partial W} \vec{B}\cdot\vec{n}dS = 0 \qquad (3.5.2)$$

Equations of this kind have not a *general name* in the physical literature. Since in the following we shall deal mainly with this kind of equation we use, tentatively, a single name and call them *equations of structure*, or *structural equations*.

We shall show that the equations of structure have a geometrical origin, more properly can be described using the elementary notions of algebraic topology. Such notions are that of "chain" and of "coboundary" that we shall explain in Section 4. There, we shall show that the "structural equations" are realization of a single *linear* process, that of forming the "coboundary" of a "chain". This will give a reason of the linearity of the structural equations.

3.6 Phenomenological equations

Configuration-kind and source-kind variables of the same physical theory are linked by *constitutive* equations, also called *material equations* or *equations of state*. They include physical constants, geometrical and material parameters.

The interaction between two phenomena is described by giving the link between the variables of one theory and the variables of the other. These are the *interaction equations* that contain coupling coefficients.

Both the constitutive equations and the interaction equations are of phenomenological kind and are usually called *phenomenological equations*.

While the structural equations are always linear, the phenomenological equations are generally nonlinear. Moreover, they can be differential or integrodifferential, as in the theory of hereditary response (hereditary materials, optical dispersion).

3.7 Other equations

The equation of structure and the phenomenological equations do not exhaust, of course, the equations arising in a physical theory. But they are, so to speak, the building blocks of all equations of a physical theory. So the Poisson equation

$$\text{div}\ [\varepsilon(-\text{grad}\ \Psi)] = \rho \tag{3.7.1}$$

arises as a mixing of two equations of structure with one phenomenological equation, i.e.

$$\text{div}\ \vec{D} = \rho, \quad \vec{D} = \varepsilon\vec{E}, \quad \vec{E} = -\text{grad}\ \Psi \tag{3.7.2}$$

In general the wave equations and the field equations are obtained by combining phenomenological and structural equations.

All what we have said can be represented in the block diagram, of Table 2.

2 - preliminary classification of physical variables of a physical theory

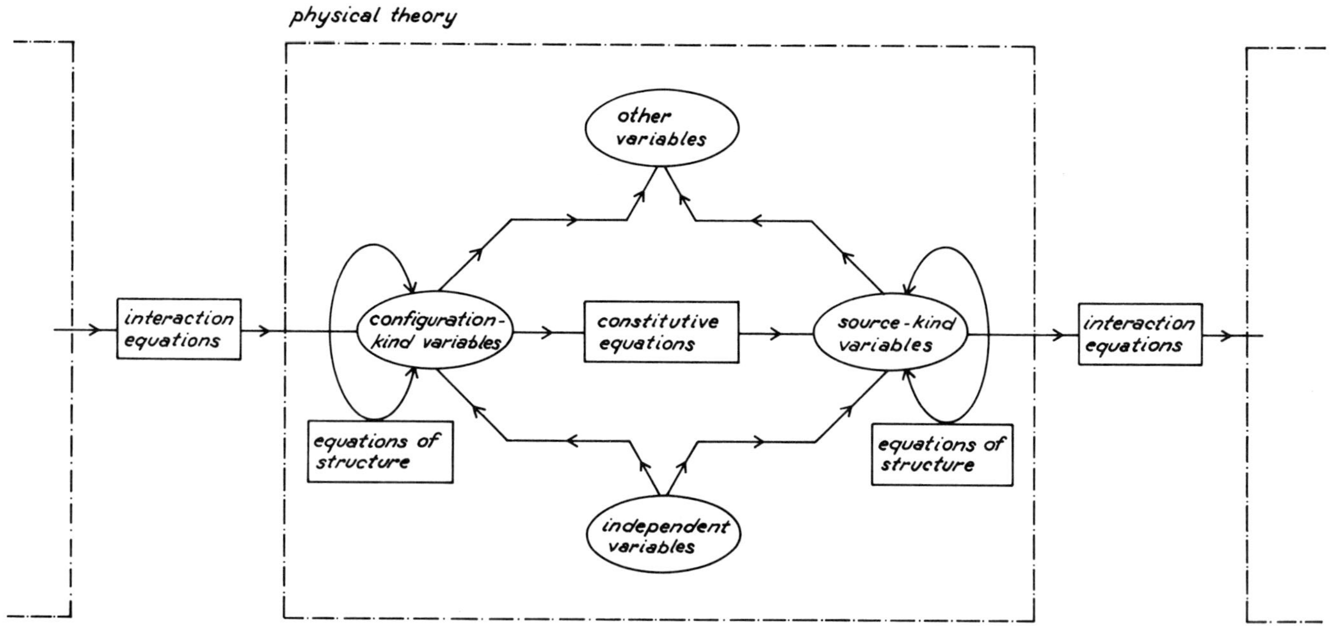

4. CHAINS AND COBOUNDARY IN PHYSICS

4.1 Cell-complex

In order to give a rational classification of the basic geometrical and chronometrical elements we shall consider a region Ω of a space R^n. This may be the three-dimensional space (n=3); the one-dimensional time axis (n=1); the four dimensional space-time (n=4) or any other space whose points are the n-tuples of coordinates formed by the independent variables used in a physical theory (see sect. 3.3). To fix our ideas we consider a region of R^3. To display the geometrical elements of R^3 it is expedient to subdivide the region Ω in three-dimensional cells whose faces are formed by the coordinate surfaces of a coordinate system x^1, x^2, x^3. Every cell is composed of vertices, edges and faces that, in turn may be considered as zero, one, two-dimensional cells respectively. Then we have four geometrical elements that are called 0-cells, 1-cells, 2-cells, 3-cells respectively.

In a time axis a cell-complex is formed by 0-cells (time-instants) and 1-cells (time-intervals). In space-time the 0-cells are the events. Since spaces with $n \leq 4$ are the most commonly used we shall use special symbols for the cells of the various dimensions:

0-cell: point, time-instant, event: P

1-cell: line segment, time interval: L

2-cell: surface segment, line segment time interval: S

3-cell: volume, surface segment time interval: V

4-cell: Hypervolume, volume time interval: H

For every point of R^3 they pass

$$\binom{3}{1} = 3 \text{ coordinate lines; } \binom{3}{2} = 3 \text{ coordinate surfaces}$$

as is shown in Fig. 1c. This permits to group the p-cells into families: all p-cells that lie on a coordinate manifold, like x^1x^2 belong to the same family. Then we have 3 families of 1-

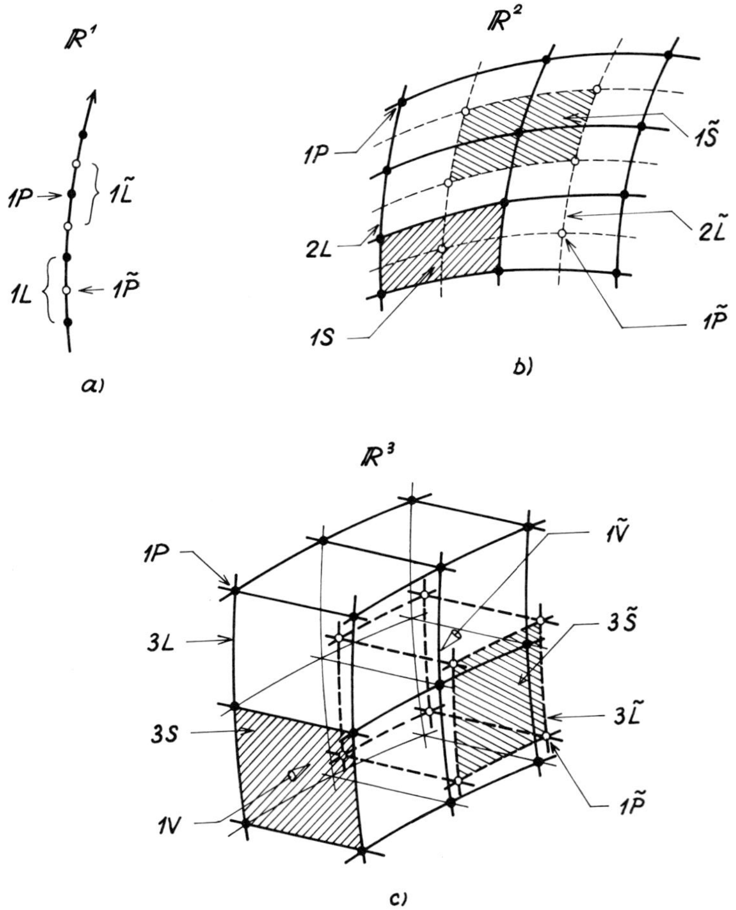

Fig. 1

cells and 3 families of 2-cells: these will be denoted with the symbols 3L and 3S. We obtain in this way a rational classification of the geometrical elements of R^n that is quoted in the left columns of Table 3. The set of all the cells of the various orders is called *cell-complex* and will be denoted K.

3 - a classification scheme of the basic geometrical elements of a cell-complex K and its dual $\tilde{K}$ in R^n

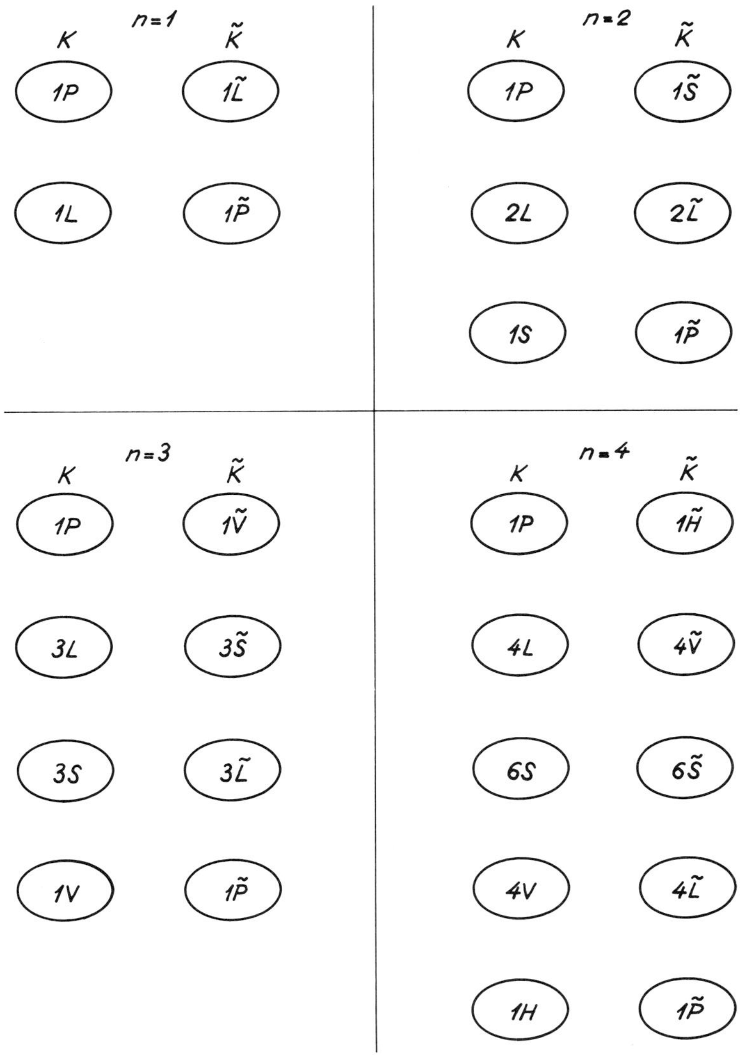

4.2 Dual cell-complex

For physical theories it will be useful to consider the centers of gravity of the 3-cells and to consider them as vertices of another cell-complex, as shown in Fig. 1. This will be called the *dual* cell-complex and we denote it by $\tilde{K}$. A characteristic of the dual cell-complex is the fact that to every p-cell of K there corresponds an (n-p)-cell of $\tilde{K}$ and vice-versa: this is shown in the right column of Table 3 that gives a classification scheme of the geometrical elements of a space R^n, for n = 1,2,3,4. Cell-complexes are one of the typical subjects of *algebraic topology*, founded by Poincaré in 1895.

4.3 Inner orientation of a p-cell

The triangle is the simplest poligon, the tetrahedron is the simplest polyhedron: they are called the *simplexes* of the spaces of dimension two and three respectively, see Fig. 2. The number of vertices of a simplex is one more than the dimension of the corresponding space. Then it is customary to denote the vertices of a simplex of R^n with the notation

$$\langle P_0, P_1, \ldots P_n \rangle \tag{4.3.1}$$

in which the index zero is included.

Let us consider a given arrangement of the vertices, say for a tetrahedron the arrangement $\langle P_2, P_3, P_0, P_1 \rangle$. All other arrangements can be grouped in two classes: those that differ for an even or an odd number of permutations from the initial one. We say that the given arrangement and all those of the same class define an *inner orientation* of the simplex. The arrangements of the other class define the opposite orientation. In this way, one evidences the combinatorial character of the notion of orientation: it is easy to realize that the inner orientation defined in this way reduces to the intuitive notion of orientation (the way to go along the perimeter of a triangle). This can be done

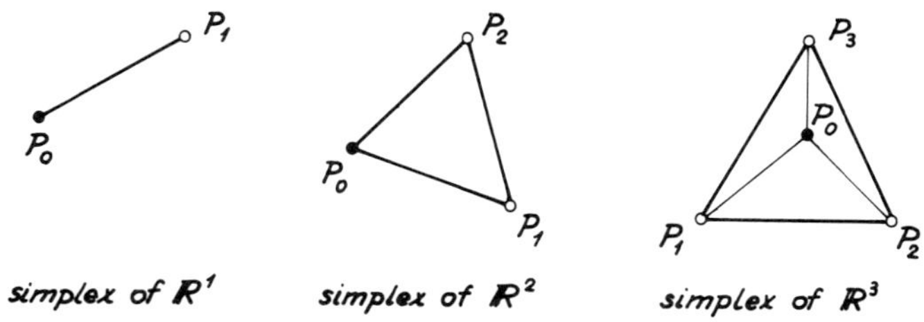

Fig. 2

using the notion of induced orientation.

4.4 Induced inner orientation

The inner orientation of a simplex induces in a natural way an inner orientation on its faces according with the rule [3, p. 30]

$$(-1)^K \langle P_0, P_1, \ldots \hat{P}_K, \ldots P_n \rangle \qquad (4.4.1)$$

where the circumflex upon P_K means that this vertex must be omitted. When this rule is applied to a 1-simplex, 2-simplex, 3-simplex respectively it gives the orientation indicated in Fig. 3.

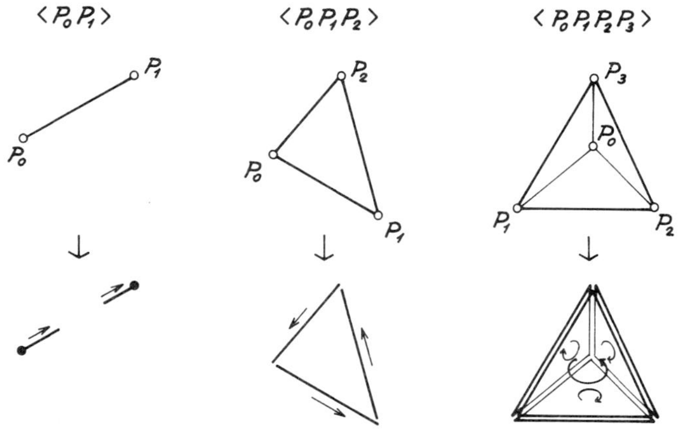

Fig. 3 Induced inner orientation

Since the set of all faces of a simplex forms its boundary we may say that the orientation of a simplex implies the orientation of its boundary.

We now show that the converse is also true. Let us observe that the orientation of two adjacent faces is such that the orientations induced on their common element are opposite. This property is known as the Möbius law of edges [1, p. 14]. If we orient one face of a simplex and then propagate the orientation according with this law, one obtains a compatible orientation of the whole boundary of the simplex, and this is equivalent to the orientation of the simplex itself, i.e. to the choice of an order of its vertices. Then to orient a simplex we may give a compatible orientation to its faces.

To give an inner orientation to a p-cell we may proceed in this way: we divide the p-cell into p-simplexes, then we orient one simplex and propagate the orientation to the adjacent simplexes using the Möbius law of edges. It is easily seen that this process is equivalent to the choice of an orientation of one face of the p-cell and to the propagation of the orientation with the edge law. Then to orient all p-cells of same family, i.e. lying on the same coordinate manifold, we need only to fix an orientation of a simplex lying on that coordinate manifold.

By definition the inner orientation of a point (0-cell) means that incoming lines are considered positive (or negative), see Fig. 4.

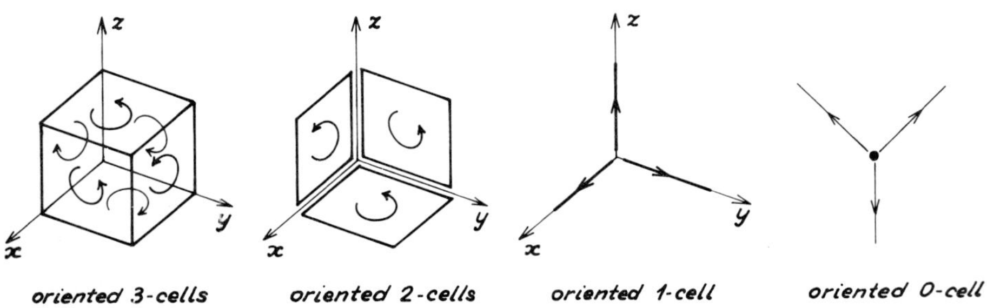

Fig. 4 - Inner orientation of the p-cells in R^3

4.5 Outer orientation of a p-cell

If we consider the dual cell-complex $\tilde{K}$ we may use the same criterion to give an inner orientation to all its cells of the various dimensions. Since every p-cell of the primal cell-complex K is crossed by an (n-p)-cell of $\tilde{K}$ we may consider the inner orientation of this (n-p)-cell as defining a new kind of orientation of the corresponding p-cell (Fig. 5).

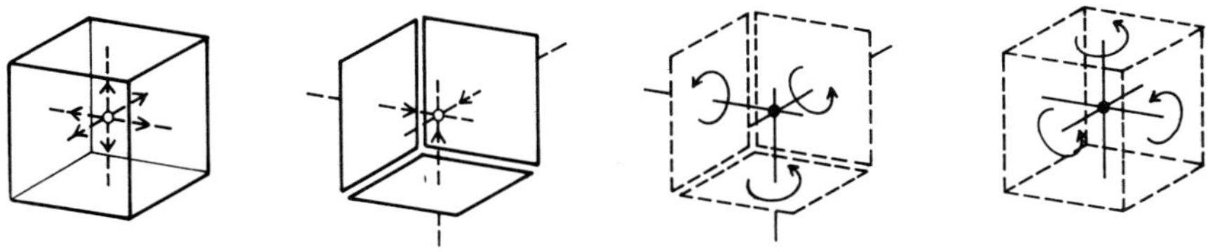

Fig. 5 - Outer orientation of the p-cells of $\mathbb{R}^3$

Precisely we call *outer orientation* of a p-cell the orientation of its dual (n-p)-cell [2, p. 56].

The outer orientation for a 3-cell amounts to the choice of the outward or inward normals, while the outer orientation of a point is defined fixing a sense of rotation around the lines that arrive or leave the point.

The link between inner and outer orientation in three-dimensional spaces may be established by means of a screw (see Fig. 6).

4.6 Induced outer orientation

The outer orientation of a p-cell induces in a natural way an outer orientation on its faces. This is described in Fig. 7.

4.7 Incidence numbers

Let us consider an oriented cell-complex, i.e. a cell-complex in which all p-cells have been equipped with an inner or outer

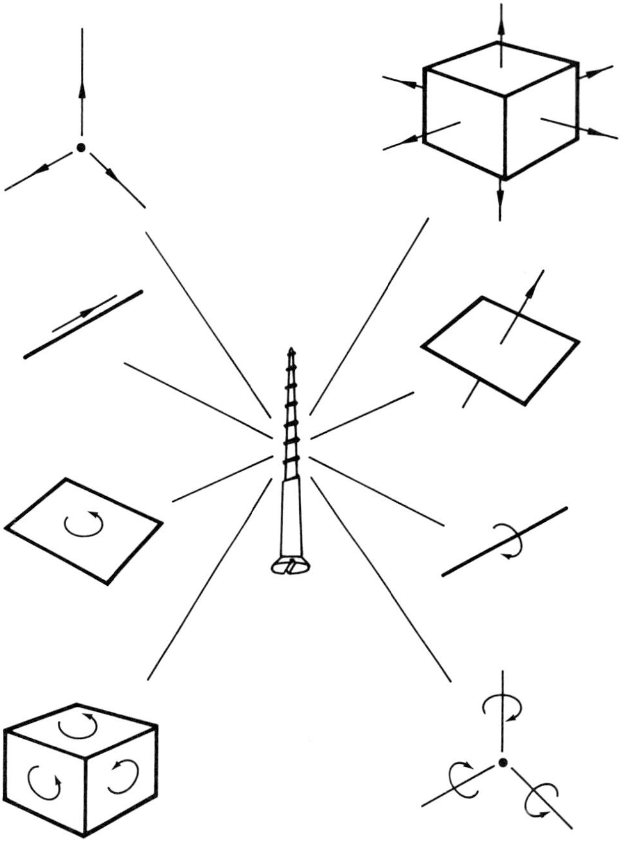

Fig. 6

orientation, and this for p = 0,1,... n.

If we consider a p-cell we may induce an orientation on all its faces: for some face the induced orientation will agree with the orientation previously assigned to the face, for others it will not. If we number all p-cells of the cell-complex according to whatever criterion and we number also all (p-1)-cells, we may consider the incidence of the r-th p-cell with the s-th (p-1)-cell. If the s-th (p-1)-cell is not a face of the r-th p-cell

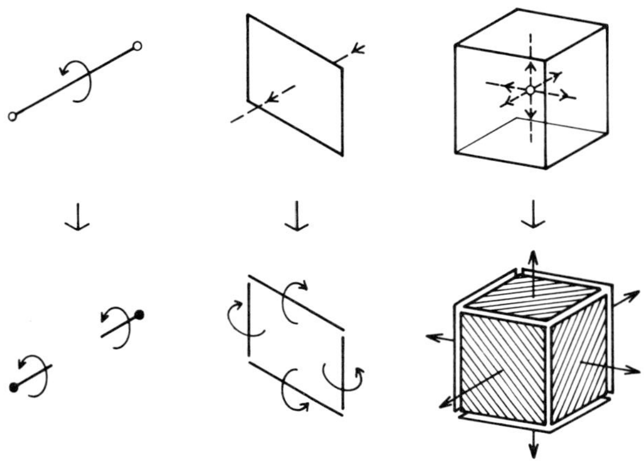

Fig. 7 - Induced outer orientation

we assign to the pair (r,s) the number 0. If the s-th (p-1)-cell is a face of the p-cell we assign to the pair (r,s) the number + 1 or - 1 according to whether the induced orientation is compatible or not with the orientation of the (p-1)-cell. Then we are lead to introduce the incidence numbers between the p-cells and the (p-1)-cells denoted by

$$e_{rs}^{(p-1,p)} = \begin{cases} +1 \\ 0 \\ -1 \end{cases} \tag{4.7.1}$$

These incidence numbers may be considered as entries of a matrix called the *incidence matrix* and denoted

$$E^{(p,p-1)}$$

So in R^3 we have three incidence matrices

$$E^{(3,2)} \qquad E^{(2,1)} \qquad E^{(1,0)} \tag{4.7.2}$$

These matrices reflect the connectivity of the various cells and they are of combinatorial nature.

4.8 Chains on a cell-complex

One of the basic notions of the homology theory is that of "chain". Let us number with an arbitrary criterion all p-cells of a cell-complex and give an orientation to every p-cell. If we associate with every p-cell a mathematical element like a number, a vector, a tensor, a matrix, etc. we have defined a *p-chain*. A p-chain is then a mapping between the oriented p-cells of a cell-complex and the elements of a set C. For later purposes we shall suppose that this set has at least the structure of an additive commutative group, i.e. that an operation of sum is defined on every pair of its elements u and v, and that $u + v = v + u$.

4.9 Chains in physical theories

It happens that in physics we use p-chains without having explicitly recognized it. So writing an equilibrium equation we use a small volume element, usually a parallelepiped in a rectangular coordinate system. This volume element is just a 3-cell of an underlying cell-complex that we usually do not explicitly mention. When we say that with every face of the parallelepiped is associated the force transmitted through it, we say that with every 2-cell of the underlying cell-complex is associated a vector, i.e. an element of a space $C = R^3$. To define the contact force transmitted through a 2-cell it is necessary to define an outer orientation of the 2-cell because we must select one of the two opposite forces transmitted through it. Then the distribution of the force on the 2-cell of a cell-complex is described by assigning with every oriented 2-cell of the cell complex a vector. This amounts to say that we have constructed a 2-chain whose "coefficients" are elements of the same linear space $C = R^3$. Let us denote by $\vec{f}_s$ the force transmitted through the s-th 2-cell: the corresponding 2-chain will denote

$$f^{(2)} = (\vec{f}_1, \vec{f}_2, \ldots \vec{f}_{\alpha_2}) \qquad (4.9.1)$$

where α_2 is the number of 2-cells of the cell-complex.

Let us consider as a second example a magnetic field in a finite region Ω of the space. If we cover this region with a cell-complex K then with every oriented 2-cell of K there is associated the amount of magnetic flux transmitted through it. If ϕ_s is the magnetic flux transmitted across the s-th oriented 2-cell the whole distribution of magnetic fluxes is described by the 2-chain

$$\phi^{(2)} = (\phi_1, \phi_2, \ldots \phi\alpha_2) \tag{4.9.2}$$

In this case the coefficients of the 2-chain i.e. the amounts ϕ_s are numbers and then it is C = R.

As a third example let us consider a Cosserat continuum: with every oriented 2-cell there is associated not only a force but also a couple. If $\vec{f}_s$ denotes the force transmitted through the s-th oriented 2-cell and $\vec{\mu}_s$ the corresponding couple we may consider the vector with complex components

$$\vec{m}_s = \vec{f}_s + i\vec{\mu}_s \tag{4.9.3}$$

where i denotes the imaginary unit. Then we can say that in a polar continuum with every oriented 2-cell is associated a complex vector. Then we obtain a 2-chain

$$m^{(2)} = (\vec{m}_1, \vec{m}_2, \ldots \vec{m}\alpha_2) \tag{4.9.4}$$

whose coefficients are elements of the space $C = C^3$.

In these three examples we have 2-chains with different space of coefficients.

We now pass to consider examples of 0-chains. Let us consider the displacement vector of continuum mechanics: with every 0-cell of a cell-complex K we may associate the displacement vector $\vec{u}$. Then the displacement field is described by a 0-chain

$$u^{(0)} = (\vec{u}_1, \vec{u}_2, \ldots \vec{u}\alpha_0) \qquad \vec{u}_s \in R^3 \tag{4.9.5}$$

where α_0 is the number of 0-cells of K.

Another example of 0-chain is that of the velocity potential in the fluid dynamics of an irrotational, perfect stationary fluid motion. Let $\Psi(x,y,z)$ denote the velocity potential, and Ψ_s its value at the s-th 0-cell. The field of the velocity potential is then described by the 0-chain

$$\Psi^{(0)} = (\Psi_1, \Psi_2, \ldots \Psi\alpha_0) \qquad \Psi_s \in R \qquad (4.9.6)$$

Let us give now some examples of 3-chains. First let us consider a continuum body that occupies a region Ω of the space in which a cell-complex has been built up. With every 3-cell we may associate the mass contained inside it. Then the mass distribution is described by the 3-chain

$$m^{(3)} = (m_1, m_2, \ldots m\alpha_3) \qquad (4.9.7)$$

We may also consider the external force acting on every 3-cell. Let $\vec{F}_s$ be the external force acting on the s-th 3-cell. The external force distribution is then described by the 3-chain

$$F^{(3)} = (\vec{F}_1, \vec{F}_2, \ldots \vec{F}\alpha_3) \qquad F_s \in R^3 \qquad (4.9.8)$$

Up to now we have considered cell-complexes in the three-dimensional space because we have considered physical quantities that depend on the three space coordinates. If we consider physical quantities that depend on a set of n independent variables we may consider the space R^n and build up a cell-complex on it. So if we consider a theory whose physical quantities depend on the space and time variables, x, y, z, t, the space is R^4. But we may consider the time t as a parameter and the three space coordinates as independent variables and then the space is R^3. On the contrary we may consider the three space coordinates as parameters and the time t as independent variable: in this case the representative space is the real axis R.

To do some examples let us consider the motion of a particle in space. The physical variables that are used in particle dynamics depend on the time t. Then we may consider a one dimensional representative space, the time axis. On it we may construct a cell-complex dividing the time axis in time intervals.

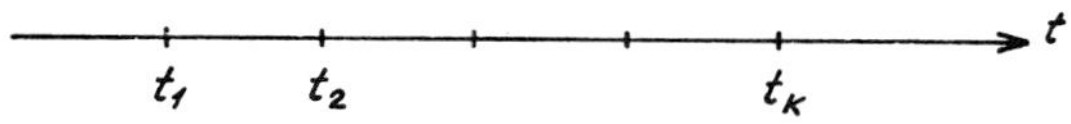

Fig. 8

The radius vector $\vec{r}$ (t) is naturally referred to time instants (0-cells) and then its variation in time is described by a 0-chain

$$r^{(0)} = (\vec{r}_1, \vec{r}_2, \ldots \vec{r}\alpha_0) \qquad r_K \in R^3 \qquad (4.9.9)$$

where $\vec{r}_K = \vec{r}(t_K)$. The displacement vector $\vec{S}$ is relative to a couple of time instants and then is naturally referred to the time intervals (1-cells). Then its variation in time is describ-ed by a 1-chain on the time axis

$$S^{(1)} = (\vec{S}_1, \vec{S}_2, \ldots \vec{S}\alpha_1) \qquad \vec{S}_K \in R^3 \qquad (4.9.10)$$

where $\vec{S}_K$ is the displacement of the particle relative to the k-th 1-cell. The impulse *h* given to the particle is referred to time intervals then its time distribution is expressed by a 1-chain

$$h^{(1)} = (\vec{h}_1, \vec{h}_2, \ldots \vec{h}\alpha_1) \qquad \vec{h}_K \in R^3 \qquad (4.9.11)$$

on the time axis. And so on.

Lastly we consider a thermal field, and assume as independent variables the three space coordinates x, y, z and the time t. Let K be a cell-complex in the region Ω of R^4 that forms the do-main of the independent variables. If we consider the heat cur-rent density $\vec{q}$ transmitted through a surface during a time inter-val we see that this physical quantity is referred to the 3-cells of K. Then its space-time distribution is given by a 3-chain

$$q^{(3)} = (\vec{q}_1, \vec{q}_2, \ldots \vec{q}\alpha_3) \qquad (4.9.12)$$

These few examples may be enough to state that *those physical quantities of a physical theory that are naturally referred to the basic geometrical elements of some space give rise to chains. The coefficients of these chains are the amounts of the physical quantity associated with every cell.*

As a matter of fact we may see that among the physical quantities of a physical theory there are some that are referred to the cells of a cell-complex K and others that are naturally referred to the cells of the dual cell-complex $\tilde{K}$. To do an example if we consider the bending of a plate subjected to a continuous load when we discretize the plate, for example dividing it in rectangular elements, with every such element associated with the corresponding load. It is then natural to consider as unknown the vertical displacement of the barycentric point of every finite element. But these barycentric points are just the 0-cells of $\tilde{K}$ while the rectangular elements are the 2-cells of K. Then when the configuration variable (here the vertical displacement) is referred to the cells of K, the corresponding source variable (here the load) is referred to the corresponding cells of $\tilde{K}$.

As a general rule if *the geometrical and kinematical variables are referred to the p-cells of K, the corresponding statical and dynamical variables are referred to the p-cells of* $\tilde{K}$.

4.10 The coboundary process

In the homology theory of cell-complexes there is one fundamental process that, starting with a p-chain leads to the construction of a (p+1)-chain: this is called the "coboundary" of the given p-chain. The process is the following: let us consider a p-chain defined on a cell-complex K. Given an arbitrary p-cell we consider all (p+1)-cells of K that are incident on it: these are called the *cofaces* on the p-cell. The mathematical entity associated with every p-cell is then transferred to every coface with the same sign or with the opposite sign provided the incidence number is +1 or -1. Since every (p+1)-cell is common coface of several p-cells, to every (p+1)-cell there comes as many values of the mathematical entity as are the *faces* of the (p+1)-cell. We then sum up all the amounts a_K coming to every (p+1)-cell: this sum is associated with the (p+1)-cell. In this way, we have obtained a (p+1)-chain that is called the *coboundary* of the given p-chain. This two-stage process is illustrated in Fig. 9. If $b^{(p+1)} = (b_1, b_2, \dots b\alpha_{p+1})$ is the coboundary of the p-chain $a^{(p)} = (a_1, a_2, \dots a\alpha_p)$ we may write

$$b^{(p+1)} = \delta a^{(p)} \qquad (4.10.1)$$

δ is called the *coboundary operator*. The coboundary operator is a linear operator that maps the p-chains into the (p+1)-chains.

Since a p-chain is essentially a vector of $R^{\alpha}p$ with coefficients in C the coboundary operator from $a^{(p)}$ to $b^{(p+1)}$ may be represented by a matrix $E^{(p+1,p)}$ that operates from $R^{\alpha}p$ and $R^{\alpha}p+1$. We may write

$$b_r = \sum_s e_{rs}^{(p+1,p)} a_s \qquad (4.10.2)$$

When the coboundary operator δ is applied in sequence two times it gives rise to the null chain:

$$\delta(\delta a^{(p)}) \equiv \theta^{(p+2)} \qquad (4.10.3)$$

where $\Theta^{(p+2)}$ denotes the chain of rank (p+2) whose coefficients are all vanishing. This property is known as the Poincaré lemma. A p-chain whose coboundary vanishes is called a *cocycle*. The Poincaré lemma states that the coboundary of a given p-chain is a cocycle.

One may also consider the p-chains on the dual cell-complex $\tilde{K}$. We denote $\beta_{\tilde{p}}$ the number of p-cells of $\tilde{K}$; $F^{(p+1,p)}$ the incidence matrices of $\tilde{K}$. Table 4 gives a classification scheme of the chains of a cell-complex K and its dual $\tilde{K}$ in a region $\Omega C R^3$.

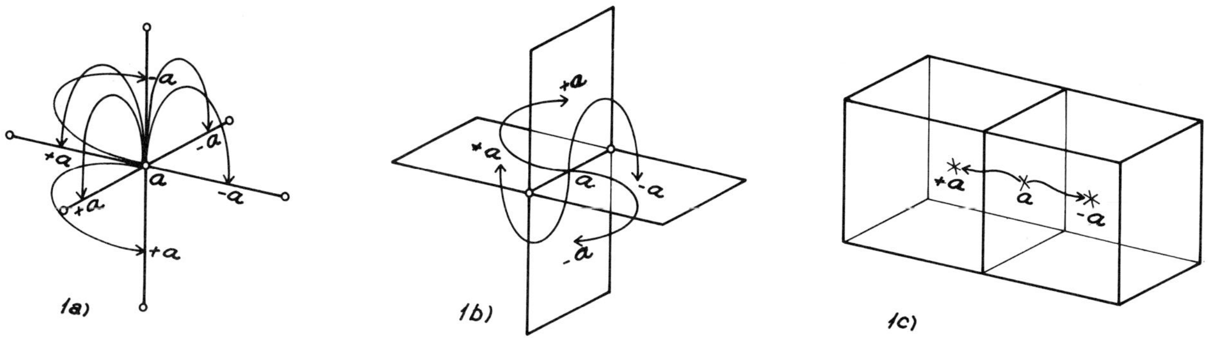

1) transferring the mathematical element to the cofaces of a p-cell

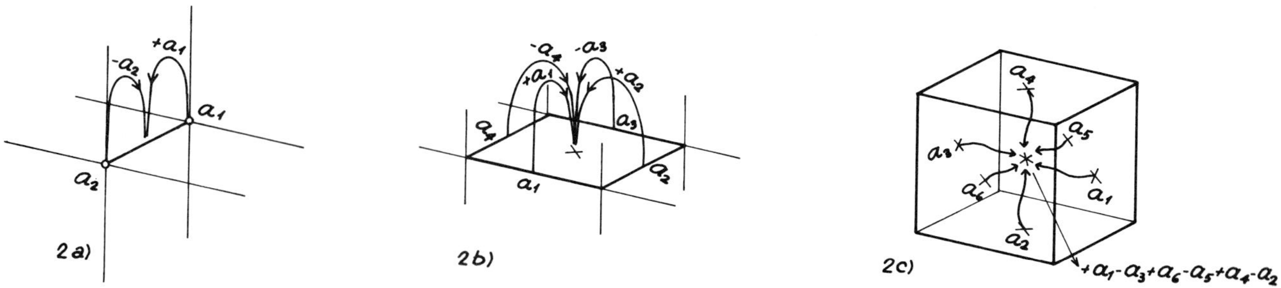

2) summing the mathematical elements assigned to every (p+1) cell

Fig. 9 - the process of forming the coboundary

4-a classification scheme for the chains of K and $\tilde{K}$ in $\mathbb{R}^3$

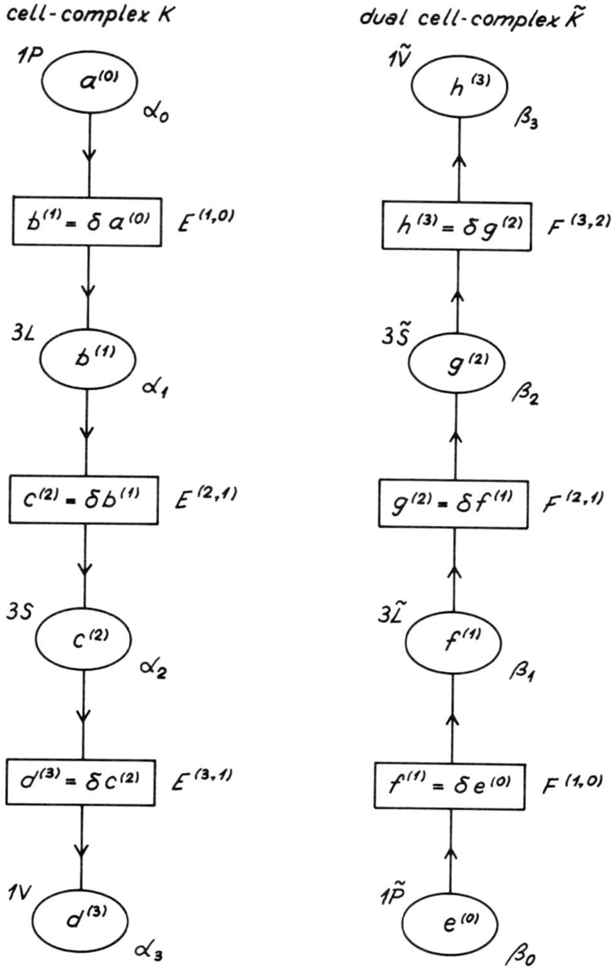

4.11 The coboundary process in physical theories

The physical interest of this process when applied to the chains formed by physical quantities comes from the fact that the equations of structure of every physical theory refer to chains in the coboundary of another. Let us give some examples.

a) *Equilibrium equations.* Let us consider a material continuum and, in it, a three-dimensional region. The equilibrium condition asserts that the sum of the contact forces transmitted through the boundary of the region and the external force acting on the matter contained in the region vanishes. If we consider as usual a 3-cell and call $\vec{F}_s$ the exterior force acting on the s-th cell, $\vec{f}_K$, the contact force acting on the K-th 2-cell, we may consider the two chains

$$f^{(2)} = (\vec{f}_1, \vec{f}_2, \ldots \vec{f}\alpha_2) \qquad F^{(3)} = (\vec{F}_1, \vec{F}_2, \ldots \vec{F}\alpha_s) \quad (4.11.1)$$

with reference to Fig. 10a the equilibrium equation can be written

$$-\vec{F}_h = (+1)\vec{f}_j + (-1)\vec{f}_i + (+1)\vec{f}_\ell + (-1)\vec{f}_m + (+1)\vec{f}_q + (-1)\vec{f}_p \quad (4.11.2)$$

or

$$-\vec{F}_h = \Sigma_K \, e_{hK}^{(3,2)} \vec{f}_K \quad (4.11.3)$$

i.e.

$$-F^{(3)} = \delta f^{(2)} \quad (4.11.4)$$

Then *an equilibrium equation states that a 3-chain is the coboundary of a 2-chain.*

b) *Circuital laws.* Let us consider a small parallelogram as indicated in Fig. 10b and a physical quantity g referred to the lines. A typical circuital law has the form

$$G = (+1)g_j + (-1)g_i + (+1)g_\ell + (-1)g_m \quad (4.11.5)$$

where G is the amount of a physical quantity associated with the 2-cell. This law can be written

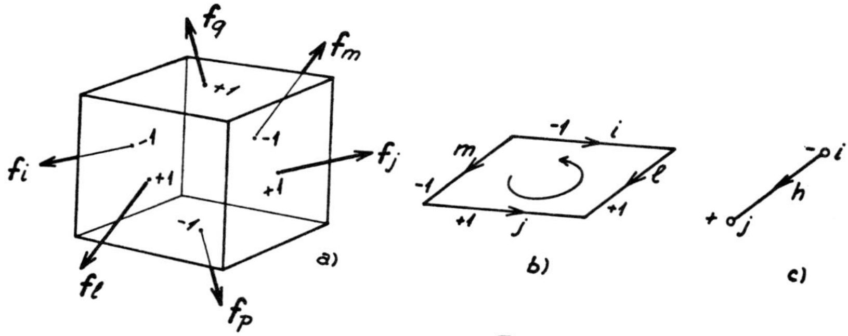

Fig. 10

$$G_h = \Sigma_K e_{hK}^{(2,1)} g_K \tag{4.11.6}$$

or

$$G^{(2)} = \delta g^{(1)} \tag{4.11.7}$$

where we have put

$$G^{(2)} = (G_1, G_2, \ldots G\alpha_2) \qquad g^{(1)} = (g_1, g_2, \ldots g\alpha_1) \tag{4.11.8}$$

Then *a circuital law is a statement that a 2-chain is the coboundary of a 1-chain.*

c) *The gradients.* Let us consider the process of forming the gradient of a scalar, vector or tensor field. The first act is that of forming the differences between the physical quantities Ψ_i and Ψ_j referred to two points P_i and P_j (see Fig. 10c. This difference

$$\phi_s = (+1)\Psi_j + (-1)\Psi_i \tag{4.11.9}$$

is associated with the line segment: we may write

$$\phi_s = \Sigma_K e_{sh}^{(1,0)} \Psi_h \tag{4.11.10}$$

or

$$\phi^{(1)} = \delta\Psi^{(0)}. \tag{4.11.11}$$

Then we see that *the construction of the gradients is equivalent to the process of forming the coboundary of a 0-chain.*

In an analogous way one can show that the compatibility equations, the circuital equations, the balance equations, the equations giving the general solution of another equation, the conservation laws are all realizations of a single statement: a (p+1)-chain is the coboundary of a p-chain.

5. A CLASSIFICATION SCHEME OF PHYSICAL VARIABLES

As we have shown in many physical theories there are physical quantities that are associated with the basic geometrical elements of the space of the independent variables chosen. This association is completed by the fact that the sign of the physical quantity depends from the orientation (inner or outer) of the geometrical element. This permits to describe the distribution of the physical quantity by a chain. Now we have seen that there exists a typical linear process, the coboundary process, that permits us to obtain a chain from a chain of one dimension lower. In turn this process reflects a typical process used in physics to link physical quantities: the corresponding equations identify with the equations of structure (see § 3.5). As we have seen if the configuration-kind variables are associated with the cells of a cell-complex K the corresponding source-kind variables are naturally associated with the dual cell-complex $\tilde{K}$.

These facts lead us towards a classification scheme for physical quantities of a physical theory in analogy with the scheme of the chains of K and $\tilde{K}$ of Table 4.

5.1 Extensive physical quantities

Let us consider those physical quantities that are referred to extended geometrical manifolds of the space of the independent variables, like lines, surfaces, volumes, hypervolumes, time intervals, etc.

By their definition these are additive on the subparts of the corresponding geometrical manifolds: this is a good reason to call them *extensive*. The name "extensive" is used here in a more general sense than in thermodynamics where it means additivity on the subparts of volumes only.

These extensive physical quantities may be scalars, vectors, matrices, operators, etc. When there are other than scalars their components with respect to a base have one or more indices. So the displacement of a particle during a time interval is associated with an extensive chronometrical element and is an extensive physical quantity (extensive in time). Since it is a vector it can be represented by its components s^A, with A = 1,2,3. The finite rotation of a rigid body with a fixed point is associated with the time intervals, then it is an extensive physical quantity. It can be described by an orthogonal matrix whose entries may be denoted R_A^B i.e. have two indices.

Examples of extensive physical quantities are the magnetic flux, the electromotive force, the electric charge, the energy, the contact force, etc.

5.2 The forming of the densities

From the extensive physical quantities one is led to introduce the corresponding specific quantities, like the flux for unit area, the contact force for unit area, the magnetomotive force for unit length, the charge for unit volume and so on. These specific quantities are the common densities. So we have the electric charge density, the mass density, the velocity (time density of the displacement), etc.

When the physical quantity is associated to lines or surfaces one must introduce a directional density or equivalently a vectorial or tensorial variable. So

magnetomotive force $\quad F_{(m)} = \int_L \vec{H}\cdot d\vec{L}$

magnetic flux $\phi = \iint_S \vec{B}\cdot\vec{n}dS$

contact force $f_A = \iint_S p_A^K n_K dS$

The vectors H_K and B_K and the stress tensor p_A^K are the densities of the corresponding extensive physical quantities. The index K refers to the geometrical element with which the extensive physical quantity is associated, and will be called *geometrical index*. On the contrary the index A, like in the contact force f_A, is referred to the mathematical nature of the extensive physical quantity and will be called *intrinsic* or *internal index*. The distinction of these two kinds of indices is of fundamental importance for the classification we are doing. Table 5 shows some extensive physical variables.

5 – Some extensive physical variable and their density 41

extensive physical variable	defining relation	corresponding density
potential difference V	$V = \int_L E_k \, dL^k$	electric field strength E_k
displacement s^A	$s^A = \int_0^\tau v^A \, dt$	velocity v^A
impulse h_A	$h_A = \int_0^\tau f_A \, dt$	force f_A
magnetic flux ϕ	$\phi = \iint_S B^k n_k \, dS$	magnetic induction B^k
contact force t_A	$t_A = \iint_S p_A{}^k n_k \, dS$	stress tensor $p_A{}^k$
mass m	$m = \iiint_V \rho \, dV$	mass density ρ
energy production Σ	$\Sigma = \iiint_V \int_0^\tau \sigma \, dV dt$	energy production density σ
fluid flow Q	$Q = \iint_S \int_0^\tau q^k n_k \, dS dt$	current density q^k

Since to every extensive physical quantity is associated a corresponding density we may classify these densities. As a classification criterion we choose the following one: *we refer a density to the same geometrical object to which the corresponding extensive physical quantity is referred.*

The density field of a physical quantity is the continuum analog of the corresponding chain. To the coboundary operator, that links two chains, there corresponds a differential operator that links the corresponding densities, as is shown in Table 6. In this way one realizes that the commonest differential operators "grad", "rot", "div" are the continuum analogs of the coboundary operator applied respectively to 0-chains, 1-chains and 2-chains.

6 - The continuum analogous of chains and coboundary

chains		integral on the s-th p-cell
$a^{(0)} = (a_1, a_2, \dots a_{\alpha_0})$	$\longrightarrow$	$a_s = a(P_s)$
$b^{(1)} = (b_1, b_2, \dots b_{\alpha_1})$	$\longrightarrow$	$b_s = \int_{L_s} B_k(P)\, dL^k$
$c^{(2)} = (c_1, c_2, \dots c_{\alpha_2})$	$\longrightarrow$	$c_s = \iint_{S_s} C^k(P)\, n_k\, dS$
$d^{(3)} = (d_1, d_2, \dots d_{\alpha_3})$	$\longrightarrow$	$d_s = \iiint_{V_s} D(P)\, dV$
coboundary		differential operators
$b^{(1)} = \delta a^{(0)}$	$\longrightarrow$	$\vec{B}(P) = \mathrm{grad}\ a(P)$
$c^{(2)} = \delta b^{(1)}$	$\longrightarrow$	$\vec{C}(P) = \mathrm{rot}\ \vec{B}(P)$
$d^{(3)} = \delta c^{(2)}$	$\longrightarrow$	$D(P) = \mathrm{div}\ \vec{C}(P)$
when the coefficients of the chain are vectors		
$t^{(2)} = (\vec{t}_1, \vec{t}_2, \dots \vec{t}_{\alpha_2})$	$\longrightarrow$	$(t_A)_s = \iint_{S_s} p_A{}^k n_k\, dS$
$F^{(3)} = (\vec{F}_1, \vec{F}_2, \dots \vec{F}_{\alpha_3})$	$\longrightarrow$	$(F_A)_s = \iiint_{V_s} f_A\, dV$
$F^{(3)} = \delta f^{(2)}$	$\longrightarrow$	$f_A = -\nabla_k p_A{}^k$
Poincaré lemma		differential identities
$\delta(\delta a^{(0)}) \equiv 0$	$\longrightarrow$	$\mathrm{rot}\ (\mathrm{grad}\ a) \equiv 0$
$\delta(\delta b^{(1)}) \equiv 0$	$\longrightarrow$	$\mathrm{div}\ (\mathrm{rot}\ \vec{B}) \equiv 0$

5.3 The standard classification scheme

The content of Table 6 can be ordered as shown in Table 7. This first column represents the chains of the various orders, the second and third columns give the continuum analogs of the first one using an intrinsic notation and a tensorial notation respectively. In both columns the physical quantity associated with the cells has the nature of a scalar. This is reflected in the fact that the mathematical entity that appears in the first box is a scalar.

When the physical quantity associated with the cells has a vector nature, this is the case of continuum mechanics, we have the fourth column. The symbol ∇_K denotes the covariant derivative that is needed to take into account the space variation of the components u_A on account of the space variation of the base vectors. The last three columns will be called *standard columns*. A pair of standard columns, one relative to a cell-complex K and the other to its dual will be called standard scheme $\tilde{K}$.

Table 8 gives the standard schemes for spaces of n = 1,2,3,4 dimensions. In the rectangular boxes that link the configuration-kind variables (K) with the source-kind variables ($\tilde{K}$) we must insert the constitutive relations.

With these rules we now construct the classification scheme of the physical quantities of some physical theories. Some of these schemes are shown in the following tables (9 to 20). All symbols and names are those recommended by the International Union of Pure and Applied Physics, S.U.N. Commission, 1965. The dimensions are those of the International System of Units (S.I. Units). The schemes for other physical theories are contained in the author's paper [5]. In the same paper the analysis we have summarized here is given in much more detail.

These classification schemes make evident the analogies between different physical theories. So the analogy among thermal conduction, fluid dynamics, electrostatics and magnetostatics is

7 - various forms of a standard column

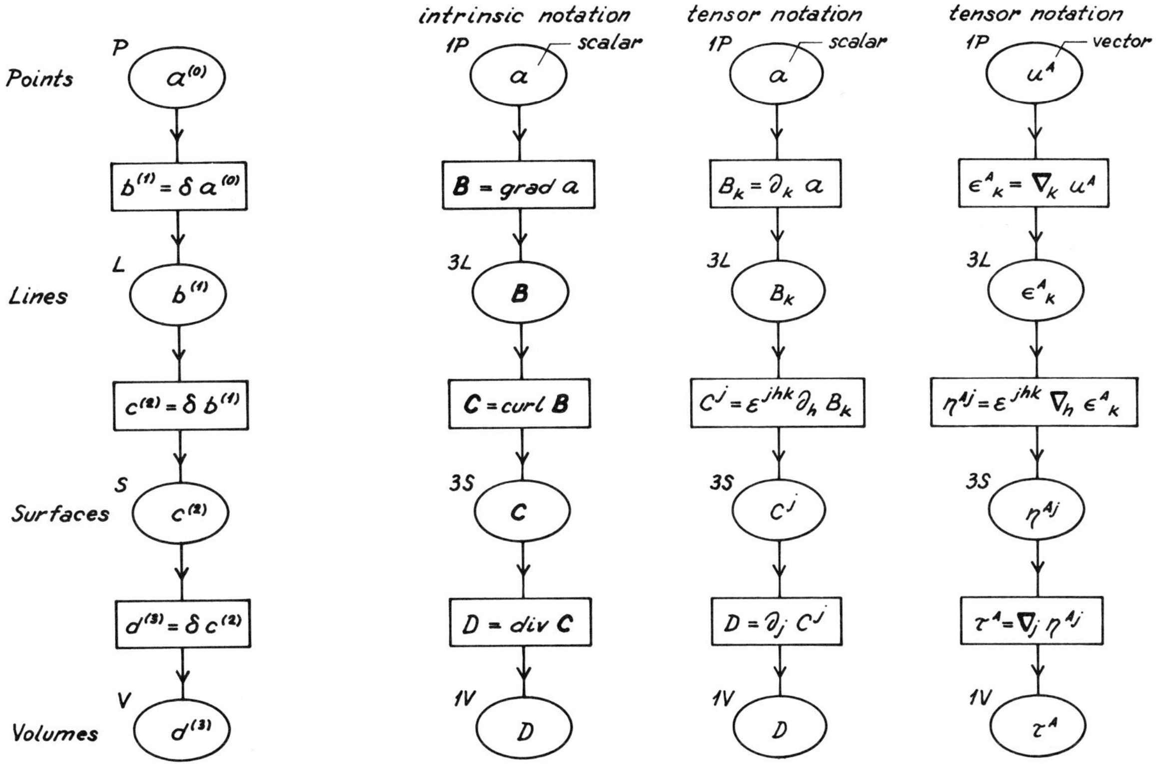

8 - the standard schemes for n = 1, 2, 3, 4.

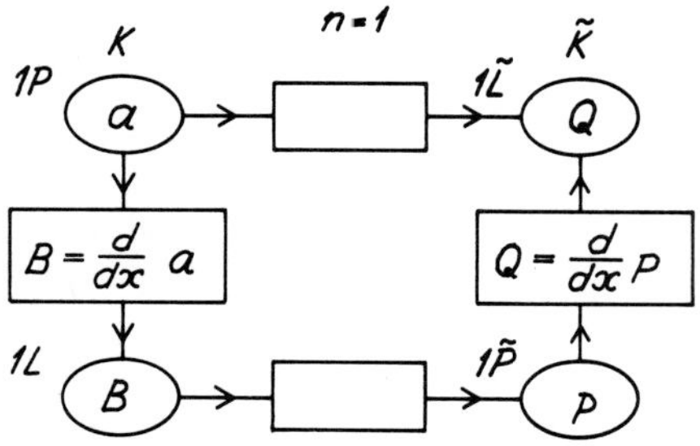

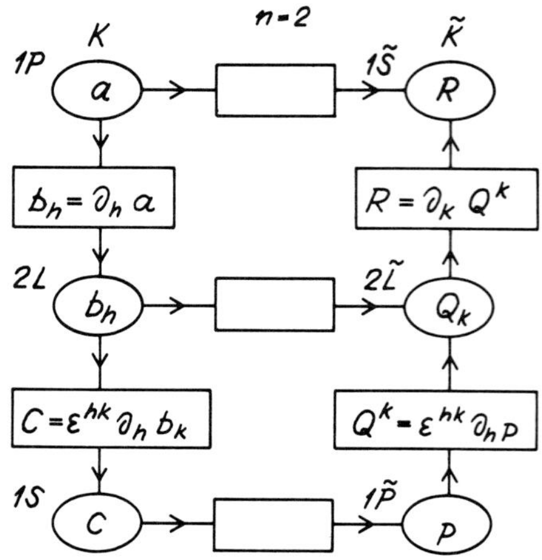

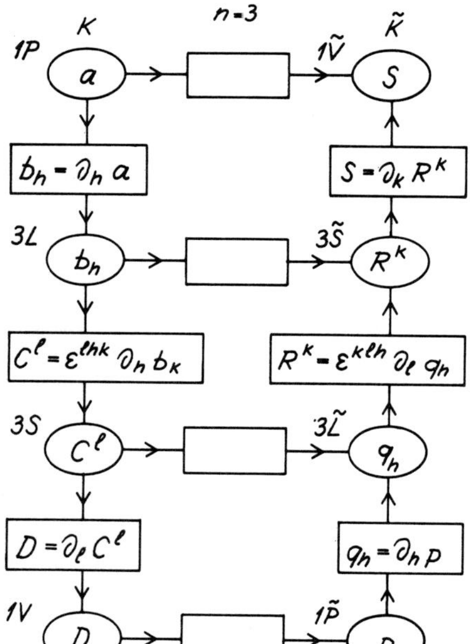

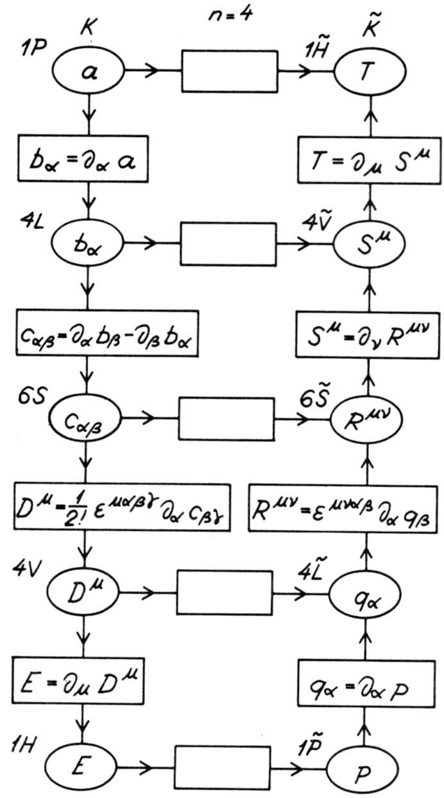

9-classical particle dynamics

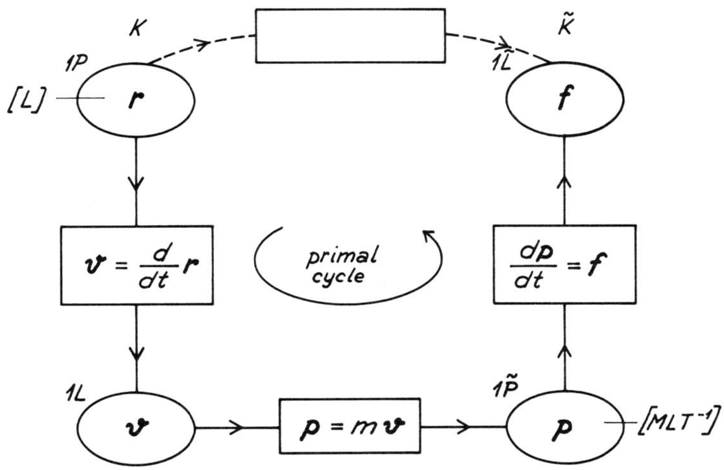

10-relativistic particle dynamics

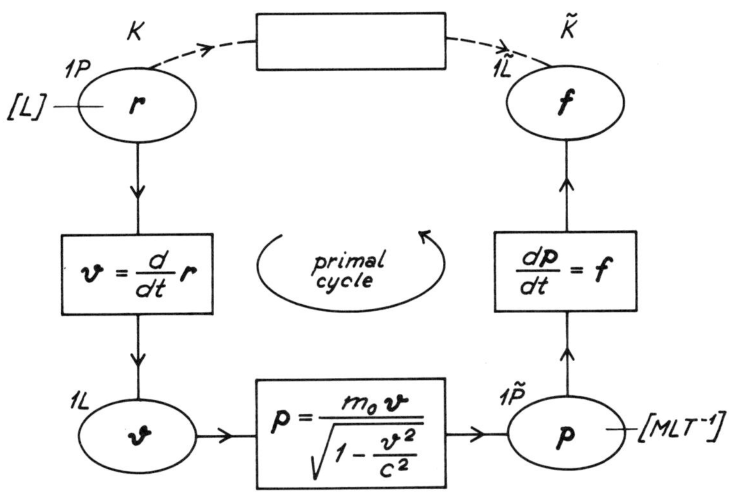

11 - longitudinal vibrations of a rod

variables t, x

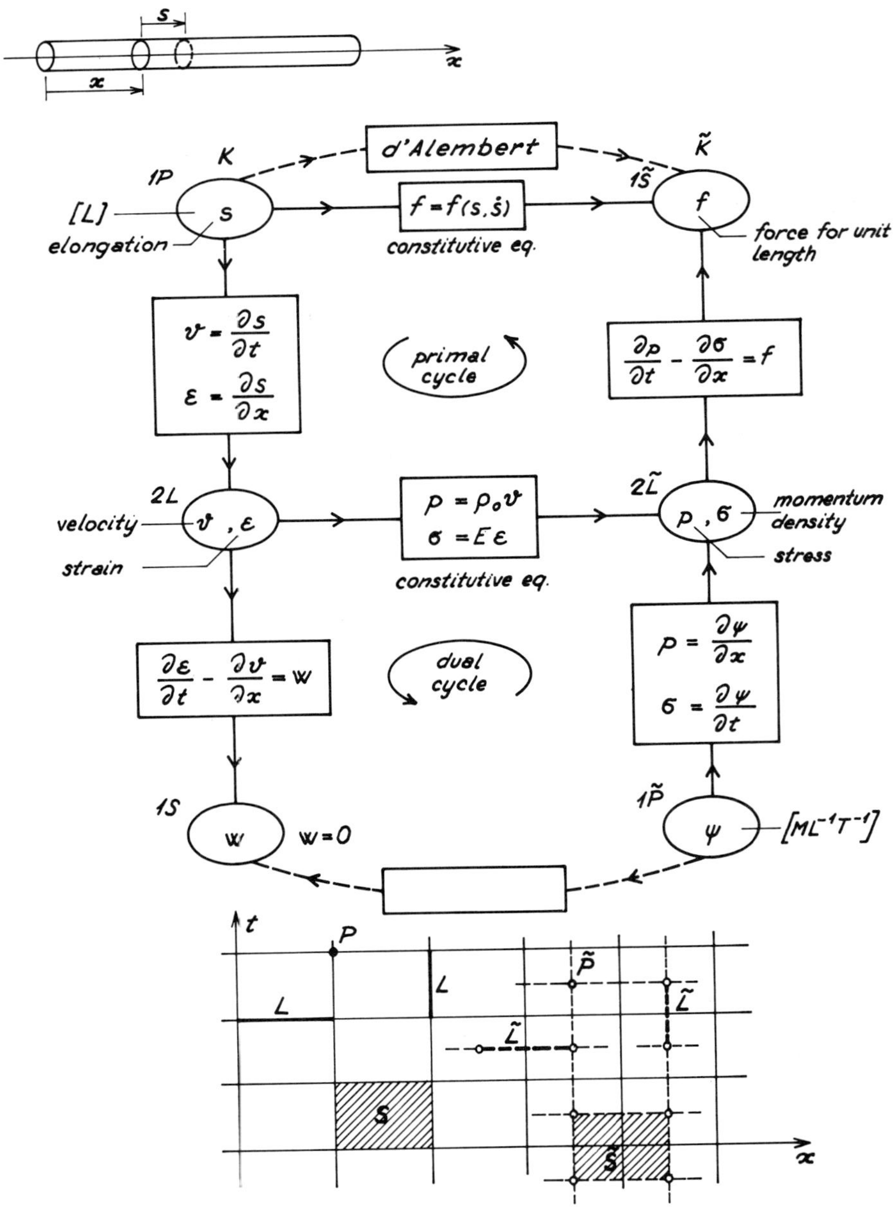

12 - thermal conduction

variables x^1, x^2, x^3; the time t is a parameter

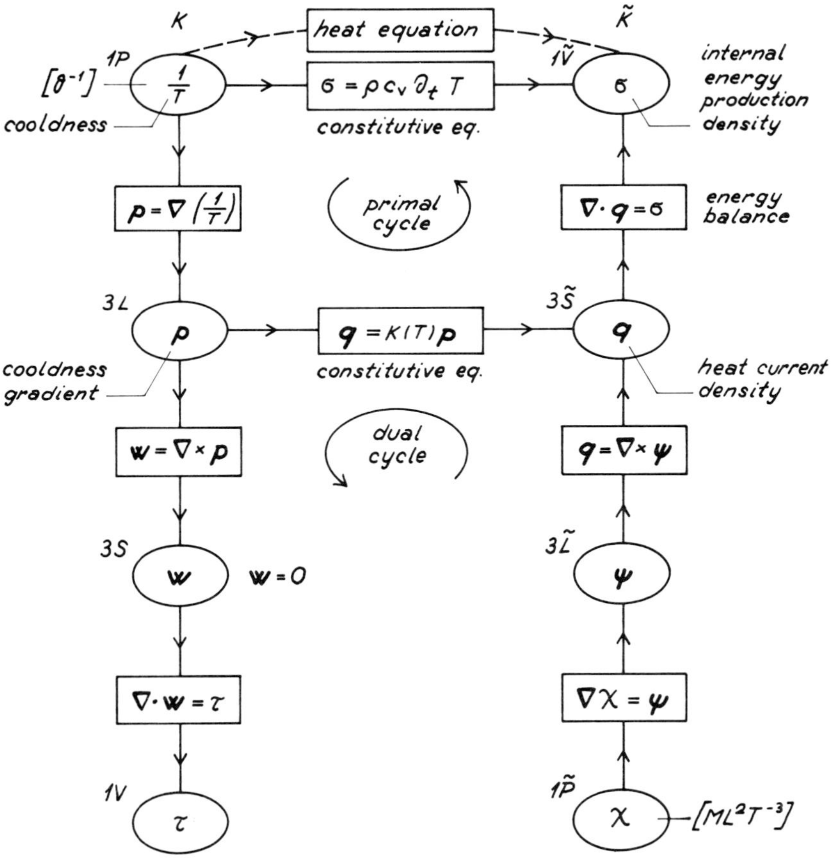

13 - thermostatics of a monoatomic ideal gas

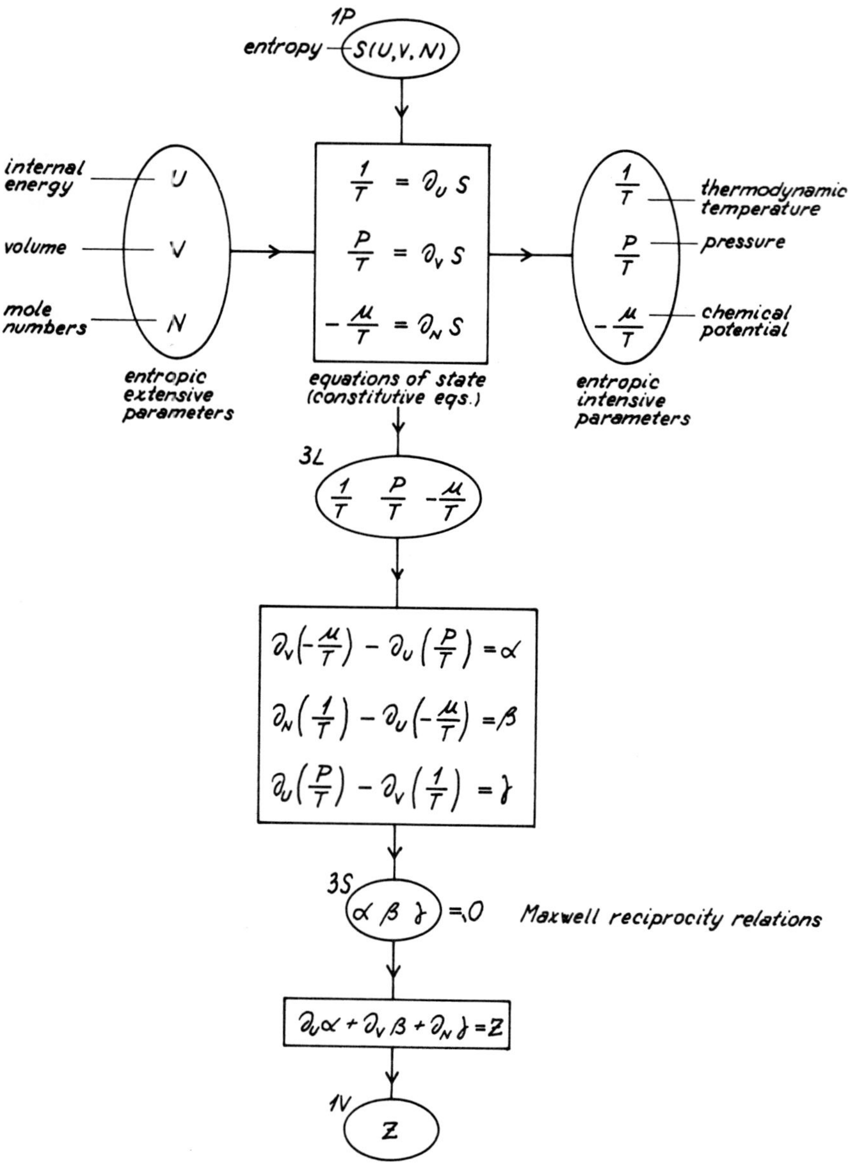

14 - irreversible thermodynamics

the time t is considered as a parameter

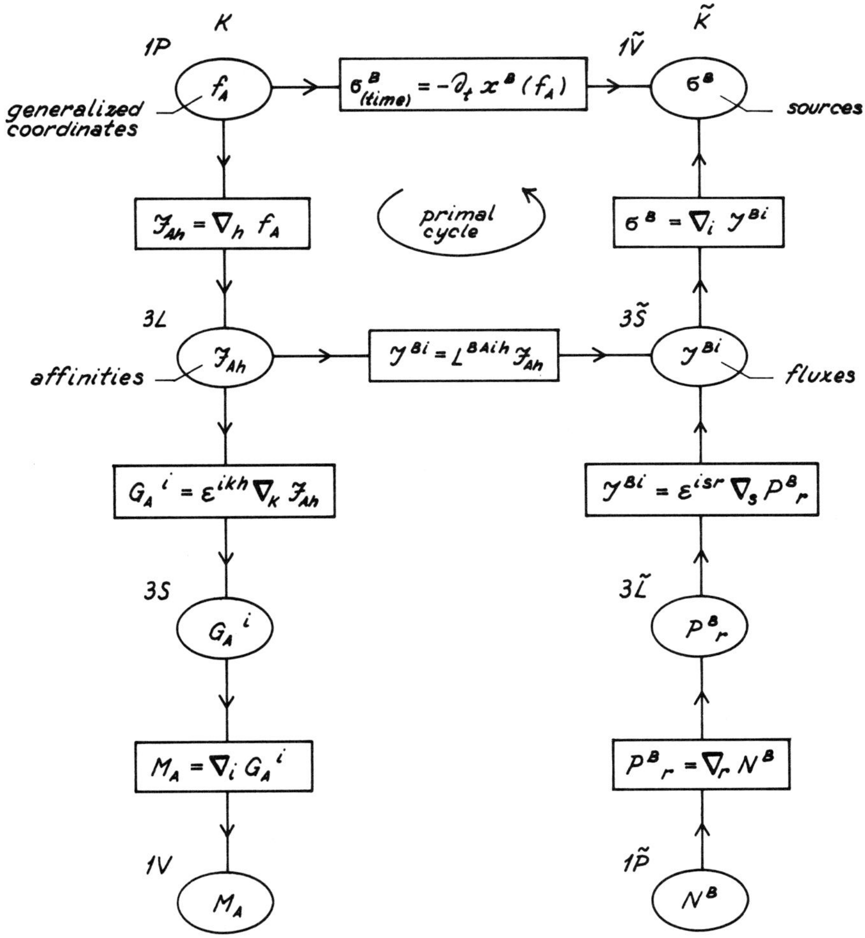

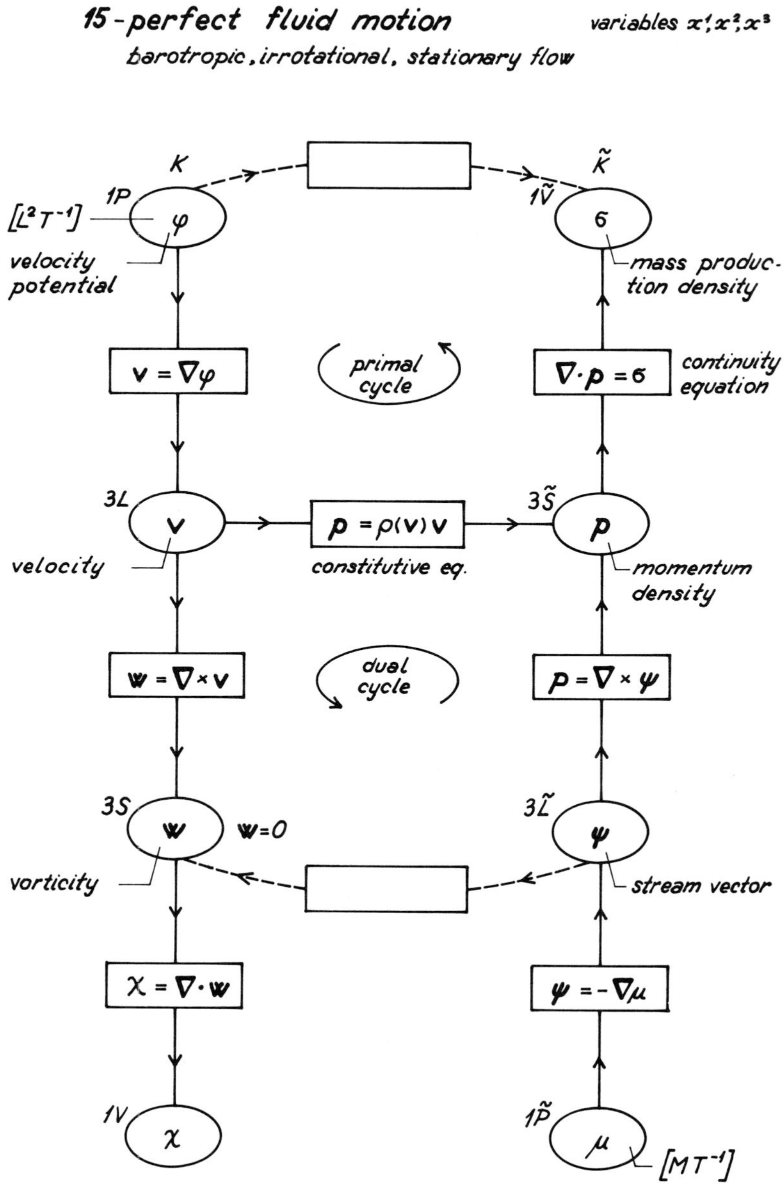
15-perfect fluid motion
variables x^1, x^2, x^3
barotropic, irrotational, stationary flow
K
$\tilde{K}$
1P
$[L^2T^{-1}]$
φ
velocity potential
$1\tilde{V}$
σ
mass production density
$\mathbf{v} = \nabla\varphi$
primal cycle
$\nabla\cdot\mathbf{p} = \sigma$
continuity equation
3L
$\mathbf{v}$
velocity
$\mathbf{p} = \rho(\mathbf{v})\mathbf{v}$
constitutive eq.
$3\tilde{S}$
$\mathbf{p}$
momentum density
$\mathbf{w} = \nabla\times\mathbf{v}$
dual cycle
$\mathbf{p} = \nabla\times\boldsymbol{\psi}$
3S
$\mathbf{w}$
$\mathbf{w}=0$
vorticity
$3\tilde{L}$
$\boldsymbol{\psi}$
stream vector
$\chi = \nabla\cdot\mathbf{w}$
$\boldsymbol{\psi} = -\nabla\mu$
1V
χ
$1\tilde{P}$
μ
$[MT^{-1}]$

16 - elastostatics

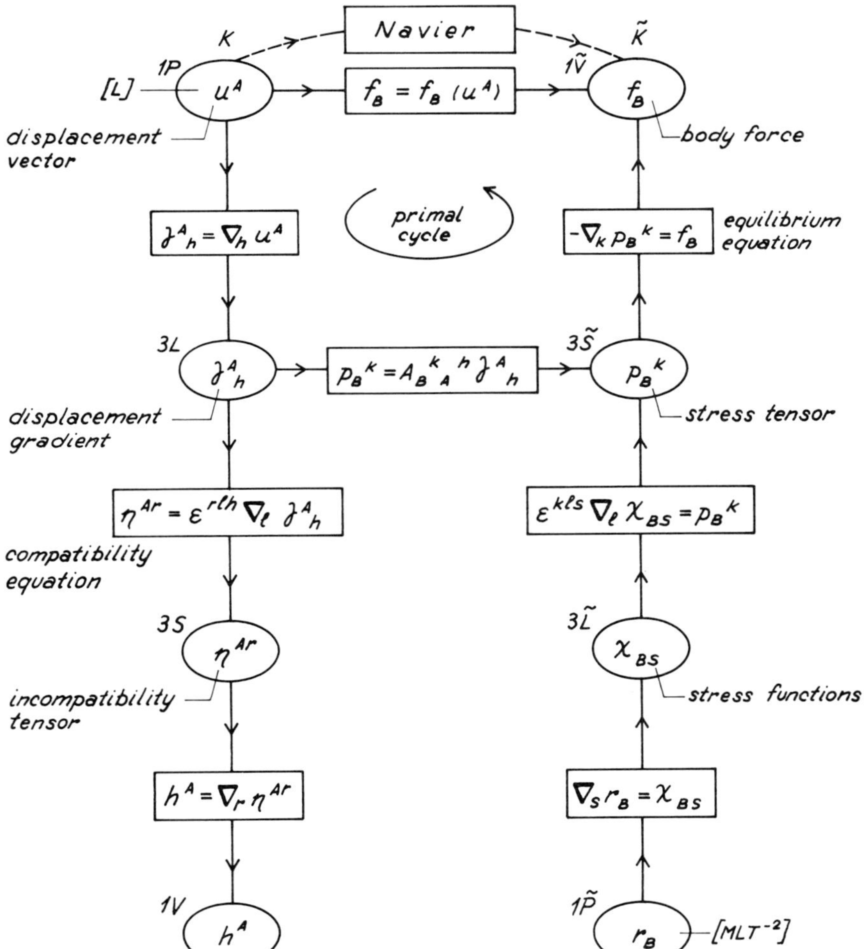

17 - fluidynamics

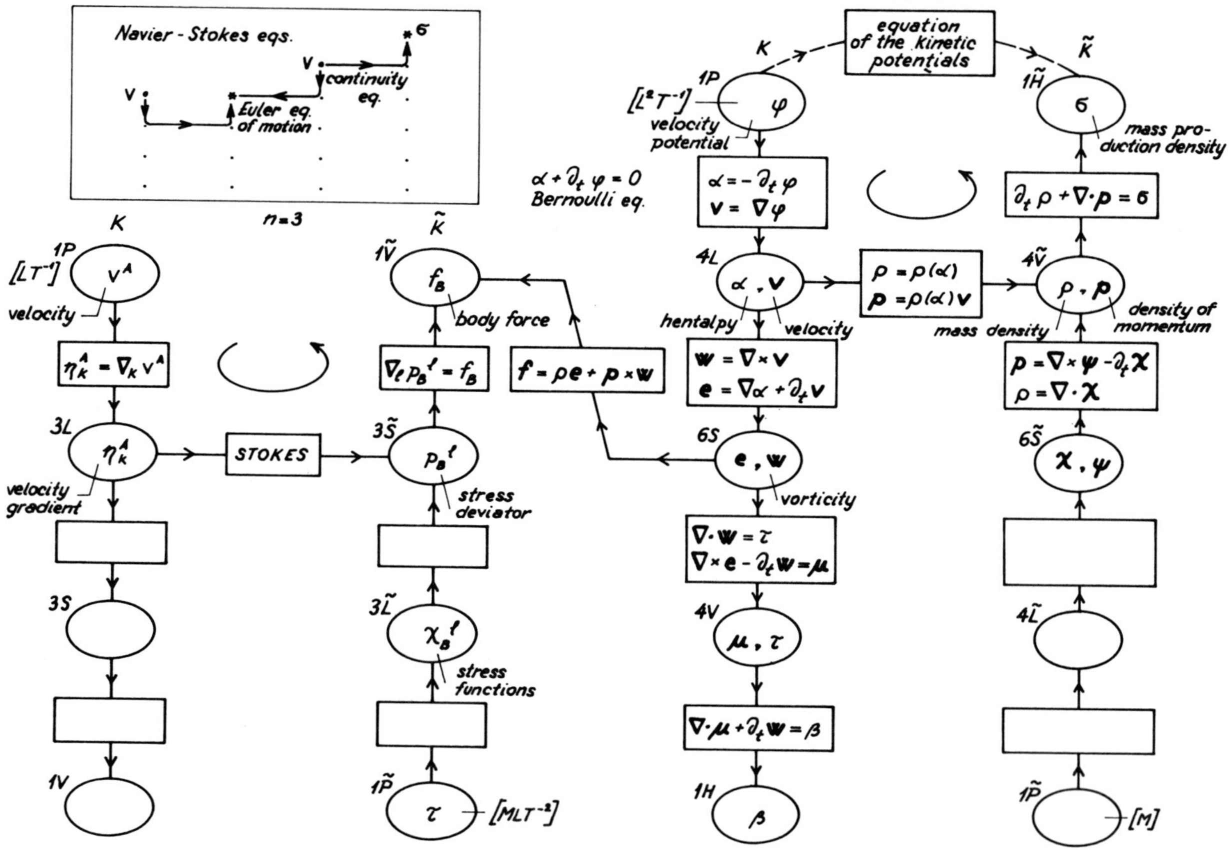

18 - electrostatics

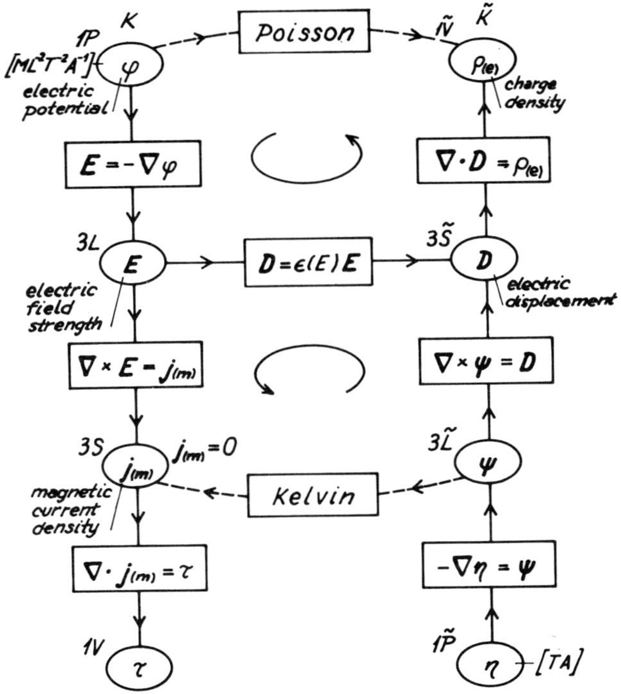

magnetostatics

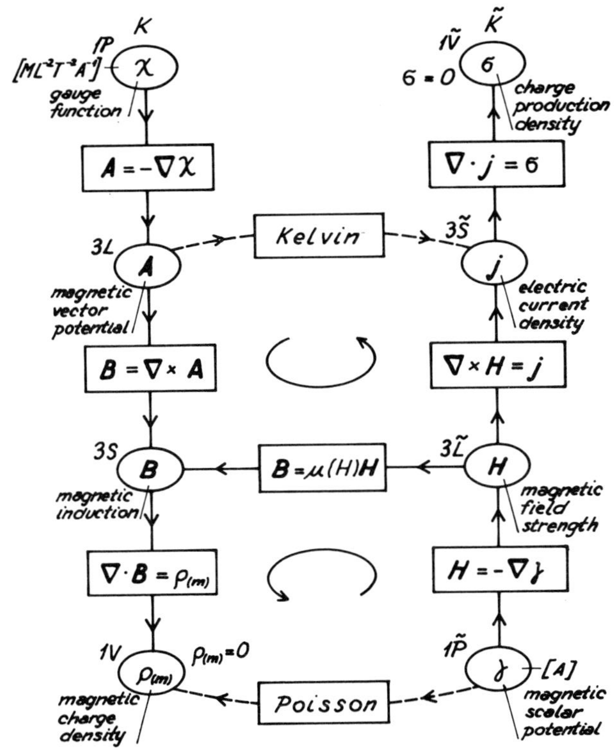

19 - eletromagnetic field

variables: t, x, y, z.

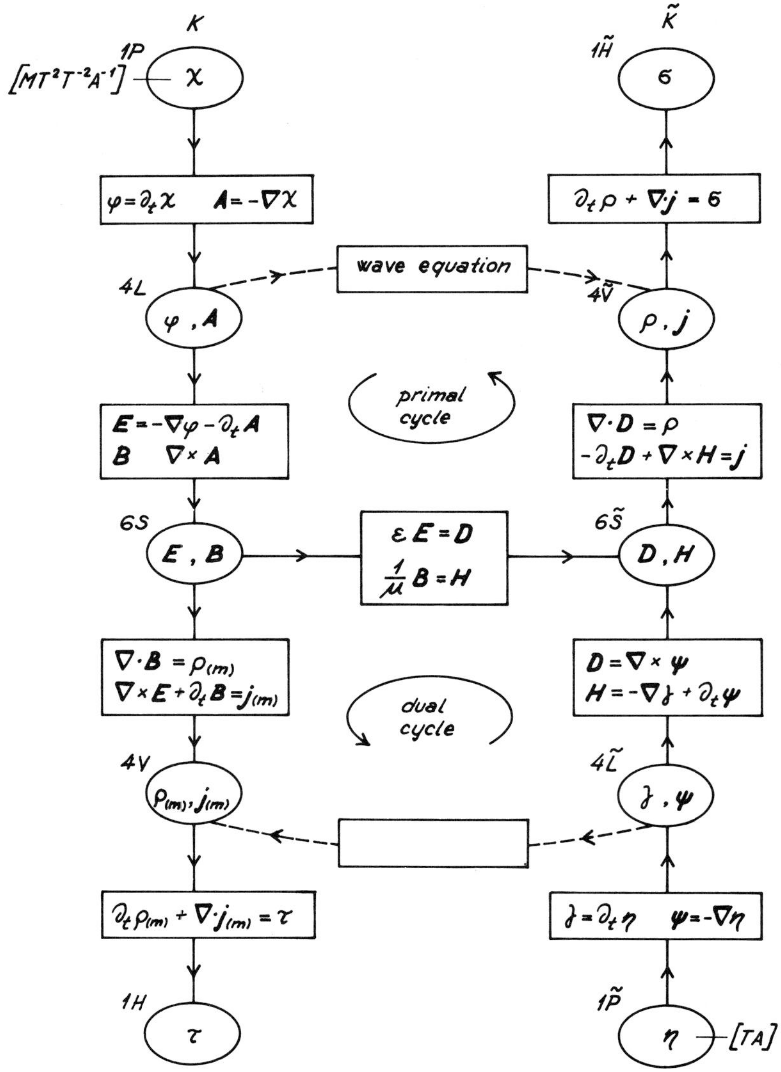

20-Schrödinger field

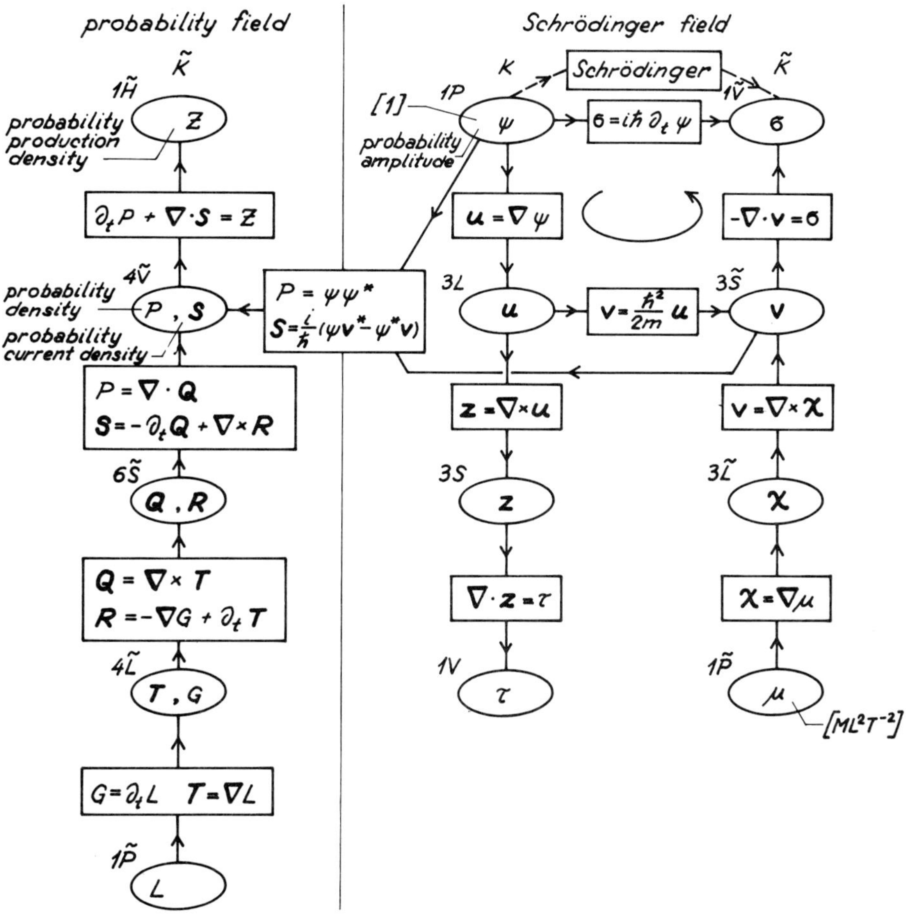

evident by a comparison of the corresponding schemes.

In order to utilize these schemes for solution purposes one must investigate the mathematical properties exhibited by the operators that appear in the structure equations and in the constitutive equations. Such investigation has been performed and the main mathematical properties have been evidentiated [6]. This has led to the construction of a mathematical model [7].

1. Klein, F., *Elementary Mathematics from an Advanced Standpoint*, Vol. II, 1948. (Translated from the German).
2. Veblen, O., Whitehead, J.H.C., *The Foundation of Differential Geometry*, Cambridge Univ. Press, 1932.
3. Franz, W., *Algebraic Topology*, Ungar Publ. Co., 1968, (Translated from the German).
4. Seifert, H., Threlfall, W., *Lehrbuch der Topologic Leipzig*, Chelsea Publ. Co., 1947.
5. Tonti, E., *On the Formal Structure of Physical Theories*, Quaderno del Consiglio Nazionale delle Ricerche, 1975.
6. Tonti, E., "On the Mathematical Structure of a Large Class of Physical Theories", *Rend. Acc. Lincei*, Vol. VII, pp. 48-56, 1972.
7. Tonti, E., "A Mathematical Model for Physical Theories", *Rend. Acc. Lincei*, Vol. VII, pp. 175-181 (first part), pp. 350-356 (second part), 1972.

A B 7 C 8 D 9 E 0 F 1 G 2 H 3 I 4 J 5